Scalable Enterprise Systems

INTEGRATED SERIES IN INFORMATION SYSTEMS

Series Editors

Professor Ramesh Sharda
Oklahoma State University

Prof. Dr. Stefan Voß
Universität Hamburg

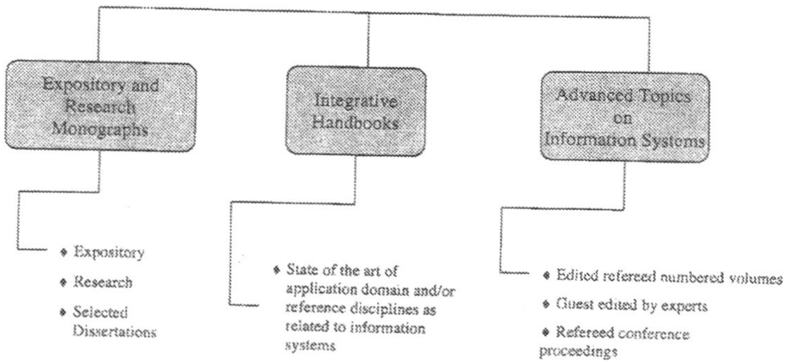

Other published titles in the series:

E-BUSINESS MANAGEMENT: *Integration of Web Technologies with Business Models*/ Michael J. Shaw

VIRTUAL CORPORATE UNIVERSITIES: *A Matrix of Knowledge and Learning for the New Digital Dawn*/ Walter R.J. Baets & Gert Van der Linden

Scalable Enterprise Systems

An Introduction to Recent Advances

Editors

VITTAL PRABHU
Marcus Department of Industrial and Manufacturing Engineering
The Pennsylvania State University
University Park, Pennsylvania

SOUNDAR KUMARA
Marcus Department of Industrial and Manufacturing Engineering
The Pennsylvania State University
University Park, Pennsylvania

MANJUNATH KAMATH
School of Industrial Engineering and Management
Oklahoma State University
Stillwater, Oklahoma

KLUWER ACADEMIC PUBLISHERS
Boston/Dordrecht/London

Distributors for North, Central and South America:
Kluwer Academic Publishers
101 Philip Drive
Assinippi Park
Norwell, Massachusetts 02061 USA
Telephone (781) 871-6600
Fax (781) 871-9045
E-Mail: kluwer@wkap.com

Distributors for all other countries:
Kluwer Academic Publishers Group
Post Office Box 322
3300 AH Dordrecht, THE NETHERLANDS
Telephone 31 786 576 000
Fax 31 786 576 254
E-mail: services@wkap.nl

Electronic Services <http://www.wkap.nl>

Library of Congress Cataloging-in-Publication Data

A C.I.P. Catalogue record for this book is available from the Library of Congress.

SCALABLE ENTERPRISE SYSTEMS: *An Introduction to Recent Advances*, edited by V. Prabhu, S. Kumara, and M. Kamath

ISBN 1-4020-7491-3

Printed on acid-free paper.

Printed in the United States of America.

TABLE OF CONTENTS

Chapter 8. PUBLISH SUBSCRIBE MIDDLEWARE.......................249

Chapter 9. EXPERIMENTAL STUDY OF SCALABILITY ENHANCEMENT FOR REVERSE LOGISTICS E-COMMERCE....287

LIST OF CONTRIBUTORS

Michael Ball is the Orkand Professor of Management Science in the Robert H. Smith School of Business at the University of Maryland. He also holds a joint appointment within the Institute for Systems Research in the Clark School of Engineering. He is Director of Research for the Smith School and is co-Director of NEXTOR, the National Center of Excellence for Aviation Operations Research. Dr. Ball received his PhD in Operations Research in 1977 from Cornell University. His research interests are in network optimization and integer programming particularly as applied to problems in supply chain management, transportation systems and manufacturing.
URL: www.isr.umd.edu/people/faculty/Ball.html

Chad Dodd obtained his B.S., and M.S., in Industrial Engineering at the Pennsylvania State University. His research work concentrated on building internet based supply chains, in specific, value nets. His interests are in e-commerce, supply chain logistics and multiagent systems. He is currently working with GE Medical Systems and is stationed at Paris, France.

Chien-Yu Chen is a Ph.D. Candidate in the Robert H. Smith School of Business at the University of Maryland. He has double majored in the Decision and Information Technologies Department and in the Logistics, Business and Public Policy Department. His research interests include supply chain management, production and operations management, and inventory control systems, especially the push-pull integration strategies by using optimization and simulation models. He taught undergraduate-level Operations Research and Operations Management courses.

Amit Chaugule is currently Senior Product Engineer for Borg Warner Morse TEC, in Ithaca, New York. Prior to Borg Warner, he worked for Mahindra and Mahindra Ltd., Bombay, India as a Product Engineer. He has over 6 years of experience in product and process development, new product launch and quality engineering. He holds a Bachelors degree in Automotive Engineering from Bombay University, India and a Masters of Science in Industrial Engineering from Oklahoma State University, Stillwater, OK, where he worked on his thesis titled "A User Oriented Enterprise Process Modeling Language" as a part of a NSF funded project. His interests include process modeling, operations management, planning and development of new product and processes particularly for the automotive industry.

Dr. Moon Jung Chung is currently a professor with Department of Computer Science and Engineering at Michigan State University. Before joining Michigan State University, he was a faculty at Rensselaer Polytechnic Institute, Troy, NY. In 1994, he received a fellowship Senior Research Associate from the National Research Council. During this tenure, he visited Army Research Lab. He has published papers in Process management in a distributed environment, Web modeling and algorithms, Parallel simulation. He has also a patent on VLSI testing. He received the B.S. degree from Seoul National University, and received the M.S. degree from Korea Advanced Institute of Science. He obtained Ph.D. degree of Computer Science from Northwestern University.

Dr. Nikunj Dalal is Associate Professor of Management Science and Information Systems at Oklahoma State University, Stillwater, OK. He has a Ph.D. in Management Information Systems from Texas Tech University. He recently completed a sabbatical at J.D. Edwards. His research interests include information systems philosophy, enterprise modeling, systems modeling, and Web perceptions. His research is published in journals such as Communications of the ACM, Decision Sciences, International Journal of Human-Computer Studies, among others.
URL: http://catt.okstate.edu/dalal

Houshang Darabi received the Ph.D. degree in Industrial and System Engineering from Rutgers University, in 2000. He is currently an Assistant Professor at the Department of Mechanical and Industrial Engineering, University of Illinois at Chicago. His research interests include the application of Discrete Event Systems control theory in modeling and analysis of business and manufacturing systems.
URL: www.uic.edu/~hdarabi/index.html

Dr. Manjunath Kamath is Associate Professor and Graduate Program Director in the School of Industrial Engineering and Management and Director of the Center for Computer Integrated Manufacturing at Oklahoma State University, Stillwater, OK. He received the B.Tech. Degree in Mechanical Engineering from the Indian Institute of Technology, Madras, India, in 1982, the M.E. degree in Automation from the Indian Institute of Science, Bangalore, India, in 1984, and the Ph.D. degree in Industrial Engineering from the University of Wisconsin-Madison, in 1989. His primary areas of interest are enterprise modeling, stochastic modeling and queueing theory, analytical performance modeling of manufacturing systems, object-oriented modeling and simulation, and Petri nets. He has served as Principal Investigator or Co-Principal Investigator on several projects funded by the National Science Foundation and the AT&T Foundation involving the enterprise process modeling, design and development of object-oriented modeling environments, integrated production management systems and queueing. Dr. Kamath is a member of IEEE, IIE and INFORMS.
URL:
http://www.okstate.edu/ceat/iem/faculty/people/faculty/kamath.html

Taioun Kim is currently an associate professor of Industrial Engineering at Kyungsung University in Busan, Korea. His educational background includes a B.S. from Seoul National University, a M.S. from KAIST in Korea and a Ph.D. from The Pennsylvania State University in Industrial and Manufacturing Engineering. His research area includes multi-agent systems, workflow management/process planning and mobile supply chain management. His current projects include intelligent process design (Korea Science and Engineering Foundation) and mobile SCM (Brain Busan 21).
URL: http://ie.ks.ac.kr/pro/twkim.asp

Soundar R. T. Kumara is currently a **Distinguished Professor** of Industrial and Manufacturing Engineering he holds joint appointments with the School of Information Sciences and Technology and the Department of Computer Science and Engineering at Penn State University. He worked as visiting faculty at several institutions including MIT, University of Tokyo, and City University of Hong Kong. His research interests span intelligent systems, sensor data fusion, process monitoring and diagnostics, Complexity theory and distributed supply chains. His current projects include Chaos Theory based situation extraction, pattern extraction for robust control of cognitive agents (DARPA), Scalable Extended Enterprises (National Science Foundation), Procurement Problem Solving Using Stochastic Programming, Game Theory and Software Agents (National Science Foundation), and OTD Process Simulation Framework using agents (General Motors). He is working mainly in developing adaptive control algorithms for large scale complex systems.
URL:
http://www.ie.psu.edu/people/faculty/kumara/Index.htm

Patrick Kwon is an assistant professor in the department of mechanical engineering at Michigan State University. His educational background includes a B.S. from University of Michigan, a S.M. from MIT and a Ph.D. from UC Berkeley all in mechanical engineering. His research area includes machining, powder processing and process management. He has published his research work in Journal of Manufacturing Science and Engineering, Journal of Mechanical Design, International Journal of Machine Tools & Manufacture and Acta Metallurgica.
URL: www.egr.msu.edu/~pkwon

Jo Min is an Associate Professor in the Department of Industrial and Manufacturing Systems Engineering, Iowa State University.He earned his Ph.D. in Operations Research at University ofCalifornia, Berkeley. His research interests are Production and Inventory Systems, Pricing and Costing, and Utilities Planning and Operation.
URL: www.imse.iastate.edu/people/min.html

Sigurdur Olafsson is an Assistant Professorin the Industrial and Manufacturing Systems Engineering Department at Iowa State University. He holds a BS in mathematics from the University of Iceland, and a MS and Ph.D. in industrial engineering from the University of Wisconsin - Madison. His research interest include optimization methods for data mining, scalability of data mining techniques, and data mining for e-commerce and production scheduling. Dr. Olafsson's work has been published in such journals as Operations Research, Management Science, and IIE Transactions,and he is a member of IIE and INFORMS.
URL: http://www.public.iastate.edu/~olafsson/

Lukasz Opyrchal is a Ph.D. student and a Graduate StudentResearch Assistant at the University of Michigan. His research interests includesecurity and content-based publish subscribe systems. In his past work,he explored the issues of security and event delivery in content-based systems. He received an MS in computer engineering from the University of Michigan, Ann Arbor.
URL: www.eecs.umich.edu/~lukasz

Brian T. Pentland is an associate professor in the Department of Accounting and Information Systems at the Eli Broad Graduate School of Management, where he teaches courses in information systems. Before coming to the Business School, he taught in the MSU School of Labor and Industrial Relations, the Anderson Graduate School of Management at UCLA and the University of Michigan School of Business Administration. His primary area of interest is in the relationship between work and technology, although he has also been developing techniques for business process modeling and the sequential analysis of qualitative data. His publications have appeared in *Administrative Science Quarterly, Academy of Management Review, Management Science, Organization Science, Accounting,Organizations and Society, Technology Studies, and Accounting, Management and Information Technologies and the Journal of Mechanical Design*. He currently serves on the editorial boards of *Organization Science, Accounting, Management and Information Technologies, and Information, Technology and People.* Pentland holds degrees from the Massachusetts Institute of Technology (S.B., 1981, Mechanical Engineering) and the MIT Sloan School of Managment (Ph.D., 1991, Organization Studies).
URL: www.msu.edu/~pentlan2

Vittal Prabhu is currently an Associate Professor of Industrial Engineering at Penn State University. Before joining Penn State, Professor Prabhu was a Research Associate (Post Doc) at the University of Wisconsin-Madison where he got his Ph.D. His research focuses on developing a rigorous, yet tractable, control theoretic framework for unified feedback control of discrete event timing and continuous variables in distributed manufacturing enterprises. Long-term outcome of this effort will be a unified analytical framework with integrated equipment level (e.g., CNC) and enterprise level control (e.g., production scheduling, inventory, supply chain). He is working with his students to develop low cost manufacturing shop floor supercomputers for distributed simulation and control. Dr. Prabhu is the Founder and Co-Director of Center for Manufacturing Enterprise Integration at Penn State.
URL: http://www.personal.psu.edu/faculty/v/x/vxp7/

Ronald L. (Ron) Rardin is Program Director for Operations Research and Service Enterprise Engineering with the National Science Foundation, on rotator assignment from the School of Industrial Engineering at Purdue University. He obtained his B.A. and M.P.A. degrees from the University of Kansas. After working in city government, consulting and distribution for five years, he continued his education with a Ph.D. at Georgia Institute of Technology. Professor Rardin's teaching and research interests center on optimization modeling and algorithms, particularly large-scale integer and combinatorial problems. He is co-author of numerous research papers in that field, as well as two comprehensive textbooks.

Sarah M. Ryan is an associate professor of Industrial & Manufacturing Systems Engineering at Iowa State University. Besides scalable enterprise systems, her research focuses on stochastic models of manufacturing and service systems for medium to long range planning. She received a National Science Foundation Early Faculty Career Development (CAREER) Award and has published articles in *Operations Research*, *IIE Transactions*, and several other journals. She teaches courses in operations research and engineering economics and serves as area editor of *The Engineering Economist*.
URL: www.public.iastate.edu/~smryan

Goutam Satapathy is a Research Engineer at Intelligent Automation Incorported at Rockville, MD. Satapathy obtained his integrated M.Tech degree in Mechanical Engineering from the Indian Institute of Technology, Bombay, India. He got his Ph.d., from the Pennsylvania State University. His research interests are in Distributed Multi Agent systems, Game Theory and Stochastic Programming. He worked on the Distributed Intelligent Agents for Logistics Project. Currently he is involved with the sector metering project funded by NASA at IAI.
URL: http://www.i-a-i.com

Lawrence M. Seiford is Professor and Chair of Industrial and Operations Engineering at the University of Michigan. Prior to joining the University of Michigan he was Program Director of the *Operations Research* and *Production Systems* programs at the National Science Foundation (1997-2000) and was a member of the faculty at the University of Massachusetts, the University of Texas at Austin, the University of Kansas, and York University.Professor Seiford's teaching and research interests are primarily in the areas of quality engineering, process improvement and performance measurement. His current research involves thedevelopment of benchmarking models for identifying best-practice in manufacturing and service systems. He has written and co-authored three books and approximately one hundred articles. Professor Seiford is Editor-in-Chief of *OMEGA, the International Journal of Management Science*, and an editor or member of the editorial board of eight other journals.
URL: http://ioe.engin.umich.edu/people/fac/seiford.html

Eswar Sivaraman is a PhD Candidate in the School of Industrial Engineering and Management at Oklahoma State University, Stillwater, OK. He has a B.S. in Manufacturing Engineering from the National Institute of Foundry and Forge Technology, India, and a M.S. in Industrial Engineering from Oklahoma State University, received, respectively, in 1996 and 1998. His dissertation research focuses on techniques for formalizing business process models and verifying the correctness of control flow and resource allocation policies. His research interests include systems modeling, construction and combinatorial problems in experimental design, and Petri nets.
URL: http://www.okstate.edu/cocim/members/eswar/

Zhenying Zhao is an Associate Research Scientist in R. H. Smith School of Business at the University of Maryland. His research interests are in the areas of mathematical modeling and optimization, supply chain management, computer simulation, production and operations management and supply chain management systems and enterprise resource planning (ERP) systems. He had worked with Toshiba, Maxtor and Compaq on business process improvement, available-to-promise decision support and supply chain.

Preface

The National Science Foundation (NSF) is the leading sponsor of basic academic research in engineering, and its influence far exceeds its budget. We think NSF is at its best when it uses that influence to focus interest within the researcher community on critical new challenges and technologies.

NSF's Scalable Enterprise Systems (SES) initiative, for which we were responsible in our successive terms in the division of Design, Manufacture and Industrial Innovation (DMII), was just such a venture. A collaborative effort spanning NSF's engineering and computer science directorates, SES sought to concentrate the energies of the academic engineering research community on developing a science base for designing, planning and controlling the extended, spatially and managerially distributed enterprises that have become the norm in the manufacture, distribution and sale of the products of U.S. industry. The issues addressed included everything from management of associated supply chains, to product design across teams of collaborating companies, to e-marketing and make-to-order manufacturing, to the information technology challenges of devising inter-operable planning and control tools that can scale with exploding enterprise size and scope. A total of 27 teams with nearly 100 investigators were selected from the 89 submitted proposals in the Phase I, exploratory part of the effort (see the list below). Seven of these were awarded larger multi-year grants to continue their research in Phase II. As the contents of this book amply illustrate, these investigations continue to flourish, with and without direct NSF support.

To provide some context for this collection of SES work, the Editors asked us to prepare this *Preface*. We have endeavoured to recount briefly some of the relevant history of the initiative and to provide some idea of its vision and scope.

GENESIS OF THE INITIATIVE

The intellectual genesis of SES was the conviction of then head of NSF's Engineering directorate, Dr. Eugene Wong, that commercial Enterprise Requirements Planning (ERP) software emerging to try to deal with extended manufacturing enterprises suffered from many of the same inadequacies Dr. Wong had seen in the early days of relational databases. Creative research was needed to advance the current state of ERP in the same way similar work had produced the user-friendly database tools now standard in all kinds of information systems.

The concept of a consistent suite of interoperable application programs to serve all major functions of a business enterprise had its origin in manufacturing. Beginning with MRP (materials requirements planning), new links to supply chain and other elements of business functions eventually led to the evolution of ERP.

In 1999 currently developed ERP products represented an evolutionary development of commercial data processing. There was little if any theoretical basis for this development and some notorious holes, such as fixed lead times and infinite capacities. Furthermore, ERP's promise of standardization of principal business functional modules to minimize customization and enhance reliability was far from being realized. The systems were also user unfriendly and placed much of the burden of feasible implementation on the potential user community.

Scalability, the problem of system behavior and performance as enterprise size, complexity and interdependence increases, was also an issue. Computer networks and the Internet are becoming a universal medium for enterprise level software deployment. This network operating environment greatly stretched the range of scalability, from a few users to millions. Such explosive growth was true not only in consumer oriented retail operations, but also in business-to-business deployments of systems in all sorts of enterprises.

FORMATIONAL WORKSHOP

Charged by director Wong with finding a way for NSF to address these critical challenges, Drs. Louis Martin-Vega and Lawrence Seiford of DMII took the lead in organizing an SES Workshop to identify researchable issues in the area. Collaborators from across NSF joined approximately 40 participants representing industry and academia at NSF on April 26-27, 1999. Four invited speakers shared their views and perspectives, and the group as a whole identified a broad set of issues needing attention. Given the wide and unstructured nature of the topic, there was a consensus that NSF should undertake an initiative in the field, and that it should begin with an exploratory first phase, followed by more in depth investigations in a second phase.

PHASE I SOLICITATION

The NSF solicitation for Scalable Enterprise Systems was released in the fall of 1999. Following the workshop guidance, it called for one-year exploratory proposals to develop concepts in Phase I, with more comprehensive grants in a Phase II to be awarded in 2001.

A total of 89 proposals were received covering the entire range of issues in SES. Peer review panels in the spring of 2000 led to the selection of 27 teams to be sponsored for Phase I beginning in fall 2000. NSF funding came mainly from DMII, but the Civil and Mechanical Systems (CMS) division of Engineering and the Computer and Information Science (CISE) directorate also provided significant support.

MIDTERM WORKSHOPS

The original SES solicitation called for a midterm workshop at the 2001 DMII grantees conference in Tampa, Florida. In a marathon session on January 7, 2001, all but three of the 27 teams took that opportunity to briefly describe the topic of their Phase I grants and their progress in the first few months. Dr. Ronald Rardin, who had by then replaced Dr. Seiford at DMII, chaired the workshop.

Probably the most visible message of all these presentations was the tremendous variety of issues and topics relevant to the SES theme. All involved with the SES initiative also took some pride in the obvious enthusiasm exhibited by so many of the investigators present.

Another SES midterm workshop was organized by the Editors of this book in connection with the national meeting of the Institute of Industrial Engineers in Dallas on May 19, 2001. Approximately ten of the Phase I SES teams offered more in-depth presentations on their plans and progress to that point in time. The organizers also used the session to set in motion the effort, which became the current book. Their resourcefulness and hard work deserves enormous credit and thanks.

PHASE II

With a second phase having been built into the structure of the original solicitation, the call for Phase II proposals in the fall of 2000 required only an email to the 27 teams eligible. Proposals were received in April of 2001, and peer reviewed at a panel held in that summer. A total of seven teams ultimately received three-year Phase II grants, again with sponsorship from DMII and substantial co-funding from partners in CMS and CISE. The main disappointment of the panel was that budget did not allow Phase II support of several other deserving teams.

CONCLUSIONS

At the time of this writing, Phase II SES grants are less than half way through their three-year durations, so much remains to be learned about the full impact of the Scalable Enterprise Systems program. However, it can already be said that dozens of Ph.D. students and a significant number of undergraduates (under the REU, research experience for undergraduates program) have been attracted to the study of systems for management of extended enterprises as a result of the SES alumni. A much larger number of scientific papers have been published, and related NSF grants have been awarded to several SES participants.

The real payoffs of the SES effort will be the genuine nuggets of discovery that begin the work of building the theory-base for research on enterprise systems envisioned when the program was undertaken. It is too early to know exactly what form they will take, but we are aware of several fundamentally new paradigms for dealing with SES challenges and a variety of software advances that may soon enrich commercial products.

It has been a great honor to have played some part in this worthy and fruitful effort.

Lawrence M. Seiford
Industrial & Operations Engineering
University of Michigan
NSF Program Director, 1997-2000

Ronald L. Rardin
Industrial Engineering
Purdue University
NSF Program Director 2000-2003

Overview of the Book

The growth in Internet and Information Technology (IT) has led to revolutionary changes in the way businesses are managed and operated. Globalisation has become the norm of the day and can be expected to increase many fold in future. Globalisation requires rapid growth and dramatic changes in the business and market environments. Enterprises must be scalable to support such changes and unprecedented growth. There is an immediate need for a single reference book that identifies the multiple dimensions of scalability and enterprise systems. Our current effort, we wish, will fulfil this need. We defined the following as our objectives in editing this book:

To create a single source of literature to address the emerging area of scalable enterprise systems.

To address the multi-disciplinary research needs stemming from engineering, computer science and business.

To address the design, modelling, analysis, implementation and experimentation needs of scalable enterprise systems.

To act as a springboard for much of the ongoing research in the field of Scalable Enterprise Systems.

We feel that this book will be of interest to both researchers and practitioners in the field of logistics, supply chain management, transportation, and enterprise integration. In addition, it will be of greatest interest to professionals in IT both at the detailed implementation level as well as strategic level.

This book addresses *six key ideas* related to scalable enterprise systems. Chapters 1, 2 and 3 address the *modelling aspects of enterprise systems*. Chapter 4 discusses *distributed control* of such systems. Chapter 5 discusses the *information systems modelling and implementation* of enterprise system with a transportation example. Given that enterprise systems produce and deal with large volumes of data, *data mining* is taken as a topic in Chapter 6. The need for a *research test-bed* is critical to study and analyse enterprise systems and this is the topic of Chapter 7. *Implementation level details* play a vital role in studying the scalability of large-scale systems. Chapters 8, 9 and 10 deal with several applied and implementation aspects, such as publish-subscribe middleware, auctions in the context of reverse logistics, and value net implementation. We briefly describe the contents of the various chapters in the following paragraphs.

In Chapter 1, Kamath, Dalal, Chaugule, Sivaraman, and Kolarik review several enterprise process modeling techniques. The authors argue that due to the distributed nature of businesses, which is a result of globalisation and IT advances, modelling, monitoring and control of business processes is critical.

They discuss several techniques for modeling with a common example. Finally they discuss the research aspects of a new process modeling framework and language.

In Chapter 2, Chung, Kwon and Pentland discuss MIDAS an XML-based scalable enterprise system. The authors discuss the design and implementation issues related to MIDAS to support collaboration among a network of suppliers engaged in the design and manufacturing of prototype parts. The discussion considers the following three key issues: (1) how to achieve the appropriate representation of process model, (2) how to model various constraints in design and manufacturing and to use them in order to find the best-of-practice process, and (3) how to maintain interdependencies within the network of vendors.

Darabi and Jafari discuss Finite State Automaton based modeling and analysis of supply chains in Chapter 3. The authors provide the framework for a new paradigm so that some aspects of business protocol defined across the supply chain need not be pre-defined. The authors propose inter and intra-business logic within the supply chain that provides for re-configurability and scalability across the network. The discussion is grounded in the theoretical aspects of discrete event control.

In Chapter 4 Prabhu discusses distributed control algorithms for scalable decision making from sensors-to-suppliers. In this chapter a review of distributed algorithms that have been developed for decision-making and control across enterprises is presented. The key barriers in scaling such decision making across multiple functions including production scheduling, maintenance scheduling, inventory management, and transportation are presented. Rest of the chapter concentrates on a class of algorithms based on distributed arrival-time control. The mathematical techniques required for modelling and analysis of these control systems are extensively reviewed.

Satapathy and Kumara discuss a multiagent infrastructure for transportation problem solving in Chapter 5. The authors present a multiagent computational approach to solve a procurement problem. In this approach, each transportation company is modeled as an entity consisting of a master and several slaves. A master represents the task negotiator and task allocator for a company. A slave represents a vehicle to which the tasks are allocated by its master. A transportation problem containing several stochastic parameters for real-time decision making are considered. Extensive treatment on the development of BDI (Belief-Desire-Intention) logical model of an agent is given. The emphasis in this chapter is on the information infrastructure for information exchange between the agents.

In Chapter 6, Olafsson discusses improving scalability of E-Commerce systems with knowledge discovery. In particular the focus is on improving scalability via dimensionality reduction and improving the information view experienced by each user. To address these issues, several common data mining problems, including feature selection, clustering, classification, and association rule discovery are presented. In addition, Olafsson presents several scalable methods and algorithms that address each of data mining methods.

Zhao, Ball and Chen discuss a scalable supply chain infrastructure research test-bed. In Chapter 7, ERP and SCM are collectively referred to as supply chain infrastructures (SCIs). The authors provide a review of various SCI architectures, describe in detail a scalable SCI research test-bed that they have developed and define an integration architecture and model. In this test-bed, middleware and a collaboration-based architecture are used to achieve "application scalability". An advanced available to promise (ATP) model and a simulation model are embedded within the test-bed to achieve "business scalability".

Opyrchal and Prakash describe Publish-Subscribe middleware in Chapter 8. Publish-Subscribe middleware has become very popular in building large-scale distributed systems. Supply-chain management, financial applications, e-commerce applications, workflow systems, and information distribution systems potentially can benefit from such a paradigm. In this chapter the authors describe methods ranging from simple subject-based implementations to complex content-based prototypes. They describe the major issues involved in designing a publish-subscribe system, such as: (1) event matching, (2) event delivery, and (3) security. The authors also discuss current research issues in event multicast and security in content-based systems.

In Chapter 9, Ryan, Min and Olafsson present the details on reverse logistics. The authors report an experimental study on the scalability enhancement for reverse logistics in e-commerce. The authors constructed an experimental prototype auction system in the context of reverse logistics for electronics products. The discussion in the chapter concentrates on the methods of improving scalability in online auctions include limiting the number of bidding opportunities, providing price information to users, and recommending auctions.

Dodd, Kumara and Kim, in Chapter 10 discuss an implementation framework for Value nets. The value net creates a network of suppliers, and distributors such that the product becomes customer centric. In order to implement the value net model, information processing and distribution need to occur in real time. The authors discuss the detailed implementation of a software agent technology based value nets implementation. In this chapter, the authors report research that adopts agent technology to handle all real-time decision processes, making the value net model a complex multi-agent network of decision makers. For the agents to properly coordinate their respective activities the authors develop Valcomm, a Web-based multi-agent language grounded in the XML and Java. A detailed discussion of Valcomm is included.

Acknowledgements

We embarked on this venture of putting a book together on this exciting and new area of extended enterprise integration during the summer of 2001. We wish to acknowledge the National Science Foundation for supporting us on our projects (DMI-9908267, DMI-0075572, DMI-075584, DMI-0075588) and starting us on this road of research. Vittal Prabhu would also like to acknowledge the support of Ben Franklin Technology Partners through the Center of Excellence grant to the Center for Manufacturing Enterprise Integration at Penn State University. Soundar Kumara would also like to acknowledge the support from DARPA under the UltraLog program (MDA972-1-1-0038). This book would not have been a success without NSF's support. We wish to thank our graduate students who helped us along the way in conducting this research, especially Atul Pandey, Karthik Ayodhiramanujan, and Muthukumar Somasundaram. Our students are our best teachers. We wish to thank our families (Radhika Prabhu; Nalini Krishnankutty, and Mahima Kumara; Surekha Kamath, Nitin Kamath, and Nikhil Kamath) for their patience and support throughout this endeavor.

Vittal Prabhu
Soundar Kumara
Manjunath Kamath

Chapter 1

A REVIEW OF ENTERPRISE PROCESS MODELLING TECHNIQUES

Manjunath Kamath[1], Nikunj P. Dalal[1], Amit Chaugule[2], Eswar Sivaraman[1], and William J. Kolarik[1]

[1]Oklahoma State University, Stillwater, OK 74078, USA; [2]Borg Warner Morse TEC, Ithaca, NY 14850, USA.

Abstract: Enterprises are becoming more distributed and collaborative as a result of rapid advances in telecommunications and Web-based technologies. The changing nature of enterprises has, in turn, led to research and development towards new techniques and products for the modelling, specification, monitoring, and operational control of business processes and enterprise functions. This chapter overviews several prominent enterprise process modeling techniques, illustrates their key features with the help of a common example, and points to the need for further process-modelling research to support the design/redesign and improvement of next-generation enterprises. This chapter ends with a brief account of recent research on the development of a new process-modelling framework and language.

Key words: Enterprise modelling, process modelling, framework, modelling language, Petri nets, XML.

1. INTRODUCTION

Today's global markets and rapid advances in Internet and Web-based technologies have enabled newer enterprise structures such as supply chains and extended/virtual enterprises. As a result, business processes have become more distributed and now require a higher degree of collaboration with customers and suppliers to satisfy business objectives of effective and efficient product/service delivery. Process modelling is useful in understanding, analysing, designing/redesigning, and controlling an enterprise's business processes. There is a need to modernize the process-modelling arsenal in order to address the requirements of emerging enterprise structures. The objective of this chapter is to give the reader a comprehensive overview of several prominent enterprise modelling techniques. Potential areas within process modelling that need new research are identified, and recent research that addresses some of these needs is described.

1.1 Enterprise modelling

An enterprise is a complex entity composed of people and processes, producing products or services for customers. According to Vernadat (1996),

> "Enterprise modelling is concerned with the representation and specification of the various aspects of an enterprise's operations, namely, *functional* aspects that describe what things are to be done and in which order; *informational* aspects that describe which objects are used or processed; *resource* aspects that describe what or who performs things and according to which policy; and *organisational* aspects that describe the organisational structure within which things are to be done."

Enterprise modelling has been an active area of research since the mid-80s when industry realised that more systematic approaches were needed to design enterprises and support them throughout their life cycle (Bernus and Nemes, 1996). Recent standardization efforts indicate the importance of enterprise modelling efforts in the design of integrated enterprises (GERAM, 1999).

1.1.1 The GERAM Framework

The Generalized Enterprise Reference Architecture and Methodology (GERAM) framework was developed by the IFAC/IFIP Task Force on "Architectures for Enterprise Integration" (GERAM, 1999). The scope of GERAM extends to all the knowledge (methods, models, and tools) needed to build and maintain the integrated enterprise, be it a part of a traditional enterprise, a single enterprise, or a network of enterprises (virtual enterprises, extended enterprises, and supply chains). GERAM is not a modelling technique or reference architecture in itself, and it does not impose any particular set of methods, models, or tools. It provides a standard set of enterprise-related generic concepts for information integration and engineering, and defines the criteria to be satisfied by any particular tool or method. Also, GERAM clearly identifies the role played by enterprise models with respect to enterprise engineering and integration.

1.1.2 Enterprise process modelling

Within the context of enterprise modelling, business process modelling and functional modelling play a central role. Business process modelling emphasizes *enterprise behaviour*, which concerns the order in which things are to be executed, i.e., the flow of control. Functional modelling emphasizes *enterprise activities*, which concern the things to be done, i.e., the transformational or decision-making functions (Vernadat, 1996). Functional models typically include the specification of inputs and outputs, which are usually material and data/information requirements for the various activities. Resources required to perform the activities and organizational units responsible are sometimes shown within these models.

According to GERAM (1999),

> "Process modelling is the activity that results in various models of management & control as well as the service and production processes, and their relationships to the resources, organisation, products etc. of the enterprise. Process modelling allows us to represent the operation of enterprise entities and entity types in all their aspects: functional, behaviour, information, resources, and organisation. This provides for operational use of the models for decision support by evaluating operational alternatives and for model-driven operation control and monitoring."

The focus of this chapter is on modelling enterprise functionality and behaviour within the context of enterprise/process improvement and redesign. The chapter also hopes to stimulate new research leading to the development of enhanced process-modelling capabilities, which offer accurate, precise, and timely decision support for improvement efforts within dynamic business/technical environments.

1.2 Significance of process modelling in the context of enterprise systems

In recent years, many large- and medium-sized enterprises have implemented enterprise systems (also called enterprise resource planning or ERP systems) with the goals of improving and/or reengineering their business processes to achieve competitive advantage. Despite some successes, reports of failed implementations, cost and time overruns, and scalability problems abound. A careful examination of field experiences reveals that a common source of difficulty lies in understanding and implementing compatible **business processes**. According to Davenport (1998), an ERP system, by its very nature, imposes its own logic on the operations of an enterprise, which may conflict with the logic of the enterprise. Keller and Detering (1996) of SAP AG state "As many companies get ready to implement standard software, they encounter the problem of how to simplify and model the enormous complexity of their business processes." Business processes hold the key to the future success of enterprise systems; they operate on business data to accomplish the enterprise's objectives. However, enterprise systems are quite inflexible with respect to process specification and implementation (Davenport, 1998) and are very difficult to change and extend because of their complex, proprietary application program interfaces (APIs) and complicated database schemata – a far cry indeed from proposed open standards of electronic commerce (Radding, 1999). Process models that can be easily created, modified, and analysed will simplify the task of extending enterprise systems since complex implementation details can be hidden.

Current trends in the evolution of enterprise systems include the extension of ERP systems to the Web environment to support the extended enterprise, which includes suppliers and customers, with modules such as Operations, Repairs and Maintenance (ORM), Customer Relationship Management (CRM), and Supply Chain Management (SCM). The next emerging phase in the evolution of enterprise systems exploits collaborative relationships among trading partners. Hence, there is an increased interest in techniques, existing and new, for modelling, specification, implementation, maintenance, and performance improvement of business processes (Knowles, 1999).

1.3 Outline of the chapter

The rest of this chapter is organized as follows. Section 2 presents an overview of several prominent enterprise modelling efforts and focuses on their process modelling capabilities. A purchasing system example adapted from Alter (2002) is used throughout this section to illustrate the process modelling features of the various techniques. Section 3 discusses the areas within process modelling that need new research and development. Section 4 summarizes the progress made on a recent research effort on the development of a new process-modelling framework and language, and Section 5 concludes the chapter.

2. REVIEW OF PROCESS MODELLING TECHNIQUES

Enterprise modelling has been an active area of research and development for nearly two decades. This section covers several prominent enterprise modelling efforts and emphasizes related process modelling techniques. The following purchasing system example, adapted from Alter (2002) is used to illustrate the process-modelling features of the enterprise modelling efforts reviewed.

When a material requirement request arrives from Material Planning, the purchasing system prepares a purchase order and sends it to the supplier. A copy of the purchase order is stored in a shared database. If the material received matches the purchase order completely, then Receiving accepts it and adds a receipt confirmation to the shared database. When an invoice arrives, Accounts Payable uses the purchase order to verify that the material was ordered and the receipt confirmation to verify that the supplier delivered it before making the payment.

This simple, intuitive example is used to guide our discussion of various techniques. A preliminary version of the review in this section was presented in Kamath *et al.* (2001).

2.1 Data flow diagrams (DFD)

Data flow diagrams (DFDs) focus on the logical flow of data between different processes in a system rather than on the flow of control (Alter, 2002; Dao and Rodjak, 1991; Whitten *et al.*, 2001). DFD was the key modelling tool within the classical Structured Analysis and Design Technique (SADT), which was popular in the 1980s, for developing information systems (Colquhoun *et al.*, 1993; Whitten *et al.*, 2001).

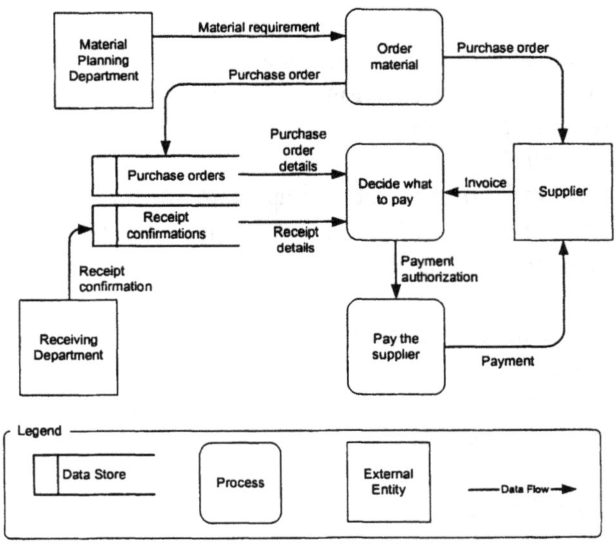

Figure 1. First-level DFD of the Purchasing System

There are four basic symbols in a DFD. In Gane & Sarson's DFD notation (Whitten *et al.*, 2001), a square box represents an external source or destination (sink) of data. A rounded rectangle represents a process. A data store is represented by an open-ended rectangle. A directed, labelled arrow represents a data flow. A DFD is usually drawn as a set of diagrams. The highest-level diagram is called the *context diagram*, and it defines the scope of the system by showing the sources and destinations of data used and generated by the system. The context diagram has only one process – the system being modelled. This diagram is then decomposed until the level of detail required is reached. The general principle in data flow diagramming is that a system can be decomposed into subsystems, which can further be decomposed into lower level subsystems, and so on. Each subsystem represents a process or activity in which data is processed. Data enters the system from the environment; data flows between processes within the system; and data is produced as output from the system. Arcs are labelled to indicate what flows between the system components. The first level DFD (an expansion of the context level diagram) of the purchasing system is shown in Figure 1. Strictly speaking, the "Receipt confirmation" data flow from the external entity "Receiving Department" to the data store "Receipt confirmations" must pass through a process, which has been omitted here for simplicity.

DFDs are probably the simplest, and yet, one of the most powerful and useful techniques available to the system analyst for modelling processes. Because of the small set of vocabulary and rules, and the focus on functional modelling, a typical enterprise user can readily understand a DFD. Breaking down a process helps the enterprise user clearly understand the detailed requirements of its different functions. One key benefit of DFDs is that they promote awareness of information sharing among different functions within an enterprise (Burch, 1992).

Timing (or control flow) is not an important consideration in a DFD, and as a result, it is not possible to completely and unambiguously capture the process behaviour within a DFD. Process numbering, if used, could mislead the reader into assuming sequence dependence among the processes. With the use of a single database, there is perhaps no need to represent different data stores in a DFD and clutter the diagram. Millet (1999) suggests the replacement of several data stores by a few databases in a DFD, in order to partly alleviate this concern of clutter.

2.2 IDEF0/IDEF3

IDEF (ICAM **DEF**INITION) was a series of techniques developed by the U.S. Air Force Program for Integrated Computer Aided Manufacturing (ICAM) in the 1970s (Bravoco and Yadav, 1985; Colquhoun *et al.*, 1993; www.IDEF.com). The techniques developed include IDEF0 for modelling functions, activities, or processes resulting in a *functional model*; IDEF1 for information modelling; and IDEF2 for modelling system dynamics. IDEF0 evolved from the Structured Analysis and Design Technique (SADT), which is a process-centred technique that relies heavily on DFDs (Bravoco and Yadav, 1985; Colquhoun *et al.*, 1993).

2.2.1 IDEF0

The basic building block of an IDEF0 model is an activity box that provides descriptive information about the activity that is taking place and data or object interfaces represented by arrows on potentially all four sides of the box (see Figure 2). The arrows entering the left side represent the inputs needed; those leaving the right side, the outputs produced; those on the top, the controls, requirements or constraints that must be satisfied; and those at the bottom, the mechanism or resources needed for the activity. To represent a complete process, these boxes can be linked with each other. For example, the output from one process block can be the input, control, or both, of the next activity block. This can be seen in Figure 2, where the purchase order becomes an input as well as a control for the second activity. The result of applying IDEF0 to a system is a model that consists of a hierarchical series of diagrams and text referenced to each other (Colquhoun *et al.*, 1993).

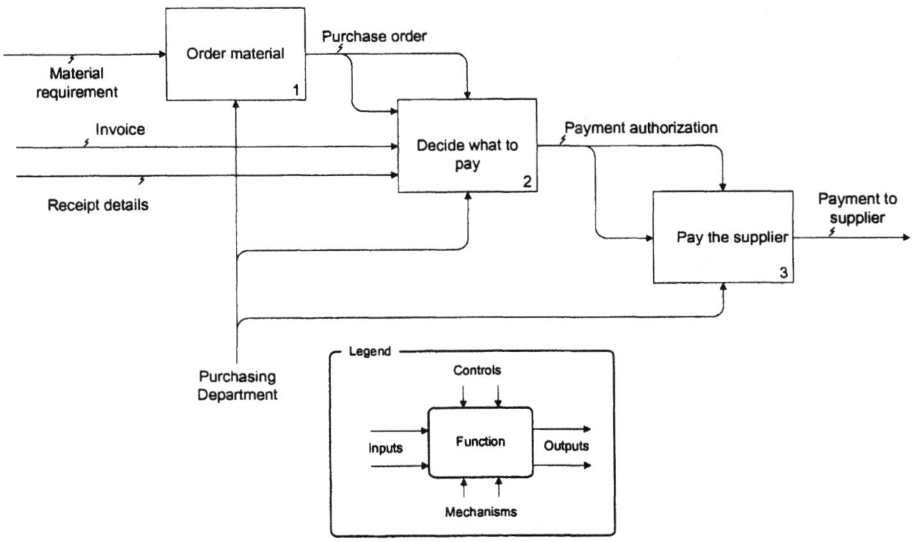

Figure 2. IDEF0 Representation of the Purchasing System

The strengths of IDEF0 are its clearly defined syntax, controlled exposition of detail, and a limited set of graphical symbols. Time is not included in IDEF0. A common misinterpretation of an IDEF0 diagram is that it represents a "sequence of activities." The modeller can arrange the activity blocks, in the proper sequence, from left to right, so as to represent the flow of control. However, it is left to the discretion of the user to model a sequence or not. Hence, we could say that IDEF0 is primarily a technique for modelling enterprise functionality. Enterprise behavior can be captured using the IDEF3

Process Description Capture Method (Benjamin *et al.*, 2000; Whitman *et al.*, 1997; www.idef.com).

2.2.2 IDEF3

IDEF3 is a structured method for capturing details of a process and has two distinct representations of the process within the boundary of the process. This boundary is termed a scenario. A process can be represented by a sequence of activities using a "process description diagram," (see Figure 3) in which each activity is represented by a rectangle termed a "unit of behaviour (UOB)." The process sequence is represented by a number of UOBs connected together by directed arcs to represent control flow. Junctions explicitly describe the logic (AND, XOR, etc.) of multiple arcs either merging or diverging. Thus precedence, causality, and logical relationships within a scenario are captured in an unambiguous manner in this process description diagram. The second representation in IDEF3 (not shown here) is termed as the Object State Transition Network (OSTN) diagram. The OSTN diagram describes what happens to an object as it passes through a sequence of activities. This representation focuses on objects involved in the process and their state changes in a single scenario or multiple scenarios.

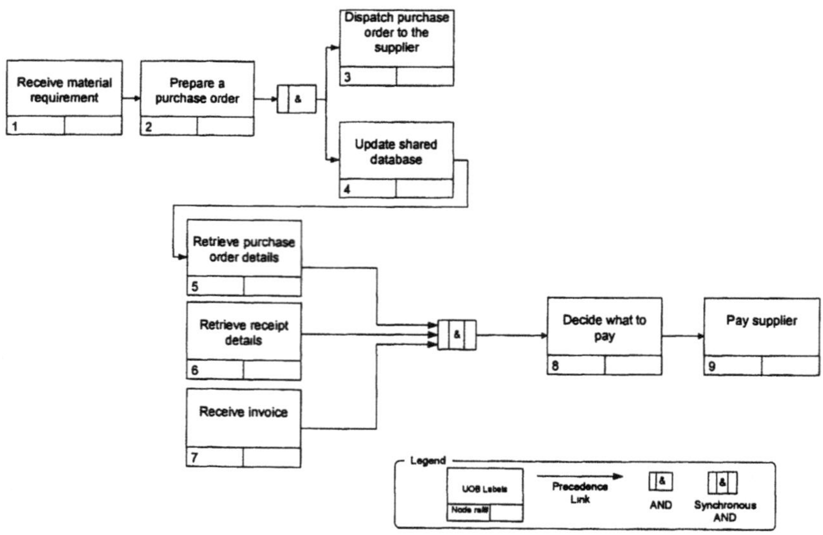

Figure 3. IDEF3 Description of the Purchasing System

IDEF3 has a clearly defined methodology for describing the details of a process. An IDEF3 description clearly captures the logic of process executions, and hence, can be used to support simulation of alternative process implementations (Benjamin *et al.*, 2000; Whitman *et al.*, 1997). By separating the process description into a process description diagram and an object centric diagram, one can construct multiple views of the same process. Multiple views are beneficial when users with different backgrounds want to have a description of the same process. However, an IDEF3 description should be supplemented by an IDEF0 representation to capture complete details of a process, such as inputs, outputs, and resource requirements of activities.

2.3 CIMOSA

CIMOSA (Open System Architecture for CIM) is the outcome of efforts undertaken in the late 1980s under the European Strategic Program for Research and Development in Information Technology (ESPRIT) project AMICE (Kosanke, 1995). CIMOSA includes an enterprise modelling language, which supports a model-driven enterprise in all life cycle phases. A well-defined framework, architecture, and methodology support the modelling language. As shown in Figure 4, the CIMOSA framework has three dimensions. Along the enterprise derivation dimension, the modelling process starts with a requirements definition model followed by the design specification model and finally, the implementation description model. Along the view dimension, there are four different views, each representing one particular aspect of the enterprise. The model creation dimension represents three instances of a model, namely, *generic, partial,* and *particular*. A *generic* model is applicable across a broad domain whereas the *particular* model is for a specific set of business processes in a specific enterprise (Zelm *et al.*, 1995).

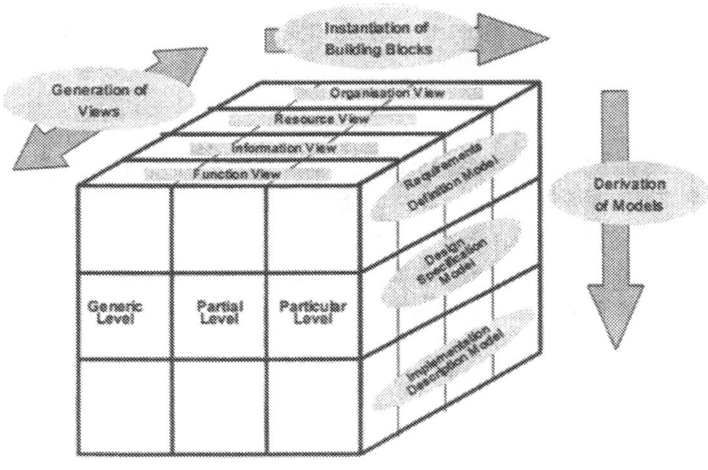

Figure 4. CIMOSA Modelling Framework (adapted from Kosanke (1995))

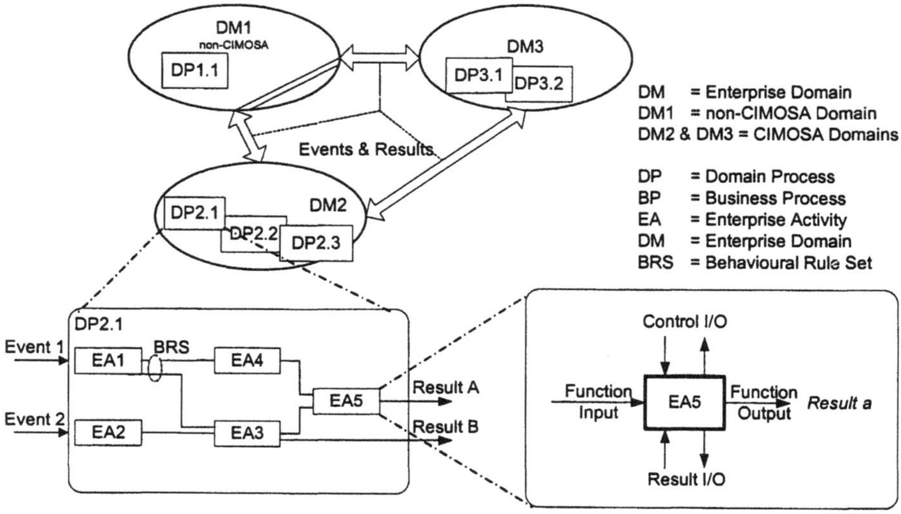

Figure 5. CIMOSA Process Modelling Constructs (adapted from Kosanke (1995))

A CIMOSA business process model (Zelm *et al.*, 1995), as shown in Figure 5, consists of a hierarchy of enterprise functions with Domain Processes (DPs) at the highest level. DPs are broken down into a set of Business Processes (BPs) at the next level, which are further broken down into Enterprise Activities (EAs). The BPs and EAs are usually linked by a set of behavioral rules, which represent control flow. An EA is represented as a set of Functional Operations (FOs). Within the CIMOSA modelling framework, FO represents an elementary unit of work, which can be executed by only one Functional Entity (resource). CIMOSA defines three types of FOs, one for manual operations, one for machine operations, and one for program (software) operations (Didic, 1994). Figure 6 shows the CIMOSA model of the purchasing system. It is a model at the highest level showing domain processes, external events, and internal events.

The main goal of CIMOSA is to support an event-driven, process-oriented modelling of manufacturing enterprises to support both the engineering and operation of enterprises. It clearly distinguishes between model engineering and model use, and defines a formal engineering release (of models) for operational support in the enterprise life cycle. The link between process modelling and model execution provides flexibility to respond to changes in the enterprise processes.

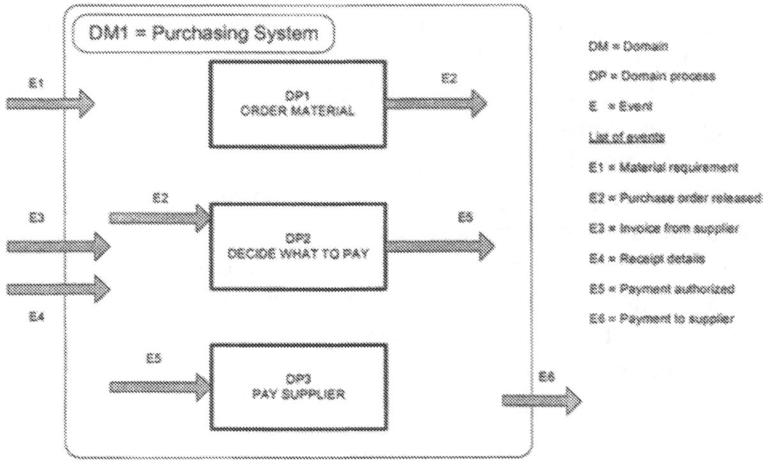

Figure 6. Domain Processes of the Purchasing System

2.4 ARIS and its event-based process chain method

Architecture for Integrated Information Systems (ARIS) (Scheer, 1992), defines an architecture for a complete, enterprise-wide information system. ARIS defines three views, namely, the functional view, the information view, and the organization view. These views are defined in all life cycle phases of the information system, namely, requirements definition, design specification, and implementation description. All three views are treated in isolation, and the relationships between the three views are represented by a control view. Business processes are described by process chain diagrams, which are event-driven. Process improvement can be done after analysing the process chain using decision-support tools. The changes/improvements in the processes can directly change the relevant information in the repository, since the process view is connected with all the other views (information, function, etc.) by the ERM (entity-relationship model). ARIS focuses on the analysis and requirements definition phases during the design of information systems, and not on the execution of business processes.

Recently, ARIS's event-driven process chain has been adapted for use within SAP's R/3 system. This is described next.

2.5 Event-driven process chain (EPC) method in SAP R/3

SAP AG (www.sap.com), a leading provider of enterprise software solutions, has developed the R/3 reference model, which is a business process model for the processes supported by the R/3 enterprise software (Keller and Detering, 1996). The SAP R/3 reference model is primarily intended for identifying possibilities for "optimising" the routines and procedures in a company. The most important addition to the R/3 model is the control logic – "when something should be done and in what order." Process descriptions are captured using *Event Driven Process Chains* (EPCs).

EPCs are made up of active components that do things (functions) and passive components (inputs, outputs, organizational units, etc.) that only come into play in response to certain business situations. The events, which are represented by hexagons, act as triggers for the functions. Identifying events in a process is not an easy task, but once they are identified, the resulting EPC acts as a control-flow logic diagram for the organization. A complex business process can be defined by connecting events with the functions using logical operators. The focus of the R/3 reference model is complete chains of processes in response to business events.

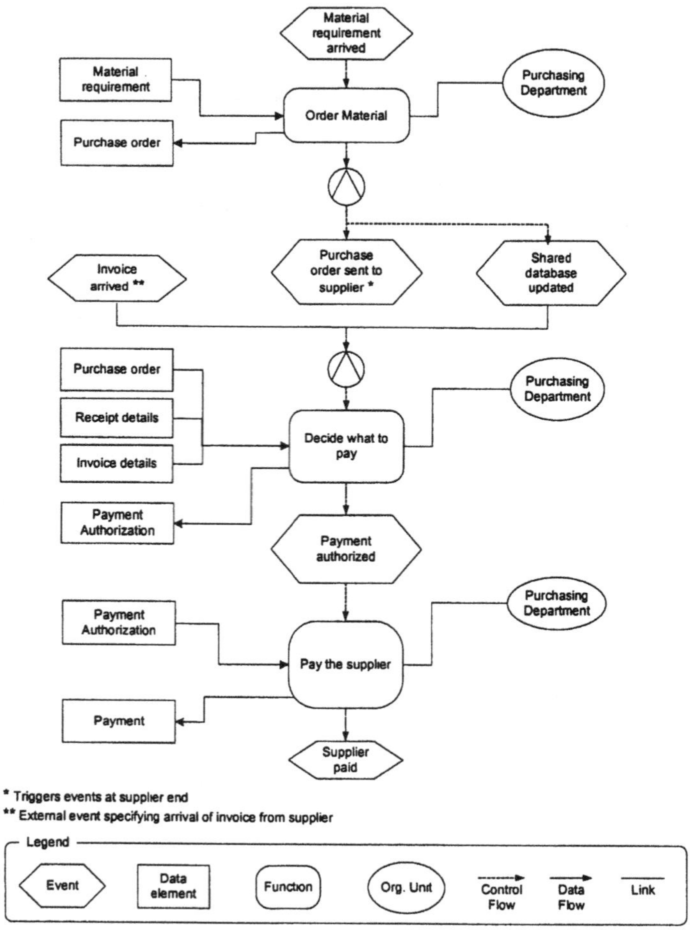

Figure 7. EPC Diagram of the Purchasing System

In addition to describing the control flow, an EPC diagram has constructs to show the information/data necessary towards the development of an information system. Departments involved in carrying out the various functions in a process chain are also shown in the EPC diagram. The process selection matrix facilitates process modelling by providing the user with standard business processes and their scenarios. The EPC model of the purchasing system is shown in Figure 7.

A distinguishing feature of the EPC diagram is the explicit modelling of events. This becomes important because one of the main functions of the EPC diagram is to specify the transaction processing logic within SAP's ERP system in response to business events. Another unique aspect is the creation of two simpler diagrams, namely, the EPC structure diagram or the lean EPC, and the EPC assignment diagram, both derived from an EPC diagram, to confront the problem of information overload (Keller and Detering, 1996). The lean EPC, by focusing on control flows, allows the user to analyse business processes independent of organizational restrictions. On the other hand, the EPC assignment diagram can be used to handle organizational and information-related questions.

2.6 Integrated enterprise modelling (IEM) method

The goal of the IEM method is to develop a unified model and to represent the different aspects of a manufacturing enterprise as views of the unified model (Mertins *et al.*, 1992; Mertins *et al.*, 1997). IEM uses the object-oriented modelling approach and exploits its three main features – encapsulation (the close relation between behaviour and data of an object), inheritance, and the class concept. The kernel or the meta-model of IEM comprises two main elements, a generic block called generic activity model (GAM), and a set of three operands or objects of a manufacturing enterprise. The two main views supported are the function view and the information view. According to Mertins *et al.* (1992) the three fundamental objects of a manufacturing enterprise are: *product*, *order*, and *resource*. This classification in IEM is derived from the three main questions within a manufacturing enterprise, what to order – *product*, how many – *order*, and what is required to manufacture the product – *resources* (Mertins *et al.*, 1992).

The basic building block of IEM's function view, GAM, is similar to that of the IDEF0 technique (see Figure 8). First, the *action* is described using a rectangular block. The input and the output objects are identified in the next step and the *function* block is created. The trigger (order) and the resources, when added to the function block, result in a complete *activity* block. The functions can be linked together to view the network of functions in an enterprise. The complete activity blocks can be linked together to yield a particular model for an enterprise. Figure 8 shows the IEM model of the purchasing system. The model shows both the flow of control and the flow of objects that are manipulated by the activities. Only limited information is available to support model creation using the IEM constructs.

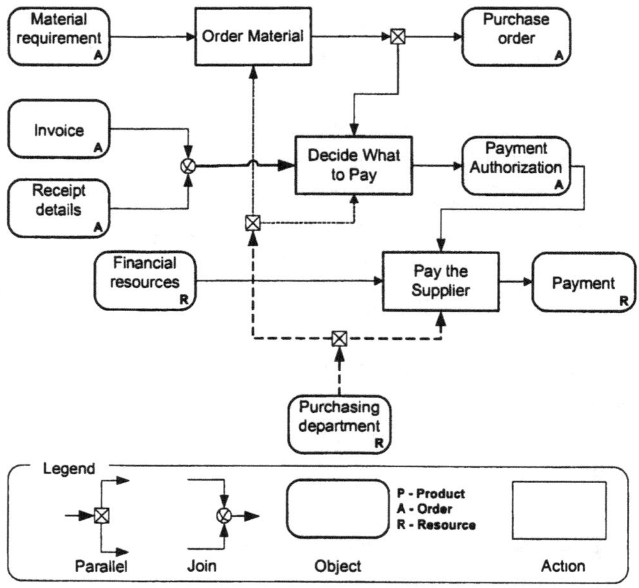

Figure 8. IEM Model of the Purchasing System

The IEM approach assumes that users are familiar with object-oriented concepts, because creation of additional subclasses of the three basic classes may become necessary. IEM also assumes that views cannot be predefined, but can be defined as the model is being developed depending on the perspective required. However, if the modelling constructs do not support the required views, then the usefulness of the model created is limited.

2.7 TOronto virtual enterprise (TOVE) method

TOVE is an enterprise modelling language with a deductive capability (Fox *et al.*, 1993; Gruniger and Fox, 1995). One of TOVE's goals is to provide a shared terminology for the enterprise that every application can jointly understand and use. The meaning of each term is defined in a precise manner using first-order logic. The semantics are implemented in a set of Prolog axioms that enable TOVE to automatically deduce the answer to many "common sense" questions about the enterprise. TOVE also defines a symbology for depicting a term or a concept in a graphical context.

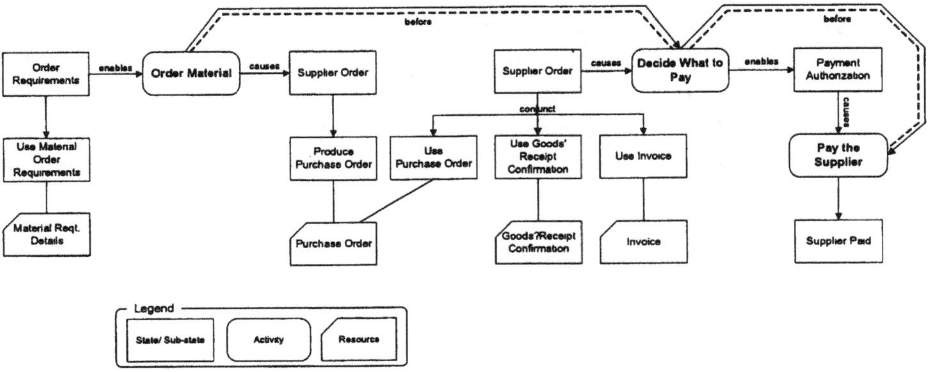

Figure 9. TOVE Activity Cluster Diagram of the Purchasing System

TOVE uses a single construct called an "Activity". An activity can be performed when something enables it. When an activity is performed, something is "caused." An activity and its "enabling" and "caused" states are called an activity cluster. Different types of states are defined; they are broadly classified into terminal and non-terminal states. "Consume" and "Release" are examples of terminal states. Non-terminal states include Boolean combinations like exclusive-OR, inclusive-OR, AND, etc. TOVE also provides the ability to model timing information and resource requirements. The strength of TOVE is that it has an underlying formal theory base. The graphical representation is limited to an activity cluster diagram that relates activities with each other. This is illustrated in Figure 9, which shows the activity cluster diagram for the purchasing system. Concepts of views and life cycle are not defined in TOVE.

2.8 Baan's dynamic enterprise modelling (DEM) technique

Baan (www.baan.com), like SAP AG, is a major enterprise software solutions provider. Baan introduced the concept of Dynamic Enterprise Modelling (DEM) in 1996. By using DEM, users can create graphical models of their organisation and process flow diagrams of individual procedures. Four models comprise the DEM approach (Cuenca et al. 1999; van der Rijst, 1997). They are the *Business Control Model*, which describes the primary processes within an enterprise and the way these are controlled; the *Function Model*, which is a set of all relevant functions that exist within an enterprise in the form of a hierarchical representation of major, mega, and main functions; the *Process Model*, which "describes how the functions represented in both the Function Model and Business Control Model are accomplished by defining the subsequent activities and events" (van der Rijst, 1997); and the

Organization Model, which shows the structure of the organization. Process modelling within DEM is accomplished using Petri nets (van der Rijst 1997). A Petri net is a bipartite graph that comprises two types of nodes called places and transitions. Places are connected to transitions and transitions to places by directed arcs (David and Alla 1991; Zurowski and Zhou, 1995). Section 4.1 contains additional information on Petri nets.

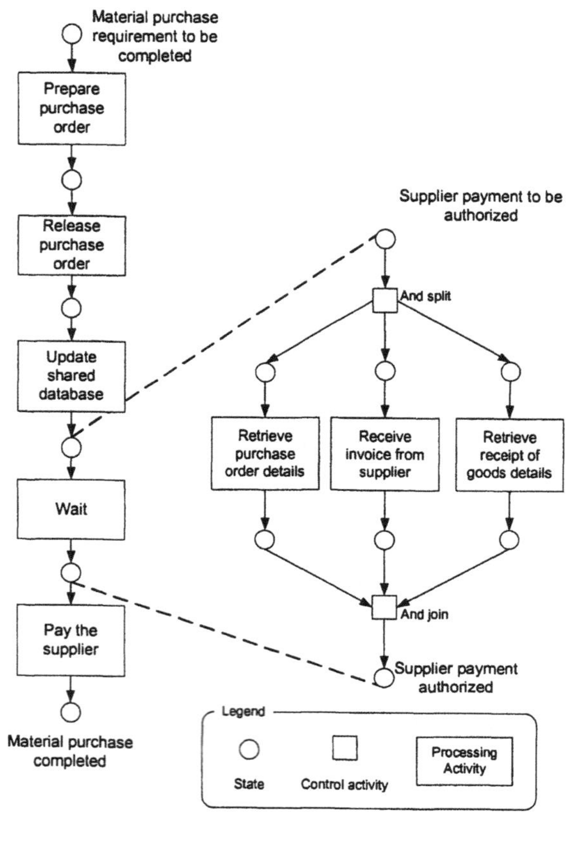

Figure 10. DEM Process Model of the Purchasing System

The Petri net constructs used within DEM are shown in Figure 10. DEM methodology provides guidelines to the user for modeling a business process. The modeling approach is event-based with an emphasis on control flow. Circles (places) represent states that contain job tokens, squares (immediate transitions) model control activities that route job tokens through a process, and rectangles (timed transitions) model manual or automated activities. The control activities are similar to logical connectors such as inclusive-OR, exclusive-OR, and AND. The type of the logical connection has to be mentioned on the graphical representation. Explicit modeling of states is similar to the detailed modeling of events with the SAP EPC (see Section

2.5). States in the Process Model link the detailed Process Model to points within the Business Control Model. The modeling of pure information (data) flows is not supported within DEM (van der Rijst, 1997).

2.9 Unified Modelling Language (UML)

Unified Modelling Language (UML) was adopted in 1997 as a standard by the Object Management Group (OMG), a consortium of over 800 software vendors and customers. "The UML is a language for specifying, visualizing, constructing, and documenting the artefacts of software systems, as well as for business modelling and non-software systems (www.omg.org)." One of the main objectives of UML is to provide a visual tool that has platform independence. UML offers several groups of diagrams to model a software system from different perspectives. A detailed explanation of these diagrams is given in the OMG UML specification (Booch *et al.*, 1999).

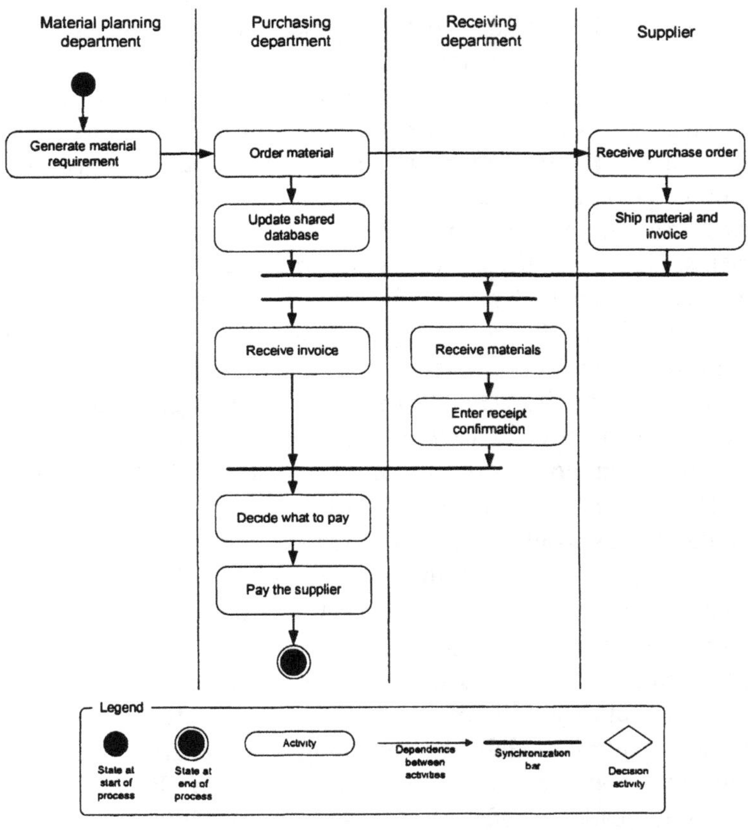

Figure 11. UML Activity Diagram of the Purchasing System

Of relevance to our discussion are activity diagrams, which could be used to graphically depict the control flow through activities of a business process. Schader and Kortahus (1998) present an approach that uses activity diagrams to represent business processes. Figure 11 shows an activity diagram model of the purchasing system. Clearly, the emphasis of activity diagrams is on control flow. However, the various diagrams within UML are interrelated, and more work is needed to better understand the role of the other UML diagrams in the context of enterprise-process modelling. Also, UML diagrams may not be appropriate for an enterprise user who is typically interested in describing process details. UML diagrams may be more appropriate for an information system analyst for developing software specifications based on the process details provided by an enterprise user.

2.10 Workflow management

Another facet of enterprise functionality and behavior is workflow management, which includes both business process specification and automated execution of business procedures. Workflow management emphasizes the increased role that information systems have come to play in today's businesses (Sheth *et al.*, 1999). Workflow Management involves two phases – (i) the *modelling phase* that abstracts from business procedures and defines computer-implementable workflow specifications, and (ii) the *execution phase* that executes instances of the workflows to meet business requirements, and both these phases are managed and coordinated by a Workflow Management System (WfMS) (Hollingsworth, 1995). The growing interest in workflow management has led to the creation of many software products, each championed by its own set of vendors, and file-formats, etc. In an effort to standardize terminology, and enable inter-operability between different vendors, the Workflow Management Coalition (WfMC), an international consortium founded in 1993, has established standard reference architectures that software vendors have to conform to, and basic terminology as would be essential for any process modelling language, amenable to computerization (Hollingsworth, 1995; WfMC, 1999).

Petri nets (see Section 2.8 and Section 4.1) have become the dominant technique for modelling workflow processes (Janssens *et al.*, 2000). There are many reasons for the popularity of Petri nets for modelling workflow. Examples are combination of mathematical foundation and graphical representation, explicit modelling of enabling and execution of tasks, support for concurrent, cooperative processes and the availability of analysis techniques. The reader is referred to Janssens *et al.* (2000) for a summary of the various Petri net classes used for workflow modelling.

2.11 Evaluation of process modelling techniques – a summary

All of the techniques reviewed in this section provide a basic set of constructs to model enterprise functionality. DFD and IDEF0 emphasize the flow of data, which was the focus of information systems development approaches in the 1970s and 1980s when these techniques were developed. In the late 1980s the focus shifted to data-centred approaches based on entity-relationship (ER) models. In the early 1990s, business process engineering became the driver for enterprise information systems development. As a result, improving the ability of process-modelling techniques to model enterprise behaviour became important. The emphasis within process modelling shifted from data flow to control flow. The following techniques – EPC (ARIS and SAP), Baan-DEM, CIMOSA, IEM, TOVE, and UML Activity Diagrams, provide detailed constructs to model control flow. In the IDEF approach, behaviour modelling is handled by a separate technique, namely, IDEF3.

Modelling of control flow, time (activity durations), and resource requirements are essential to support any formal quantitative analysis needed for business process improvement initiatives. It appears that perhaps with the possible exception of CIMOSA, none of the techniques attempts to develop process models that can directly link description with formal qualitative or quantitative *analyses* in an integrated manner. All of the techniques support the description of enterprise operations with varying degrees of detail. Linking this description to formal analysis is typically not an objective. Even within CIMOSA, the details pertaining to the use of the process model in enterprise execution and control are not completely defined (Didic, 1994). Integrating descriptive modelling and quantitative/qualitative analyses is only one of the challenges facing the process-modelling field. The newer enterprise structures have brought about a host of other challenges as described in the next section.

3. MODELLING NEXT-GENERATION ENTERPRISES

In this section we discuss the challenges faced while trying to model the next-generation of enterprises, which are distributed and Web-based, and operate in a collaborative environment. All of the big five ERP vendors, JD Edwards, Baan, Oracle, Peoplesoft, and SAP, have Web-enabled their software. By extending existing ERP systems to support Internet-based supply chain management (SCM), Internet-based customer relations management

(CRM) and electronic commerce, organizations can leverage their investment in their existing ERP solutions while speeding up efforts to develop electronic commerce capabilities. Hence, it is not surprising that information technology professionals have strongly endorsed efforts to extend ERP to a Web environment. "The Web is exactly what we want, and beyond the Web, the whole e-commerce thing," states Dan Spaulding, director of Halliburton, a worldwide corporation running a single instance of SAP (Radding, 1999).

In addition to the newer enterprise structures mentioned above, other issues such as integrating process description and analysis, bridging the gap between business and engineering processes, and supporting modern cost and resource management concepts like activity based management (ABM), continue to pose new challenges in the field of enterprise process modelling (Dalal *et al.*, 2003).

3.1 Modelling and incorporating distributed computing

Existing process modelling techniques do not explicitly incorporate the distributed computing paradigm of the Internet and are hence unable to capture the reality of newer enterprise systems that operate in an Internet-based environment. There is a need for techniques that explicitly address the distributed infrastructure of the Internet. Also, users of large-scale and medium-scale ERP systems are spread geographically, making it imperative that users be able to create, modify, and analyse process models from any location on the globe. According to Sarjoughian *et al.* (2000), current collaborative modelling systems do not "allow multiple distinct modelling sessions to collaborate with one another at a higher level." With next-generation extended enterprises and virtual corporations, the need for distributed access to process models will be even greater.

3.2 Integrating process description and analysis

Process modelling techniques primarily utilize notation, terminology, and theory from the field of software and information systems development. On the other hand, techniques and models such as simulation, queueing, and optimization, used for process analysis and improvement belong to the field of operations research. There is a need to establish a common theoretical base that can support the descriptive process modelling desired by business users and the rigorous, quantitative analysis of business processes desired by technical users. Rigorous quantitative analyses (simulation and/or analytic) are essential to link changes in process logic and parameters to changes in performance measures or metrics of the enterprise. Furthermore, an underlying formalism would enable one to generate multiple views of the enterprise at various levels of abstraction in a consistent and reliable manner.

These capabilities are essential for efficient and effective management of the enterprise.

3.3 Linking engineering and business processes to support process improvement initiatives

Conventional wisdom points to the fact that business results are tied to physical results, i.e., resources are converted to products and products satisfy market demands. Many companies use terms such as "line of sight" and "value chain" to describe and seek out connections between business and physical results. In most attempts to model value chains, we observe inputs and outputs, as well as unitary disconnects between layers in the enterprise structure (Kolarik, 1999). In general, contemporary knowledge in the process field is disjoint in terms of focus. Engineering approaches typically tend to focus on physical conversion in the process spectrum, while business approaches tend to focus on market and financial strategies at the other end of the spectrum. Typically, engineering approaches describe cause-effect relationships in subprocesses and focus design efforts on production processes made up of machines and people. The business approaches tend to focus on business strategies, which address market and financial structures. More effective and efficient process design, control, and improvement require modelling technologies, which possess scalable and dynamic properties, and provide a seamless linkage between business and technical process issues (Kolarik 1995; Sanders *et al.*, 1998).

3.4 Incorporating activity-based management approaches

Information on cost drivers and process performance measures (e.g., response time, product or service quality, and productivity) are not readily captured within existing process modelling techniques. This makes it difficult to utilize the operational improvement capabilities of activity-based management (ABM) (Cooper and Slagmulder, 1999a,b). It is critical that the next-generation enterprise systems provide timely financial and performance information for strategic, tactical, and operational decision-making. Process-modelling techniques and methodologies must be designed to capture (perhaps in an automated way) the detailed cost and performance data necessary to support ABM efforts.

4. THE DISTRIBUTED INTEGRATED MODELLING OF ENTERPRISES (DIME) FRAMEWORK

Taken together, the four categories identified in the previous section clearly define the case for an integrated modelling framework that seeks to meet enterprise management needs in a holistic manner. This section provides an overview of a new framework called the Distribute Integrated Modelling of Enterprises (DIME) Framework. The material contained in this section is based on Kamath *et al.* (2002a,b). Our vision for the DIME framework is a Web-based environment in which enterprise users will define, modify, and analyse business processes using user-oriented modelling applications resident on Web clients. Graphical business process models will be translated into formal models using the underlying Petri net-based constructs. The formal representation of the enterprise's business processes will be stored on enterprise servers. Server applications will store and manipulate the formal representations and also perform qualitative and quantitative analyses using formal models to support process improvement initiatives. The Web-based clients will address the needs of diverse enterprise users, ranging from managerial to technical.

4.1 Petri nets as a theoretical base

The formal representation scheme (back-end) in the DIME framework is based on Petri nets. Petri nets model a business process by capturing the relationship between conditions (e.g., order waiting and request arrived) and events/activities that affect these conditions (e.g., accept order and execute the request). These conditions and events/activities constitute the two types of nodes in the graphical model. The nodes that represent the conditions are called places, and are represented graphically by circles. The nodes that represent events/activities are called transitions, and are represented graphically by bars or rectangles. Directed arcs, which capture relationships, connect the places and transitions. No two nodes of the same type can be connected with each other. The truth of a condition represented by a place is identified by the presence of a token in that place. If places are used to model the availability of resources, then the tokens indicate the number of units available. Graphically, tokens are represented by black dots inside the places.

Based on this basic theory, a number of extensions have been proposed (David and Alla 1991; Desrochers and Al-Jaar 1994; Kamath and Viswanadham 1986; Zurawski and Zhou 1994). Some of the important extensions include (i) *Stochastic Petri nets*, which capture timed activities/events by using (probabilistic) timed transitions – these Petri net models can then be translated into a Markov chain for further analysis, and

(ii) *Colored Petri nets*, where the tokens can be colored (or carry attribute information) to differentiate the entities that flow through the system.

The arguments in favor of using Petri nets as formal representations of business processes are:

- They represent a hierarchical modeling tool with a well-developed mathematical foundation (Murata 1989).
- Petri nets support both qualitative and quantitative analyses and many computerized tools are available for the same (Desrochers and Al-Jaar 1994).
- They lend themselves very well to simulation/execution on distributed/parallel architectures which is essential for representations of realistically-sized enterprise systems (Ferscha 1998)
- They have been used by themselves, or in conjunction with other techniques to model business processes (Sagoo and Boardman 1998; van der Aalst 1999a,b).

Petri nets support advanced and formal analyses such as the reachability of various process states, detection of deadlocks, and behavioural ambiguities. According to Curtis *et al.* (1992) such analyses are critical to permit evaluation of a proposed process for syntactic correctness, completeness, consistency, risks, and opportunities for improvements.

4.2 DIME framework development

The front-end of the DIME framework is based on the development of a graphical process modeling language that builds upon the strengths of existing languages (Chaugule, 2001). The language, called the DIME Descriptive Modelling Language (DDL), has the following two main objectives – (i) ease of use for both modellers and readers and (ii) support for formalization of the process model into an equivalent Petri net description. Toward achieving objective (i), the DDL uses a parsimonius set of building blocks that intends to completely capture both *data flow* (i.e., inputs and outputs to an activity) and *control flow* (i.e., transfer of control from one activity to another) within the process while trading off the ability to describe finer organizational support details. The logical operands available within DDL allow for the representation of a wide variety of transitioning conditions, and modeling time-based, or control-based entry and exit criteria for the various activities comprising a process. Figure 12 describes the basic building blocks used in the DDL (Chaugule, 2001). Figure 13 shows the (previous) purchasing system example drawn using the DDL.

More importantly, the simplicity of the modeling language is best exploited in developing formalizations of the process model. *Formalization* replaces the semantics of context-specific, conceptual descriptions of the process, with more abstract descriptions rich in support for detailed analysis and in answering situational what-if questions. Petri nets were chosen as the medium of choice for developing formalizations of the business process model to support analysis questions related to reachability, deadlock, etc. (Curtis *et al.*, 1992; Janssens *et al.*, 2000; van der Aalst, 2002). Figure 14 depicts the mapping scheme used to translate business process models into Petri net models for subsequent analyses.

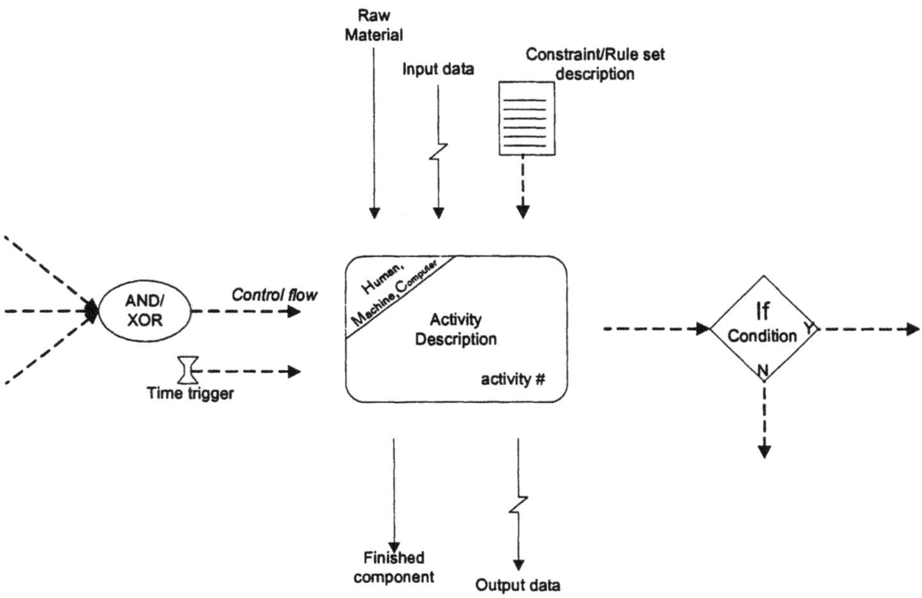

Figure 12. DDL Building Blocks

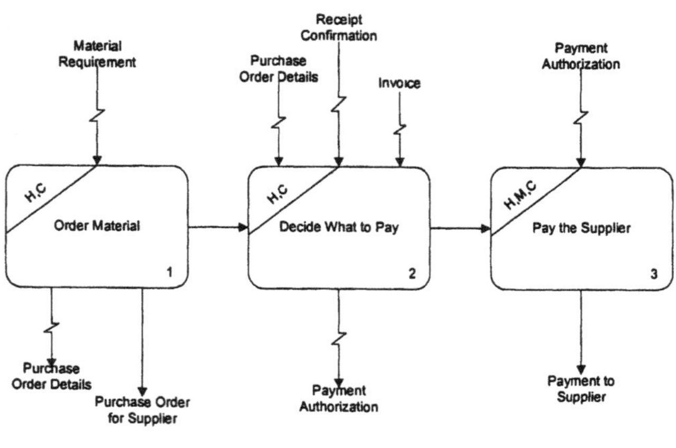

Figure 13. DDL Representation of the Purchasing System

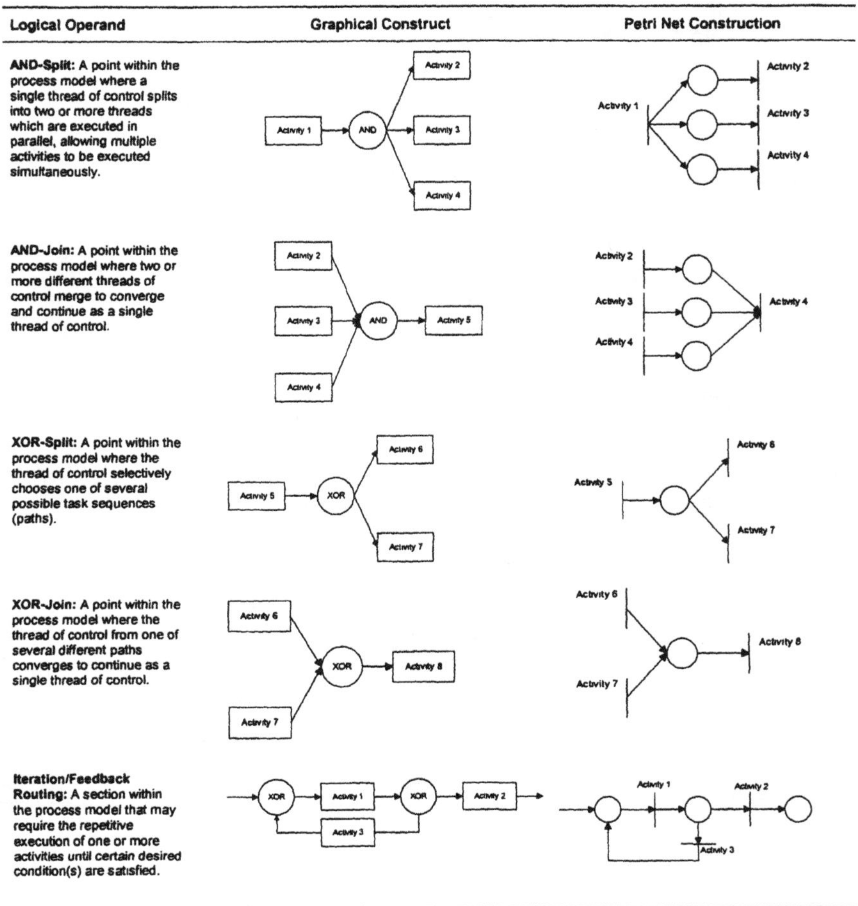

Figure 14. DIME Mapping Scheme

The linkage between Petri net theory, and the DDL constructs in a Web context represents an important step in the establishment of a scalable theoretical foundation for user-oriented enterprise process modelling. A variation of the previous purchasing system example is used to illustrate the different models and representations. These are shown in Figure 15.

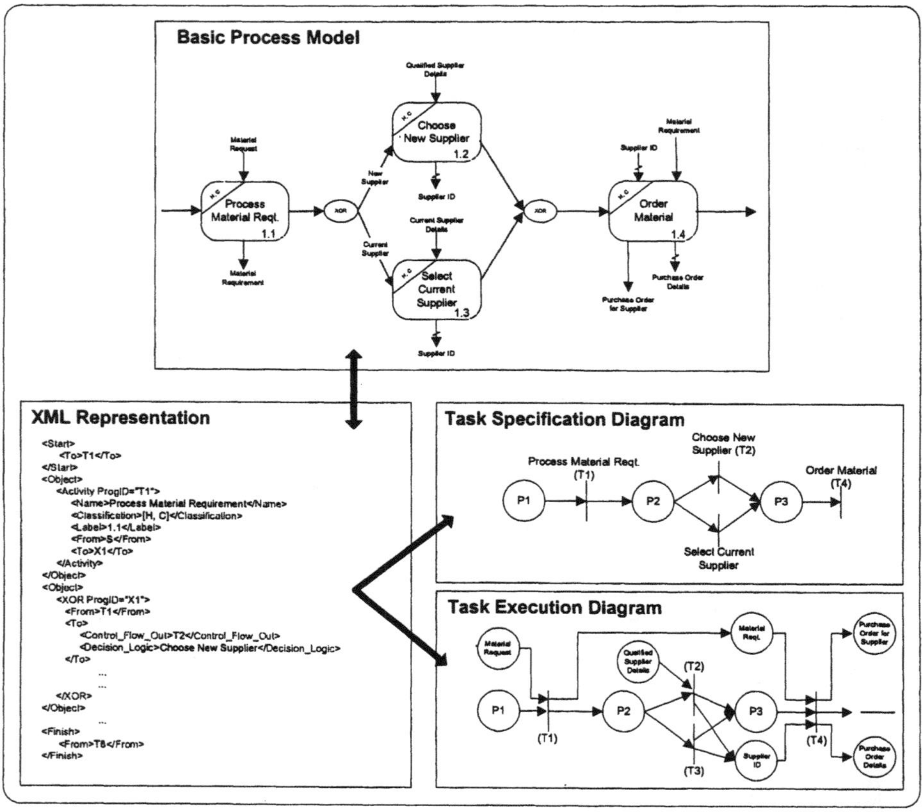

Figure 15. Different DIME Representations

4.2.1 Enterprise analysis using Petri net models

The task specification diagram or TSD (see Figure 15) is a representation of the structure of the process definition, devoid of specifics related to resource requirements, etc., and allows the modeller to study what is *going to happen, and in what order*. This information allows the modeller to step back and verify "structurally" what the process is expected to do. It also allows the modeller to study the process definition for structural inconsistencies. For example, is it possible for a task to wait "forever" for some predecessor task(s) to relinquish control? Is it possible that there are some tasks that are never realized/implemented?

We can check for ambiguous process definitions, i.e., inconsistencies arising from syntactic misrepresentations and logical couplings. To illustrate – suppose an XOR operand is used to bring together control flows that are created by a preceding AND operand; the chance for multiple repetitions of the task (following the XOR) is very real, which may or may not have been intended. At a stage like this, the TSD (or the modelling tool) should query the modeller to clarify his/her intent. The TSD is ideal for identifying possible problems in syntactic usage and for semantic checks useful in process verification. In other words, we can *identify potential sources for conflicts and ambiguities in decision choices*. The task execution diagram or TED (see Figure 15) is essentially the TSD enriched with operational specifics of data and resource allocation requirements. TED supports more advanced and formal analyses such as the reachability of various process states and detection of deadlocks. These additional complexities arise from the sharing of scarce resources by multiple activities.

If activity durations are specified, then time can be associated with transitions, which will enable deterministic cycle time calculations (David and Alla 1991; Murata 1989; Zurawski and Zhou 1994). In a similar manner, cost elements can also be aggregated to more accurately estimate cost related measures. Petri nets have been expanded to incorporate the stochastic nature of many real-time systems, integrating queueing network theory and Markovian analysis, to provide a more complete framework for performance evaluation and structural verification of systems, existing, or proposed (Ciardo *et al.* 1989; David and Alla 1991; Zurawski and Zhou 1994). Simulation is a possibility with the availability of commercial languages such as AlphaSim (www.Alphatek.com), which is based on Petri net models.

4.2.2 Current and future work on the DIME framework

The DIME framework comprises three distinct layers – a descriptive layer, a scalable representation layer, and an enterprise analysis layer. The descriptive (modeling) layer includes the graphical front-end language, the DIME Descriptive Modeling Language (DDL) described in the beginning of this section. DDL incorporates many user-oriented features and uses a tiered approach for enterprise process modelling (Chaugule, 2001).

The scalable representation layer is so named because it contains computer-processable representations that enable internal user groups as well as supply-chain partners to share process information. This layer includes the XML (www.w3.org) representations of the descriptive and formal models as well as the DIME mappings that are achieved between these representations. The XML representation of the descriptive model is captured using the DIME Markup Language (DML). DML is automatically created from the DDL using a browser-resident program. Each time the user creates or modifies a DDL model, it is automatically validated for consistency of data structures against the DML-schema before it can be automatically mapped to the DML representation. The XML parser module present in the Web browser flags all errors. A new XML schema can be made from libraries of existing schema. Hence, the need to build new XML documents from "scratch" is reduced.

This feature is useful when the size and complexity of the enterprise system increases over time.

The other XML representations within this layer are for (1) Petri net models, (2) other formal models such as queueing and simulation, and (3) formal views, such as cost, resource, and productivity views. Petri net models provide the DIME theory base. The two-way mapping between the DML and the Petri net representation is fundamental to our approach in DIME. Our mapping scheme has polynomial complexity and in many cases generates a Petri net model that has a one-to-one correspondence with the descriptive model. The mapping from DML to PNML, the Petri net Markup Language (and other similar DIME mappings from DML to other markup representations) is achieved by means of XSL transformations. The transformation code specifies the set of rules that produces the output document from the input XML document. A transformation written in XSLT is called a style sheet. The style sheet is made up of template rules, which have two parts, a 'pattern' that is matched against the nodes in the source document and a 'template' that can be instantiated to form the result document.

The DIME framework is scalable, flexible, and robust for many reasons. First, the Web-based creation, management, and use of models, enables users from different locations to work independently on parts of the same enterprise process model. The existing model is not affected by the addition of a new sub-model or by the addition of a new group of users assigned to modelling a different set of processes. Once the new sub-model is created, it can be linked to the existing model via an appropriate parent process by the owner of the parent process. Second, the XML-based scalable representation layer is flexible, yet powerful. With the help of the DIME mappings, this layer will 1) enable models created by different users to be aggregated automatically in a platform and machine independent manner, 2) allow other members of the supply-chain to share and access the enterprise model as needed, and 3) enable the use of Petri net theory, queueing theory, and other formalisms to access, analyse, and modify the enterprise process model. Third, the DIME descriptive modelling language (DDL) will enable us to compactly capture the semantics of all business processes for a variety of enterprises, whether they be traditional manufacturing firms, extended/virtual enterprises, or parts of a supply chain.

5. CONCLUSIONS

This chapter reviewed several prominent enterprise process modelling techniques and made a case for further research and development in process modelling to support the requirements of next-generation enterprises. The progress on the development of a new process-modelling framework, DIME, was described. The work described in this chapter was conducted as part of a research project funded by the National Science Foundation under Phase I of the Scalable Enterprises Initiative. Significant further research is needed to establish an enduring theory base and associated modelling techniques and methodologies to support the engineering of integrated enterprises.

6. ACKNOWLEDGEMENTS

We gratefully acknowledge the funding support provided by the National Science Foundation through grant # DMI-0075588 under the Scalable Enterprise Systems Initiative. We also thank Mr. Karthik Ayodhiramanujan Mr. Ananth Krishnamoorthy, Mr. Vinil Komalan, and Mr. Srikant Vinakota for their assistance during the preparation of this paper.

7. REFERENCES

Alpha/Sim ®, *A General Purpose Discrete Event Simulation Tool*, http://www.alphatek.com.

Alter, S., 2002, *Information Systems – The Foundation of E-Business,* Prentice Hall, NJ.

Benjamin, P., Delen, D., Mayer, R., and O'Brien, T., 2000, "A Model-Based Approach for Component Simulation Development," *Proceedings of the 2000 Winter Simulation Conference,* J.A. Joines, R.R. Barton, K. Kang, and P.A. Fishwick (Eds.), IEEE, 1831-1839.

Bernus, P. and Nemes, L. (Eds.), 1996, *Modeling and Methodologies for Enterprise Integration*, Chapman & Hall, UK.

Bernus, P., Nemes, L., and Morris, R., 1996, "The Meaning of an Enterprise Model," in *Modeling and Methodologies for Enterprise Integration*, Bernus, P. and Nemes, L. (eds.), Chapman & Hall, UK, 183-200.

Bond, B., Genovese, Y., Miklovic, D., Rayner, N., Wood, N. and Zrimsek, B., 2000, "ERP is Dead – Long Live ERP II," Research Note, Gartner Group, October 4, 2000.

Booch, G. J. Rumbaugh, and I. Jacobson, 1999, *The Unified Modeling Language User Guide*, Addison-Wesley Publishing Company.

Bravoco, R.R. and Yadav, S.B., 1985, "Requirement Definition Architecture – An Overview," *Computers in Industry*, 6, 237 – 251.

Burch, J.G., 1992, *Systems Analysis, Design, and Implementation*, Boyd & Fraser, Boston, MA.

Chaugule, A., 2001, "*A User-Oriented Enterprise Process Modeling Language*," MS Thesis, Oklahoma State University, Stillwater, OK.

Ciardo, G., Muppala, J.K., and Trivedi, K.S., 1989, "SPNP: Stochastic Petri Net Package," in *Proc. of the International Workshop on Petri nets and Performance Models*, IEEE, 142-150.

Colquhoun, G.J., Baines, R.W., and Crossley, R., 1993, "A State of the Art Review of IDEF0," *International Journal of Computer Integrated Manufacturing*, 6, 252 – 264.

Cooper, R., and Slagmulder, R., 1999a, "Designing ABC Systems for Strategic Costing and Operational Improvement," *Strategic Finance*, August, 18–20.

Cooper, R., and Slagmulder, R., 1999b, "Activity-Based Cost Management System Architecture–Part 1," *Strategic Finance*, October, 12–14.

Cuenca, L., Hawa, M., and Ortiz, A., 1999, "Analysis of Architectural Issues in Enterprise Modelling: A Compared Analysis of DEM (Dynamic Enterprise Modelling) of Baan and CIMOSA (CIM – Open System Architecture)," *Baan Brothers Foundation 2nd Roundtable Proceedings*, Utrecht, The Netherlands.

Curtis, B., Kellner, M.I., and Over, J., 1992, "Process Modeling," *Communications of the ACM*, 35(9), 75–90.

Dalal, N.P., Kamath, M., Kolarik, W.J., and Sivaraman, E. (2003), "Toward an Integrated Framework for Modeling Enterprise Processes," *Comm. of the ACM* (forthcoming).

Dao, K.L. and Rodjak, D., 1991, "CIM Flows from Data Flow Diagrams," *Manufacturing Systems*, 67-72.

Davenport, T.H., 1998, "Putting the Enterprise into the Enterprise System," *Harvard Business Review,* July-August, 121-131.

David, R. and Alla, H., 1991, "Petri Nets for Modeling Dynamic Systems" *Automatica*, 30(2), 175-202.

Desrochers, A.A., and Al-Jaar, R.Y., 1994, *Applications of Petri Nets in Manufacturing Systems,* IEEE Control System Society: NY.

Didic, M., 1994, "CIMOSA Model Creation and Execution for a Casting Process and a Manufacturing Cell," *Computers in Industry,* 24, 237-247.

Ferscha, A., 1998, "Optimistic Distributed Execution of Business Process Models," *Proceedings of 31st Annual Hawaii International Conference on System Sciences*, 723–732.

Fox, M.S., Chionglo, J.F., and Fadel, F.G., 1993, "A Common-Sense Model of the Enterprise," *Proceedings of the 2nd Industrial Engineering Research Conference*, 425-429.

GERAM: Generalized Enterprise Reference Architecture and Methodology, Version 1.6.2, Annex to ISO WD15704, IFIP-IFAC Task Force.

Gruniger, M., and Fox, M., 1995, "The Logic of Enterprise Modeling," In *Proceedings of the Fourth Industrial Engineering Research Conference*, 747-758.

Hollingsworth, D., 1995, "The Workflow Reference Model," published by the Workflow Management Coalition, Document TC00-1003, http://www.wfmc.org/standards/docs.htm.

Janssens, G.K., Verelst, J., and Weyn, B., 2000, "Techniques for Modeling Workflows and Their Support of Reuse," in *Business Process Management – Models Techniques and Empirical Studies,* W. van der Aalst, J. Desel, and A. Oberweis (Eds.), Springer, Berlin.

Kamath, M., Dalal, N., Kolarik, W., Chaugule, A., Sivaraman, E., and Lau, A., 2001, "Process-Modeling Techniques For Enterprise Analysis And Design: A Comparative Evaluation," *IERC 2001 Proceedings.*

Kamath, M., Dalal, N., Kolarik, W., Lau, A., Sivaraman, E., Chaugule, A., Choudhury, S., Gupta, A., and Channahalli, R., 2002a, "An Integrated Framework For Process And Performance Modeling Of Next-Generation Enterprise Systems: Design And Development Issues" *Proc. of the University Synergy Program (USP) Conf.*

Kamath, M., Dalal, N., and Chinnanchetty, R., 2002b, "The Application of XML-Based Markup Languages in Enterprise Process Modeling," *IERC 2002 Proceedings.*

Kamath, M., and Viswanadham, N., 1986, "Applications of Petri Net Based Models in the Modeling and Analysis of Flexible Manufacturing Systems," in *Modeling and Analysis of Flexible Manufacturing Systems*, A. Desrochers, (eds), IEEE, 262–267.

Kateel, G., Kamath, M., and Pratt, D., 1996, "An Overview Of CIM Enterprise Modeling Methodologies," *Proceedings of the 1996 Winter Simulation Conference*, IEEE, 1000-1007.

Keller, G. and Detering, S., 1996, "Process-Oriented Modeling and Analysis of Business Processes Using the R/3 Reference Model," in *Modelling and Methodologies for Enterprise Integration*, Bernus, P. and Nemes, L. (eds.), Chapman & Hall, 69-87.

Knowles, G., 1999, "Next Generation Enterprise Modeling: A Step Beyond Baan Application Modeling," *Baan Brothers Foundation 2nd Roundtable Proc.*, The Netherlands, 95-115.

Kolarik W., 1995, *Creating Quality: Concepts, Systems, Strategies, and Tools*, McGraw-Hill, NY.

Kolarik, W., 1999, *Creating Quality: Process Design for Results*, McGraw-Hill, NY.

Kosanke, K., 1995, "CIMOSA – Overview and Status," *Computers in Industry*, 27, 101 – 109.

Mertins, K., Jochem, R., and Jakel, F.-W., 1997, "A Tool for Object-oriented Modeling and Analysis of Business Processes," *Computers in Industry*, 33, 345-356.

Mertins, T., Susseguth, W. and Jochem, R., 1992, "An Object Oriented Method for Integrated Enterprise Modeling as a Basis for Enterprise Co-ordination," *Enterprise Integration Modeling: Proceedings of the First International Conference*, MIT Press, MA, 249-258.

Millet, I., 1999, "A Proposal to Simplify Data Flow Diagrams," *IBM Sys.Journal*, 38, 118–121.

Murata, T. 1989, "Petri Nets: Properties, Analysis and Applications," *Proc.eedings of the IEEE*, 77, 541-580.

Peterson, J.L., 1981, *Petri Net Theory and the Modeling of Systems*, Prentice-Hall, Inc., NJ.

Radding, A., 1999, "ERP, Componentization and E-commerce," available at http://itmanagement.earthweb.com/erp/print/0,,11981_616621,00.html.

Sagoo, J.S. and Boardman, J.T., 1998, "Towards the Formalisation of Soft Systems Models using Petri Net Theory," *IEE Proc. on Control Theory & Applications*, Vol. 145, 463 – 471.

Sanders, M., Patro, S. and Kolarik, W., 1998, "Process Redesign/Redefinition in Small Enterprises," *IE Research 98 Conference Proceedings*, CD.

Sarjoughian, H.S., Zeigler, B.P., Park, S., 2000, "Collaborative Distributed Network System: A Lightweight Middleware Supporting Collaborative DEVS Modeling," *Future Generation Computer Systems*, 17, 89-105.

Schader, M. and Korthaus, A., 1998, "Modeling Business Processes As Part Of The Booster Approach To Business Object-Oriented System Development Based On UML," *IEEE Proc. of the Second International Workshop on Enterprise Distributed Object Computing*, 56-67.

Scheer, A.-W., 1992, "*Architecture of Integrated Information Systems: Foundations of Enterprise Modelling*," Springer-Verlag, Berlin.

Sheth, A., Aalst, W.M.P., and Arpinar, I. (1999), "Processes Driving the Networked Economy," *IEEE Concurrency*, 7(3), 18–31.

van der Aalst, W.M.P., 1997, "Verification of Workflow Nets," in *Application and Theory of Petri Nets 1997*, volume 1248 of Lecture Notes in Computer Science, P. Azema, and G, Balbo (eds.), Springer– Verlag, Berlin, Germany, 407–426.

van der Aalst, W.M.P., 1998, "The Application of Petri Nets to Workflow Management," *The Journal of Circuits, Systems and Computers*, 8(1), 21–66.

van der Aalst, W.M.P., 1999a, "Formalization and Verification of Event-driven Process Chains,"*Information and Software Technology*, 41(10), 639–650.

van der Aalst, W.M.P., 1999b, "Interorganizational Workflows," *Systems Analysis, Modeling, Simulation*, 34(3), 335–374.

van der Aalst, W.M.P. and Hee, K., 2002, *Workflow Management: Models, Methods, and Systems*, MIT Press, Cambridge, MA.

van der Rijst, N., 1997, "Dynamic Enterprise Modeling Development Method: Guidelines and Conventions," User Manual, Baan Business Innovation, The Netherlands.

Vernadat, F.B., 1996, "CIM Business Process and Enterprise Activity Modeling," in *Modeling and Methodologies for Enterprise Integration*, P. Bernus, and L. Nemes, (eds.), Chapman & Hall, UK, 170-182.

WfMC, 1999, "Terminology & Glossary," published by the Workflow Management Coalition, Document WFMC-TC-1011, http://www.wfmc.org/standards/docs.htm.

Whitman, L., Huff, B., and Presley, A., 1997, "Structured Models and Dynamic Systems Analysis: The Integration of IDEF)/IDEF3 Modeling Methods and Discrete Event Simulation," *Proceedings of the 1997 Winter Simulation Conference,* S. Andradottir, K.J. Healy, D.H. Withers, and B.L. Nelson (Eds.), IEEE, 518-524.

Whitten, J.L., Bentley, L.D., and Dittman, K.C., 2001, *Systems Analysis and Design Methods*, Irwin/McGraw-Hill, Boston, MA.

Zelm, M., Vernadat, F.B., and Kosanke, K., 1995, "The CIMOSA Business Modeling Process," *Computers in Industry*, 27, 123–142.

Zurawski, R. and Zhou, M.C., 1994, "Petri Nets and Industrial Applications: A Tutorial," *IEEE Transactions on Industrial Electronics*, 41 (6), 567-583.

Chapter 2

DESIGN AND MANUFACTURING PROCESS MANAGEMENT IN A NETWORK OF DISTRIBUTED SUPPLIERS

Moon Jung Chung
Department of Computer Science and Engineering

Patrick Kwon
Department of Mechanical Engineering

Brian Pentland
Department of Accounting and Information Technology

Michigan State University, East Lansing, Michigan 48824

Abstract

In this paper, we presented the basic structure of an XML-based scalable enterprise system called Manufacturing Integration and Design Automation System (MIDAS) to support collaboration among a network of suppliers engaged in the design and manufacturing of prototype parts. MIDAS provides a new paradigm where design and manufacturing processes can be represented and integrated within each vendor and with other venders working together to build a prototype. The framework is developed based on Java and XML to provide the efficient and secure interchange of data among the vendors involved in designing and building a prototype. While the quantity of parts is one aspect of the problem, scalability also includes the ability of a system to adapt to a changing environment, including new products, processes, technology, markets, and business partners. The key issues are (1) how to achieve the appropriate representation of process model, (2) how to model various constraints in design and manufacturing and to use them in order to find the best-of-practice process, and (3) how to maintain interdependencies within the network of vendors.

In order to resolve these issues, the process logic, which deals with the problems such as "how to do each subtask in detail, and what methods should I use," is separated from the structural information of process such as "what subtasks are there, and what are the necessary sequence of processes?" Otherwise, excessive information demises the effectiveness of the decision makers in business, design and manufacturing tasks. Consequently, MIDAS was implemented using a layered approach where process information is separated into syntactic layer and execution layer. Process grammar captures the syntactic layer of process information: the relationships between tasks and subtasks, the alternatives of process decompositions, data dependencies among tasks. The execution layer provides the detail on how each task can be executed in a particular situation by choosing appropriate alternatives and executing the enter and exit criteria for executing each subtask.

Finally, we will present a simple example for service processes in cooperation with the companies in the Jackson Smart Park as a part of our modeling efforts. While the model we create will focus on prototyping services, we expect that the basic concepts for process configuration and constraint management will be applicable to a broader range of manufacturing industries. Such an enterprise system will be scalable in terms of not only volume but also variations, and will evolve into a powerful tool to model, configure, and reconfigure a wide variety of processes in relation to e-commerce activities in a virtual community.

1. INTRODUCTION

Existing systems to support supply chain management naturally tend to focus on transactions involving standard parts and products among networks of suppliers. Whether operated as "reverse auctions" or some other scheme for placing and filling orders, these systems provide a market-based mechanism for coordinating the activities of a widely distributed set of enterprises. Furthermore, they provide an information infrastructure that can be used to facilitate the optimization of the supply chain, and a great deal of practical value has been added by this line of research [Yin and Zhang, 1996; Artiba and Elmaghraby, 1997; Cox and Spencer, 1998; Wu, 2000; Shaw, 2001].

In this chapter, even though the general concepts behind MIDAS can be applied to a traditional supply chain management, we focus on a different and much narrower segment of the overall manufacturing sector: the creation of prototype parts and products by networks of suppliers. This focus arises from our collaboration with a group of medium to small sized companies in Jackson, Michigan, that we will call the Jackson Smart Park (JSP). The problem of coordinating distributed manufacturing processes among this network of small firms poses an entirely new and different set of theoretical and practical challenges. As a result of our research, we have identified a novel class of problems and successfully implemented a promising grammar-based framework for dealing with them [Chung, Kwon and Pentland, 2001], but much work remains to be done.

Our basic question that needs to be addressed here is: how can we provide computational support for the dynamic enactment of design and manufacturing processes by distributed participants? In particular, how can we minimize the need for iteration, which results in delays and wasted work? And when iteration is required, how can we minimize the coordination costs? The answers to these questions entail designing and exploring a new category of process management systems with the following distinctive properties:

(1) information concerning the adequacy of the process is only available during process execution; and
(2) the knowledge necessary to generate a plan is distributed (not localized) and some of this knowledge may be proprietary (secret).

These characteristics pose especially challenging problems as the scale of the system increases. The issue is not just transaction volume, but also the number of participating organizations, the number of interdependent parts that are being created, the number of alternative manufacturing processes involved, and so on. These factors increase the complexity of the prototyping process and exacerbate the problems of providing computational support. Providing good support for two companies creating a single prototype part is challenging; supporting a dozen vendors engaged in prototyping a complex sub-assembly is considerably harder.

This chapter is organized as follows. First, we begin by providing some background on how to conceptualize this class of problems and review some related approaches to process representation and process planning. Secondly, we describe the architecture and operation of the MIDAS system and the underlying process grammar framework. Finally, we provide a simple example of how MIDAS can be applied, and discuss the implications of our research and ideas for future work.

2. BACKGROUND

In theory, web technology allows the formation of "virtual" organizations to handle a wide range of design and manufacturing processes. In this new environment, "the winners will be those companies that can effectively leverage the Internet to redesign, automate, and integrate all business operations - from demand capture, production planning, and purchasing to deliver customer service, and new product development" [Aberdeen Group White Paper, 2000].

While this is a compelling vision, the reality is not here yet. To take full advantage of this technology, however, work processes must be flexible and dynamically reconfigurable. The whole design and manufacturing (D&M) process -- not just the pieces – needs to be configured in response to changes in technical considerations such as features, methods, materials, costs, and other critical decision parameters. This means planning and executing a process that extends across formal organizational boundaries.

Consider a typical scenario: a firm wants to create a prototype for a new product. The product may involve a mix of standard and custom components created from various materials and technologies. Each component requires a process to create it. Within each process, engineers should be able to select from a menu of possible technologies and possible vendors who offer competing technologies. And given that the process may be performed by many people in many locations, they should be able to coordinate these dispersed activities in a timely manner. Ideally, an engineer should be able to select optimally a methodology, design parameters and resources to perform a task. Of course, such selections are highly interdependent on each other and may interact with choices made for the other components. As the process proceeds, everything must be checked in real time to redefine and iterate the process. The execution of tasks should be coordinated and any conflicts between tasks should be resolved. Engineers should also be able to understand, revise, archive and reuse a process flow.

2.1 Why not just optimize?

The optimization of manufacturing processes using formal techniques, such as network flow analyses, has a long history and has produced a solid base of practical knowledge. The assumption of most optimization techniques is that the process, once planned, will be carried out more or less according to plan. At a deeper level, there is an assumption that the process model corresponds to reality in a well-known, reliable way [Ghosh & Tanniru,1997; Janusz, 1997].

In the world of prototyping, however, things are not so simple. There are several aspects to the prototyping problem that make it distinctive. First, a process to create a prototype does not always work out. This is especially true when innovative material or technologies are being tested. Second, the prototyping process involves business details, not just technical details. Even when the technology is entirely routine and predictable, there may be business considerations that dominate the decision-making (such as pricing, scheduling, etc.). Third, the knowledge base from which a plan can be generated is distributed among a set of participants who may or may not be known at the outset of the planning process. For example, when one is attempting to use a new material, one may need to involve the vendor for that material in the process. Finally, the information that each participant brings to bear on the prototyping process may, in some cases, involve trade secrets that cannot be shared with other participants, or shared only selectively. Such information cannot simply be put on a centralized server, regardless of the sophistication of the access control scheme.

2.2 Abstracting the key dimensions of the problem: What kind of problem is this?

In an effort to come to grips with these difficulties, we began to ask, how can we characterize this kind of process planning and select an appropriate process planning method? Chang et. al [1991] distinguish between "variant" and "generative" methods of process planning. In a variant-based approach, one has a library of plans that one can pull off the shelf and adapt incrementally to fit the current situation. Variant planning is very effective in situations where the requirements change incrementally and can be closely matched to an existing plan (for example, DCLASS [DCLASS, 1985], Granville [Charles, 1986]). In a generative approach, one has a knowledge-based representation of the possibilities and generates a new plan for a part. Generative planning is more appropriate in situations where the requirements are novel (for example, GENPLAN [Tulkoff, 1981], BYUPLAN [BYU, 1985], Wong and Siu [1995]). Many commercial process planning systems support both approaches, allowing users to generate new plans that can be stored, retrieved, and modified.

In our preliminary work with the Jackson Smart Park, we have identified a third major category of process planning methods: *enacted processes*. These involve processes which are designed "on the fly", by the participants, while the process is being executed. It is more like improvisation than planning and for the reasons just mentioned, conventional optimization techniques do not seem readily applicable. This does not, however, imply the absence of knowledge or the lack of need for computational support. Rather, it means that a different set of tools and technologies may be required depending on the constraints in hand.

	Variant	Generative	Enacted
	Off the shelf	Knowledge-based	Improvised
When is plan created?	Prior to execution	Prior to execution	Prior to and During execution
How is knowledge distributed?	Localized	Localized or distributed	Distributed
Knowledge base	Technical	Technical	Technical and Business

Table 1: Characteristics of major process planning methods

2.3 Adaptive planning systems

There has been a great deal of research on adaptive planning systems [Whiteside et al., 1997; Miura and Shirai, 2000; Melis and Meier, 2000; Dengler, 1994; Steinhage and Schoner 1997; Goel, 1994; Rea et al., 1998; vom Ende et al., 2000]. The basic idea is to revise a plan based on new information that arrives during plan execution. For example, an agent might generate a plan to travel from point A to point B over a series of connecting paths. When a chosen path is blocked, the plan must be revised. Manufacturing processes in building a prototype is analogous, except that (a) the designers may choose to revise the design requirements (change the planned destination) rather than simply revise the plan; and (b) there are many participants involved, acting in parallel. Thus, while prototyping definitely involves what might be called adaptive planning, scalability is a huge concern. In particular, scaling up from a single agent with a fixed, unidimensional goal, to a large number of agents with a variable, multi-dimensional goal is not simple.

2.4 Problem space perspective on process planning

In considering how to approach this problem, it is helpful to think in terms of the problem space that underlies a typical process planning problem. A problem space consists of initial state, goal state, intermediate states, and operators [Newell and Simon, 1972; Simon, 1973]. In terms of a manufacturing process, a typical initial state would consist of some raw materials and a set of design specifications. The goal state consists of a finished product that meets the specifications. The operators consist of the various manufacturing operations that one might apply to the raw materials to create the finished product. One might also introduce constraints on cost, quality, cycle time, and so on. Typical process planning problems may be exceedingly complex, but relatively well structured. One can use a variant approach (pull a plan off the shelf), or a generative approach (create a new plan using a knowledge base of some sort). The initial and goal states are relatively fixed, the operators can be specified, and the constraints are well known.

But in prototyping, none of these factors can be taken for granted. In his classic piece on the nature of ill-structured problems, Simon [1973] identified several characteristics that tend to make design problems (such as designing a house) ill-structured. One particularly vexing issue is that new information that changes the nature of the operators, the objectives, or the states of the problem space can be added to the problem at any time. If one tries something that doesn't work out, then one must try something else.

Also, in prototyping, there is constantly new technology becoming available. New operators are being added. The details of how some of these new technologies perform (what their exact capabilities are) may be considered an important trade secret. Hence, the knowledge base is not only distributed, it is constantly changing, and it is at least partially secret.

Also, each major aspect of the process plan involves a business decision as well as a technical decision. Interviews with the entrepreneurs at the Jackson Smart Park suggest that pricing, scheduling and other contract terms are often "adjusted" based on a variety of considerations that are not necessarily driven by the details of the current project. For example, a vendor may offer favorable terms to attract a new client. These decisions are critical to the creation of a process plan, but they are completely outside the scope of traditional, machine-oriented process planning systems. The difference is that the prototyping process crosses organizational boundaries between firms that have independent decision-making and cost accounting. When decision-making and accounting are centralized, it becomes reasonable to "factor them out" of the process planning exercise. In the case of a distributed network of independent suppliers, this is simply not realistic.

From the perspective of automated problem-solving and optimization, these are features of the problem that are annoying at best, and intractable, at worst. There is no question that they can be quite inefficient and frustrating for the participants. From the perspective of creativity and innovation, however, this is precisely the kind of flexibility and adaptability that need to be encouraged. If everything could be specified in advance, why bother prototyping? This is inherently – and deliberately – an iterative, "cut and try" approach to engineering design.

Nevertheless, even an entirely novel design may be composed of relatively familiar steps. One can generate, in advance, a rough outline of a reasonable course of action. And many of the steps within the plan may be quite routine (e.g,. turning a diameter, cutting a slot, drilling a hole, etc.). Thus, our work naturally builds on and incorporates the insights from many existing systems, which we discuss next.

3. PROCESS MODELING: A BRIEF REVIEW

There are a large number of tools and techniques available for process modeling and specification. For example, in engineering design and manufacturing, many systems have been proposed to generate design processes [Knapp and Parker, 1991] and manage workflow [Dellen et al., 1997; Lavana et al., 1997; Schurmann and Altmeyer, 1997; Sutton and Director, 1998]. Many of them use the web technology to coordinate various activities in business [Andreoli et al., 1998], manufacturing [Berners-Lee et al., 1994; Cutkosky et al., 1996; Erkes et. al. 1996], and micro-electronic design [Chan et al., 1998; Chung and Kim 1999]. Either they use a rule-based approach or use a hard-coded process flow, requiring source code modification for any change in process. Most of them are weak in at least one of the following criteria suggested by NIST [Schlenoff et al, 1996]: process abstraction, alternative tasks, complex groups of tasks, and complex sequences.

Because a task can be accomplished in various ways, representing alternatives for a task is one of the requirements for process specification. With such functions, designers can access alternative methodologies or tools and select the best one based on what-if analysis. A blackboard approach was sometimes used [Lander et. al, 1995] to generate a particular flow suitable for a given task. In Nelsis [Bosch et. al 1991] and IDEF [1998], alternative workflow is explicitly represented as branches using 'or' nodes and 'merge' nodes. The most important aspect missing from most process management systems is the abstraction mechanism [Schlenoff et. al 1996].

IDEF

Much research has also been done on formal process modeling and generating processes dynamically. For example, IDEF [1998] has been used to represent a wide range of processes. Even though several IDEF methods have been proposed to model various processes, IDEF is a typical example of over-specification, requiring such information as control mechanisms, resources, and constraints at a specification time. Therefore, the workflow represented by IDEF is difficult to understand, not suitable for sharing and improving. IDEF is mainly targeted to model the static process in a flat level, and not suitable for modeling the dynamic process in which an engineer generates the workflow incrementally during the process execution.

The Process Specification Language (PSL)

By the initiative of NIST, PSL has been proposed PSL as an interchange format designed to help exchange process information automatically among a wide variety of manufacturing applications such as workflow, process modeling, and process planning. The goal is to allow that any system would be able to automatically exchange process information with any other system via PSL. [Schlenoff 2000]. The main focus of PSL is to provide semantics of such *temporal* constructs such as "duration", activity occurrence etc.

RosettaNet

RosettaNet is a consortium of major Information Technology, Electronic Components and Semiconductor Manufacturing companies such as Hewlett-Packard Co., IBM, Cisco Systems Inc., and Ingram Micro Inc. Its goal is to create and implement industry-wide, open e-business process standards. These standards form a common e-business language, aligning processes between supply chain partners on a global basis, and speeding development and production implementation of RosettaNet Partner Interface Process(TM)

(PIP(TM)) standards. RosettaNet programs are focused on e-business processes, such as supply and demand management and B2B integration with mid-tier partners [http://www.rosettanet.org/]. Other standards for e-commerce are also emerging, such as ebXML (www.ebXML.org). These standards focus primarily on the data required to perform transactions (e.g., accept orders, confirm orders, provide invoices, pay invoices, and so on). They do not embody the domain knowledge required to represent design and manufacturing processes.

And/OR Graph and Petri nets

And/Or graphs and Petri nets have also been used to model processes and their behavior [Kamath, 2001]. They represent the alternative paths explicitly using join and fork nodes, or transitions. The flow represents all possible cases of the process before the start of process execution. That is, the process flow itself is determined in advance, and the flow of tokens or the nodes executed are determined dynamically. These approaches are not suitable when there are many tasks, each of which has many alternatives. In this situation, the process flow graph may grow exponentially. Moreover, it is not easy to represent interdependencies between tasks with these approaches. In traditional systems analysis, flow charts, data flow diagrams, and other modeling techniques provide important supports for any system implementation [Yourdon, 1989]. While unquestionably useful, these techniques describe particular systems, and provide little insight into potential alternatives [Malone et al, 1999].

The Process Handbook

The "Process Handbook" [Malone et al, 1999] is an electronic database of process descriptions that emphasizes the representation of closely related alternative processes. The Process Handbook catalogs and indexes alternative processes at varying levels of abstraction, using a generalization-specialization hierarchy. It provides a facility for process comparison, as well, but the comparison must be based on a human evaluation of the processes. Dependencies between tasks must be entered manually. Couretas and Zeigler [2001] used a similar approach to model and simulate process behavior using the Devs framework. While these frameworks are useful as repositories of process models, they are not especially helpful in creating an adaptive process.

Optimization techniques and constraint programming

When a process is already defined, constraint programming can be used to find an optimal combination of parameters [Mehra, 1994; Petrie et al, 1996]. However, constraint programming generally requires that process information be known in advance. Our preliminary research indicates that this is frequently not the case. Likewise, there are a wide variety of optimization techniques that can be applied to various aspects of process configuration and analysis. Either they use a rule-based approach or use a hard-coded process flow, frequently requiring source code modification for any change in process. Moreover, there are too many parameters to represent the processes and constraints affected all in advance. Most of the proposed approaches are weak in at least one of the following criteria suggested by NIST [Schlenoff et al. 98]: process abstraction, alternative tasks, complex groups of tasks, and complex sequences. Many systems require complete details of D&M tasks and flow before a process plan can be generated. They are rather rigid and difficult to handle exceptions. However, design is inherently iterative. When engineers are not satisfied with the design result, the process has to be iterated with new parameters and/or processes. The new process itself may be similar to the previous one with minor changes. We need to coordinate the dynamic nature of D&M processes and a systematic way of generating permissible process flows interactively.

4. FUNCTIONAL REQUIREMENTS OF PROCESS MANAGEMENT: SPECIFICATION AND EXECUTION

Designing a complex product involves a collaborative process, where many different engineering activities have to be coordinated among engineers, their tasks, and the design results. A process management environment should provide the definition, presentation, execution, and control of design methodology in a flexible and configurable way [Kleinfeldt et al. 1994]. It must be able to:

- Represent acceptable methodologies, workflow, processes, and design tools together with the information such as when and how they may be used.
- Guide the designer in selecting suitable methodologies and in constructing a workflow to complete the design under given constraints.
- Support collaboration by sharing process and facilitating interaction.

- Accommodate rapidly changing technologies and emerging new tools.
- Keep the relationships among various design specifications and maintain the consistency among designs.

During design, a methodology or workflow must be constructed based on company practice, local expertise, and individual preference that must be reflected on available tools, resources and design data. During design, the process itself can be modified because of new tools and methodologies. Life cycle support includes updating the process dynamically, and testing/validating the design process, version history and configuration management of the design process. Tests and validations of the design processes, the simulation of processes, and impact analysis may be necessary tools.

The process management should not constrain the designers, rather free them from routine tasks so that they just guide the execution of workflow. User interaction is especially important. When exceptions are encountered during the execution of flows, a user can modify the workflow locally. The system must support such activities through 'controlled interactions' with users. Process management can be divided into two parts:

(1) A formal specification of supported methodologies and tools.
(2) An execution environment that helps designers to construct and execute workflow.

4.1 Process Specification

The framework must provide a facility to represent a process for design methodology. It must show tasks involved in the workflow, data, and their relationship. Workflow defined by methodology developers (design synthesis) enforces the user to follow the flows. Processes captured as a workflow must be encapsulated and presented to designers in a usable way. Designers should be able to define the tasks, organize them by hiding the unnecessary details, and retrieve and reuse them. They should also be able to browse related processes, and compare, analyze, and modify them for their own purpose. For an extensive study on the requirement for process specification, readers are referred to the NIST report [Schlenoff et. al, 1996].

The most important aspect of the process specification is the abstraction mechanism [Schlenoff et. al 1996]. The abstraction mechanism allows a complex activity to be represented more concisely with a small number of higher-level tasks hiding the details of the tasks. Abstraction may be accomplished by simply defining a set of tasks in terms of its higher-level function without the details of their internal implementations. Using the generalization–specialization hierarchy, called class hierarchy, similar tasks can be grouped together and catalogued as a higher level task in the class hierarchy. In class hierarchy, a node (representing a task) in a lower level inherits its attributes from its higher level task so that software modules and workflow of parent classes can be shared by their descendants.

There are various approaches to bind a specific tool to a smallest unit of design task. A tool can be bound to a task statically at the compile time or dynamically during execution based on available resources and constraints. When a new tool is installed, designers should be able to modify the existing bindings. The simplest approach is to modify the source code or write a script file and recompile the system. The ideal case is 'plug and play', meaning that CAD vendors address the need of tool interoperability, e.g., the Tool Encapsulation Specification (TES) proposed by CFI [1995].

4.2 Execution Environment

The execution environment of the process management is the supporting software tools that binds data to tools and provides the execution of tasks and tools either manually or automatically. The execution environment requires the information such as the control mechanism for a task, the capability to handle exception, and the priority of the task to be executed. Execution environment may have two stages: flow generation, and execution. Only a few frameworks separate the execution environment from the specification of design process. A task can be executed in three different ways: manual flow execution, automatic flow execution, and automatic flow generation. In the manual flow execution, the environment executes a task in the context of a flow. Automatic flow execution environments execute the flow automatically based on the order specified in the flow graph. In automatic flow generation, the framework generates and executes workflow dynamically without the guidance of designers.

When a task involves many subtasks, and each subtask has more than one alternative, generating a workflow for the task that would successfully accomplish the task is not easy. In addition, design process is typically iterative. That is, the first workflow generated may not produce the output specifications that would meet the post-evaluation criteria set by the designers. Then, other alternatives and parameters should be tried. Searching through the alternatives, generating the workflow, and executing them is not trivial. Most frameworks are inadequate in this dynamic process flow generation. Typically, they use forward chaining mechanisms to generate one particular process flow [Knapp and Parker, 1991], or use knowledge-based approaches [Jacomb and Director, 1996]. However, designers need to analyze the created workflow. That is, repeatability and predictability are important factors if a framework supports dynamic creation of workflow. The major achievement in this paper is that the process grammar facilitates the mechanism of process flow generation.

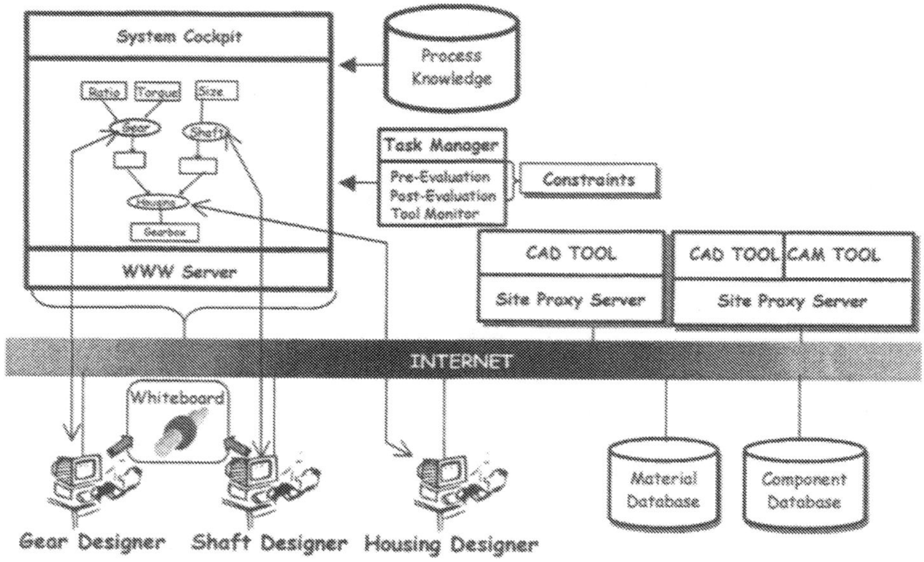

Figure 1. The Architecture of MIDAS

5. DESCRIPTION OF MIDAS SYSTEM

MIDAS is a collaborative engineering framework that coordinates design activities and facilitates the flow of process knowledge and multi-media design data among the organizational units of a virtual enterprise. Figure 1 illustrates how MIDAS links tools and sites to support design and manufacturing processes. The design activities can be invoked manually by engineers or executed automatically by the framework.

5.1 Process grammar

The core of the MIDAS system is a process grammar, which provides the theoretical foundation to represent, manipulate and execute design and manufacturing processes [Baldwin and Chung, 1995a]. In the MIDAS system,

D&M processes are modeled as a collection of tasks. Each task can be decomposed into a set of subtasks. Our process flow graph depicts tasks, data and the relationships among them, describing the sequence of tasks for a larger D&M activity. Four basic symbols are used to represent the process flow: *Logical Tasks* (represented using oval), *Atomic Tasks* (two-concentric oval nodes), *Specifications* (rectangular nodes) and *Selectors* (diamonds nodes). A logical task can be decomposed into subtasks, while an atomic task cannot. A selector selects a specification or a parameter for a task. An output specification produced by a task can be consumed by subsequent tasks as an input specification. These elements can be combined into a process flow graph using directed arcs to indicate the specifications used and produced by each task. Specifications with no incoming arcs are the first inputs to the process flow.

The process grammar provides the mechanism for transforming a high-level process flow graph into progressively more detailed process flow graphs. The grammar consists of a set of production rules. A production is a substitution that permits the replacement of a logical task node with a flow graph that represents a possible way of performing the task. If there are several production rules with the same left side flow graph, it implies that there are alternative production rules for the logical task. Several production rules for a logical task imply that there are alternative production rules for that task. Casting and machining are, for example, two alternative productions for the task Manufacturing. This capability is critical to maintaining the usability and effectiveness of the overall framework. Therefore, the process grammar naturally captures the hierarchical design methodology and allows systematic exploration of D&M space.

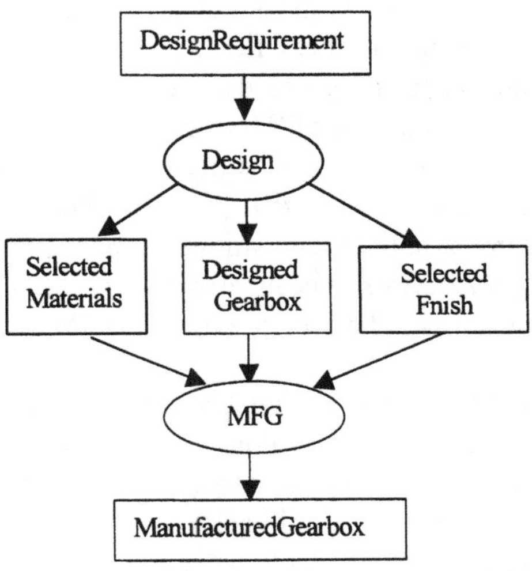

Figure 2. A high-level process flow of a D&M process for a Gearbox

Using these graphical elements, a process flow graph can be created in a top-down fashion. Fig. 2 shows a high-level process flow of a design and manufacturing process of a gearbox. Figure 3 shows a production of "Design" task in Figure 2. Process flow graphs can describe processes in varying degrees of detail as the task called 'Design' has expanded progressively into Figs. 3 and 4. A graph containing many logical nodes describes what should be done without describing how it should be done (for example, specifying which tools to use). Conversely, a graph in which all task nodes are either atomic or selectors fully describes a complete methodology for a design as shown in Fig. 4 with the exception of 'HousingDesign.' With the abstraction mechanism, the overall process can be seen better after the detail tasks are hidden from the overall process.

6. PROCESS FLOW GENERATION AND EXECUTION

The Execution environment of MIDAS keeps track of the current state of the design process, generates a detailed process flow, and executes tasks and tools. Given a process flow, a detailed flow can be obtained by applying a production to a logical task. Fig. 3 shows one production for Design task. With the exception of 'HousingDesign,' Fig. 4 shows the fully expanded flow that is obtained after applying the productions for 'StaticAnalysis' and 'DynamicAnalysis' that were shown in Fig. 3. In this case, 'DynamicAnalysis' is simply expanded into a single atomic task called 'FatigueAnalysis' because there was no alternative production or any other atomic tasks. This elaboration may continue until all tasks become atomic tasks.

When a task is finished after executing all subtasks of the selected production, a post-evaluation function is performed to see if the output specification meets the constraints. When the output specifications of a task are not satisfactory, rollback must occur. For example, consider a mechanical design process that includes a stress analysis task. If the results indicate that stresses will exceed the yield strength of the material, then a rollback must occur. Either a new material or a new geometry must be selected. Post evaluation functions are like design review steps; they provide checkpoints in the execution of the process.

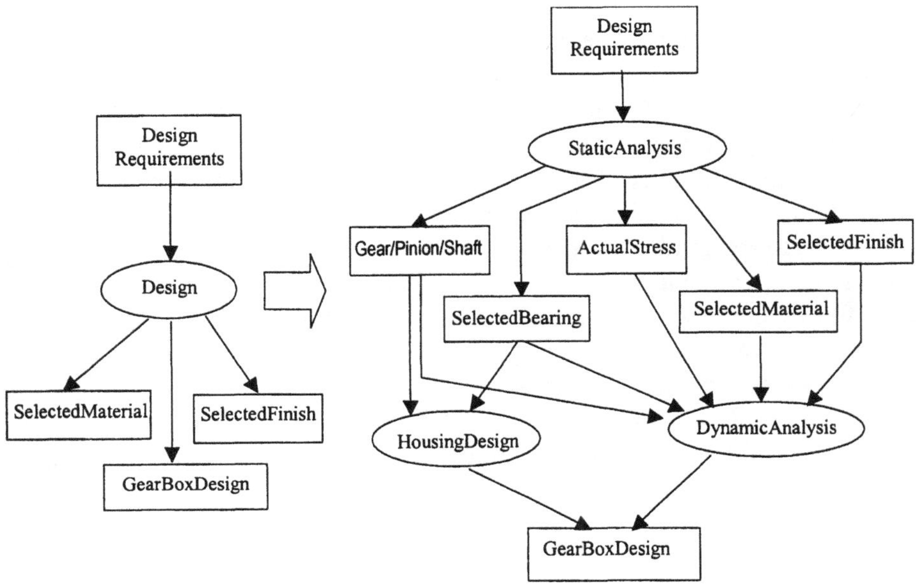

Figure 3: A Production for Designing a Gearbox

The Execution environment of MIDAS keeps track of the current state of the design process, generates a detailed process flow, and executes tasks and tools. Given a process flow, a detailed flow can be obtained by applying a production to a logical task. Fig. 3 shows one production for Design task. With the exception of 'HousingDesign,' Fig. 4 shows the fully expanded flow that is obtained after applying the *productions* for 'StaticAnalysis' and 'DynamicAnalysis' that were shown in Fig. 3. In this case, 'DynamicAnalysis' is simply expanded into a single atomic task called 'FatigueAnalysis' because there was no alternative production or any other atomic tasks. This elaboration may continue until all tasks become atomic tasks.

When a task is finished after executing all subtasks of the selected *production*, a post-evaluation function is performed to see if the output specification meets the constraints. When the output specifications of a task are not satisfactory, rollback must occur. For example, consider a mechanical design process that includes a stress analysis task. If the results indicate that stresses will exceed the yield strength of the material, then a rollback must occur. Either a new material or a new geometry must be selected. Post evaluation functions are like design review steps; they provide checkpoints in the execution of the process.

Rollback is the basic mechanism for handling iteration in MIDAS. When a rollback is necessary, different tools or parameters must be chosen or a task must be decomposed in an entirely different way. The capability of generating an alternative process in mid-stream is a critical aspect of the functionality provided by a grammar-based approach. Rather than simply searching for alternative design parameters, MIDAS supports searching for alternative processes with the update constraints. Rollback can be executed automatically or guided by manual intervention where a designer explicitly selects a new alternative. As with any design process, iteration continues until the final output specifications meet the design objectives. In this way, the process grammar naturally captures the hierarchical design methodology and allows systematic exploration of D&M space. Note that the iterative nature of the design process is not explicitly represented in the grammar. Rather, pre- and post-evaluations are embedded with each task to provide the necessary feedback loop.

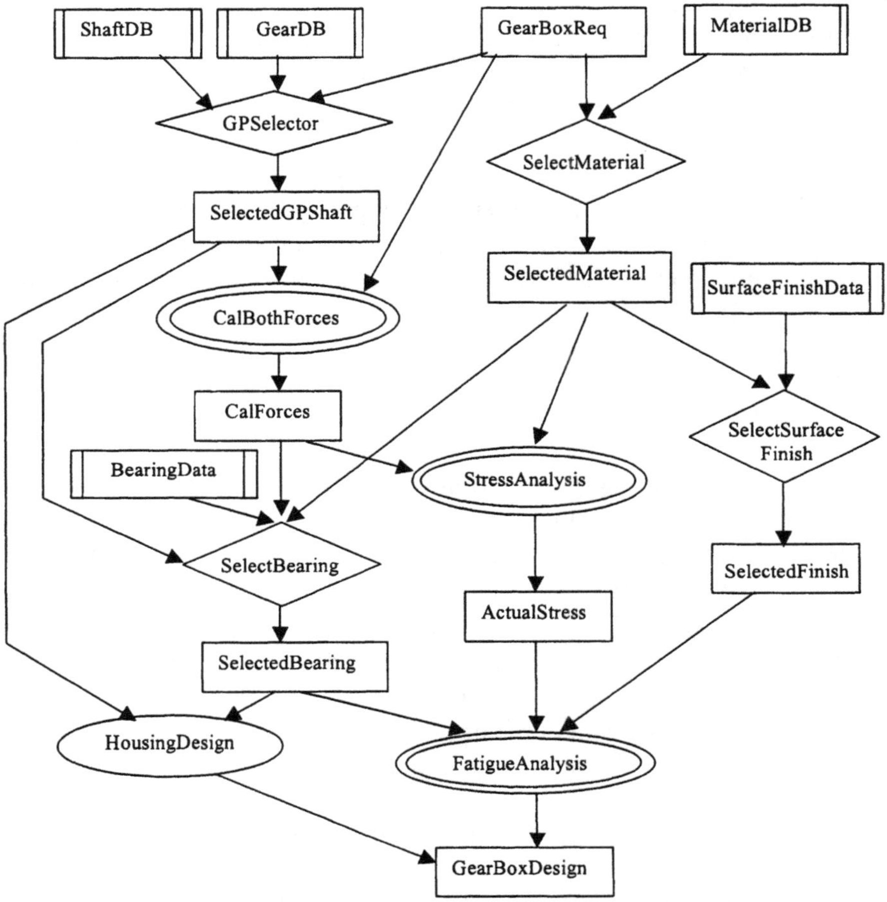

Figure 4. Expanded Flow Diagram for Design

MIDAS can be used in either interactive or automatic mode. In interactive mode, a designer directs the design process by interacting with Cockpit. Cockpit's display indicates to the user what design tasks have been done so far and what other tasks remain. Designers select tasks to be executed and the Cockpit determines what productions are available for remaining non-terminal tasks. It sends a message to the corresponding manager program requesting the rating of each production. The manager computes the rating of each production in the current situation based on the encoded knowledge in the manager program. This information is displayed to the user. In automatic mode, Cockpit selects the production with the highest rating. In interactive mode, Cockpit waits for the user to select a production. When a production for a given task is selected, Cockpit displays the refined process graph after applying the production. Cockpit also sends a message to the corresponding manager program to invoke the task. In automatic mode, it may be advantageous to explore several alternatives simultaneously. While the system explores one alternative automatically, the designer can explore another. The ability to explore several methodologies in parallel is absent in other systems and it would be very difficult to add this feature to them. During auto-mode execution, MIDAS executes tasks in a topological order [Brassard and Bradley, 1995] so that, for each task, its input specifications are available before its execution.

6.1 XML-based scalability

Enterprise systems are constantly evolving: new processes and products are added to the system, and new technology makes old constraints obsolete. In a truly scalable enterprise system, processes, tools, machines should be added in a modular way. To demonstrate the proof of concept, we will develop a prototype of an Extensible Markup Language (XML) based framework. XML is becoming the standard for Internet-based data interchange. It has been applied to workflow interoperability [WFMC, 1998] as a repository for Step Modules[Lubell 2000] and as a markup language of the machining and foundry manufacturing features [Veeramani and Garikapati, 2000].

XML is used to encode the process grammar, constraints and other business logic. From a processing point of view, XML promotes scalability because it can alleviate the processing load on the server side. As we separate process information into two layers: a syntactic layer and an execution layer, the processing of the syntactic layer can be moved to the client side if it is equipped with XML capability (such as Internet Explorer 5 or Netscape). XML tags provide common semantics used in the framework. The process logic encoded in XML is transformed into Java programs consisting of Java tag libraries. XML is also used as the data interchange element of MIDAS. While providing a common interchange format, XML still requires the specification of an appropriate Schema for a task and effective mapping of relational and object-oriented data to a hierarchical XML representation.

In some cases, selected processes are business transactions between partner companies. Processes in MIDAS have well-defined, generic interface for input and output. Therefore, their input and output data can be in arbitrary format and there is not restriction on transaction protocol. MIDAS can provide the interface with transaction management of ebXML [Riemer,

2001], meaning MIDAS can be integrated with industrial standard tools easily in terms of transactions between companies.

7. PERCOLATION AND SENSITIVITY ANALYSIS: PROCESS EXPANSION

The capability of generating alternative processes in mid-stream is a critical aspect of the functionality provided by a grammar-based approach. Rather than simply searching for alternative design parameters, MIDAS supports searching for alternative processes as well as parameters in a dynamic environment as the constraints are being constantly changed and updated. Design choice affect manufacturing options and vice versa, but the question is, how? And if any change occurs, which tasks are affected and who is responsible?

To address this system, MIDAS incorporates two key functions called Percolation and Sensitivity analysis. Percolation allows manufacturing constraints to propagate backwards to design; sensitivity analysis works the opposite way. Even with incomplete design data, constraints like manufacturing cost, manufacturing setting and machine shop resources can be propagated to the design tasks that are directly affected. Likewise, these constraints lead the designer to choose better processes and parameters in earlier stages of design. These functions are discussed here in the context of the gearbox example, along with the use of evaluation functions that control the execution of logical task.

7.1 Productions for Design Task

Percolation and sensitivity analysis allows engineers to foresee the consequences of their choices and hopefully avoid mistakes. This can be accomplished by using the grammar while expanding logical tasks into more specific or atomic tasks. Through this action, potential problems become apparent to engineers.

Let us return to our gearbox example at the design stage. As shown in Fig. 2, the objective is to design a gearbox that meets the design objectives and to manufacture it efficiently at a minimum cost. The design objectives (shown as GearBoxReq in Fig. 2) may be given in terms of operating condition and life. As shown in Figs. 3 and 4, we must select a pair of Gear/Pinion and Bearings from the component database. In this design (one way to design a gearbox), we assumed that the shafts be made from off-the-shelf round stocks and, by machining them, we can attain the specified dimension of a design. The selected shaft must be reduced to fit the gear and bearings based on the selected gear and bearings. Thus, the main design task for a shaft is to design the exact dimensions of a shaft with two fillet radii, the diameter of a shaft and the diameter of the bearing seat. The designed geometry has to withstand the loading condition and life provided in the design objectives. For an initial geometry, static analysis must first be carried out. The fillet radii, surface finish and the statically satisfied geometry are then analyzed under the fatigue cycle for operating life. If it does not meet the fatigue requirement even with the best possible finish, it is necessary to select another material or to increase the diameter. In some situations, the important design objectives may be to minimize the weight of the shaft as much as possible, the criteria for selecting a material should be based on strength per density (weight) as shown in the first task, SelectMaterial, in Fig. 4. Such information is well tabulated in Ashby [1992].

As will be discussed later, if we wanted to manufacture the shaft by casting, we cannot simply select the reasonable stock size but must design the dimension based on the analysis discussed earlier. In such a situation, an alternative design process diagram must be introduced, which will allow us to design the exact dimension, rather than searching though the database for a stock. This exact dimension information is needed to make a mold for casting. If the manufacturing resource is limited to casting, then this alternative design production must be selected, since casting is the only choice. There are some limitations in terms of fine details such as fillet radius, which cannot be achieved by casting. Either a fine detail cannot be specified or additional machining or grinding process becomes necessary. This is why the function provided by percolation is important.

With only the selected material, MIDAS can help to determine the best manufacturing method. Let us first consider casting process. The shape of a designed shaft is rather simple, which does not particularly eliminate any casting process among the possible casting methods such as Sand, Shell, Plaster, Investment, Permanent Mold, Dies and Centrifugal casting methods. For example, sand and permanent mod are the preferred methods for the magnesium alloy, EZ33A, while any casting method can be utilized for AZ91B [Schey, 1988]. For AZ91B, if dimensional accuracy and surface finish are important, Investment and Die casting methods are reasonable [Kalpakjian, 1994]. Between the two, die-casting achieves a shaft with less porosity while investment casting provides more flexibility in terms of the size and production rate of the part. Identifying these differences is a good motivation for introducing sensitivity analysis. Both sensitivity and percolation are illustrated in more detail later.

7.2 Productions for Manufacturing Task

For simplicity, we only consider the manufacturing processes of a shaft. When examining for process planning, it can be manufactured either by casting, machining and casting/machining as shown in Fig. 5. Three important data, SelectedFinish, SelectedMaterial, ShaftDesign, are required to execute the task called 'Manufacturing'. If the required finish can be met directly with a chosen material and cast method, casting will be the final logical task as shown in Fig. 5(b). Otherwise, if the required surface finish cannot be met, grinding (or machining) may be necessary as shown in Fig. 5(c). For many materials, deformation processing and heat treatment would be required after casting to enhance their mechanical strength. However, they have been omitted here again for simplicity. For a variety of materials, machining from a round bar (off-the-shelf stock) is the preferred method as shown in Fig. 5(d) with an additional process of Grinding.

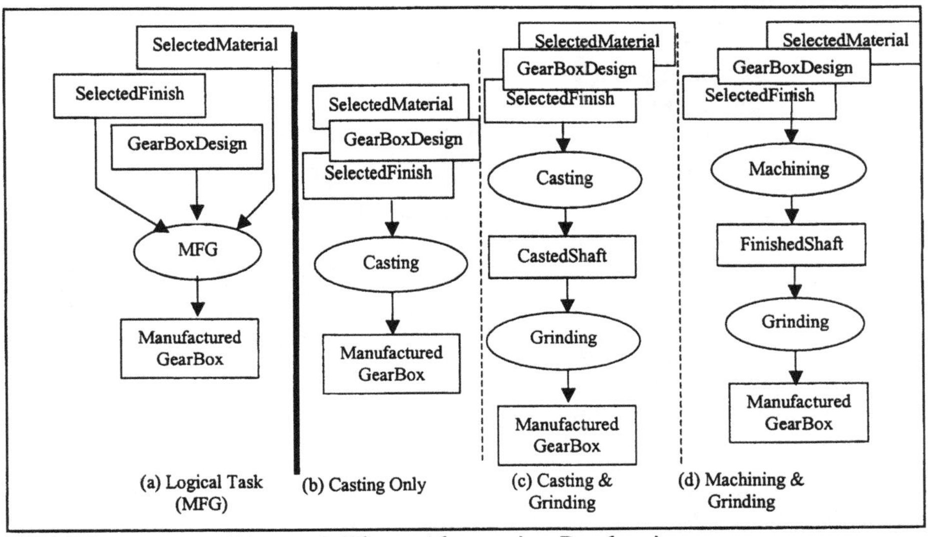

Figure 5. Three Alternative Productions

In machining, RoughCut is used to get rid of possible oxidation layers from a larger diameter round stock. After RoughCut, FineCut and Grinding are required to obtain the desired shape of the shaft as shown in Fig. 6. However, depending on the exact capability of the machine tool available, the manufacturing process plan will be drastically changed as shown in Fig. 7. For a manual machine, a skilled machinist must be assigned (Fig. 7(b)). For a

CNC turning machine, an appropriate CNC code can be developed (Fig. 7(c)). However, the tool must be changed when we progress through each stage of RoughCut, FineFinish and Grinding. If a machining center is available, even changing tools can be done automatically and the flow diagram shown in Fig. 7(b) can be used.

When a CNC turning machine is used (Fig. 7(c)), however, all the details must be worked out. For example, the amount of excess materials removed by RoughCut must be determined. Based on that, the insert must be determined for the best performance. The rest of the excess materials must be removed by FineCut and Grinding. For each operation, appropriate tools or a grinding wheel must be selected based on the capabilities of the machines and the material chosen for design. With a machining center (Fig. 7(d)), both RoughCut and FineFinish can be carried out in one operation by using a turret where multiple tools can be loaded before machining process begins. After these operations, Grinding can take place.

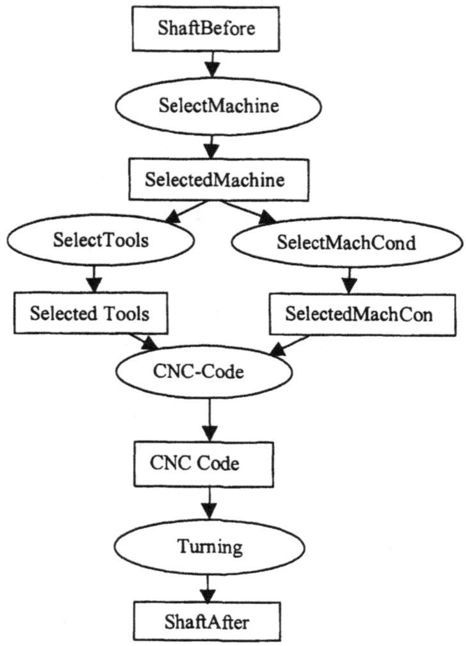

Figure 6 Tasks in Rough Cut or Fine Cut

Even though three independent logical tasks, RoughCut, FineCut and Grinding, are specified, a skilled machinist can perform all three tasks. RoughCut and FineCut include the following atomic tasks, selecting machining condition, selecting tools, assigning a machine and generating CNC. RoughCut and FineCut are generically the same process; therefore, the required tasks can be represented in one flow diagram as shown in Fig. 6. Grinding also has similar atomic tasks.

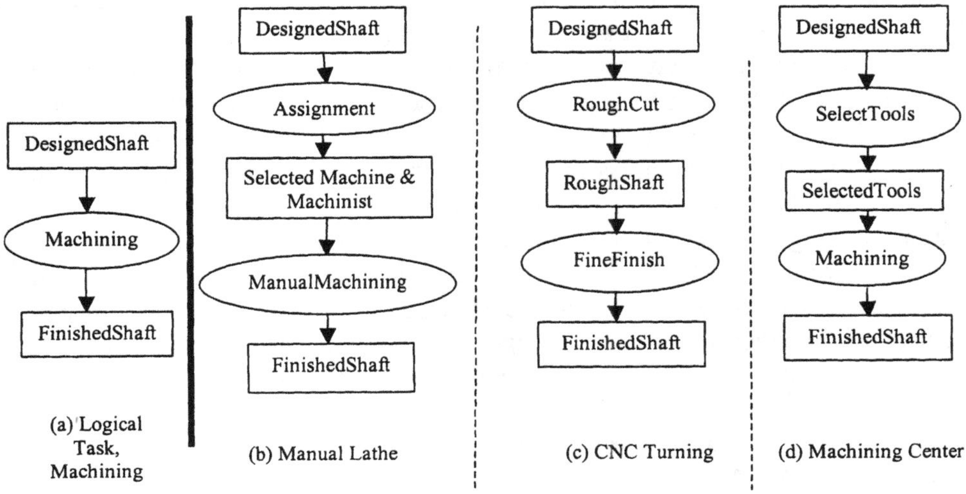

Figure 7: Three Different Productions for Machining

Productions for Machining	Requirements	General Constraints		Specific Constraints	
		Volume (V)	Set-up Cost (S)	Tolerance	Surface Finish
Manual Lathe	Skilled Machinist Manual Lathe	1	1	5-10	5-8
CNC Lathe	CNC Programmer CNC Lathe	9	5	9	5-9
Machining Center	CNC Programmer Machining Center	10	10	5	7-9

Table 2: Criteria for Selecting a Production for the task Machining

When many alternative productions are available for a task, pre-evaluation function is used to select the most promising production. Table 2 shows the criteria of selecting a particular production. The criteria are based on either the inherent capability limitation or the cost of each production. As shown in Fig 7(b), the machining plan for a manual lathe requires assigning a skilled machinist and a lathe. A skilled machinist must select a tool and a set of appropriate machining conditions for a given work material. These decisions may not have to be conveyed to any other person in the organization and they do not need to be represented as individual tasks. A skilled machinist can take care of these details. The specific constraints depend on the skill level and machine types etc. that are specific to aparticular location. To select the best production, users have to assign the 'weight' for each criteria category to determine the fitness of each production. The fitness is a fuzzy value, which are similarly used in many systems. It can be a spread-sheet type that provides the fitness of a production. Using this table, the pre-evaluation will weight the final fitness based on the weight of each factor provided by user, and total sum the weight as the fitness. For example, the pre-evaluation function can be in a form of Total Cost = V*S. The most-fit production will be the production with the minimum Total Cost. This information is used not only to select a production but also to manage interdependent processes. Because of different processes involved, the cost evaluation function may be different for each production. There are, however, many generic costs that are similar in most of the productions. Many different scenarios as have been discussed can be represented in MIDAS.

8. A SIMPLE EXAMPLE

Figure 8 shows a simple outsourcing arrangement, where several companies are engaged in manufacturing and delivering the finished parts to the customer. This example is drawn from our partners in the Jackson Smart Park.

In this scenario, Company A designs the basic part. Company B performs some specialized analytical work to confirm that the design will meet the requirements. Company C is the die maker, who prepares dies for both casting and trimming. Finally, Company D actually produces and finishes the parts once the dies are ready. Each step along the way requires a "checklist" both before the work is done and after. Before the work is done, the vendor can provide useful feedback about the feasibility of the proposed work. After the work is done, it is necessary to confirm that the requirements have been met.

Note that there may be many alternatives for each of these process steps: different technologies, different vendors, and so on. And at each step, there is a "make or buy" decision, as well. Also, for the overall process to function smoothly, all the participants need access to critical information (such as design changes). Ideally, they need to be able to provide feedback early in the process concerning the feasibility and schedule for their part of the work.

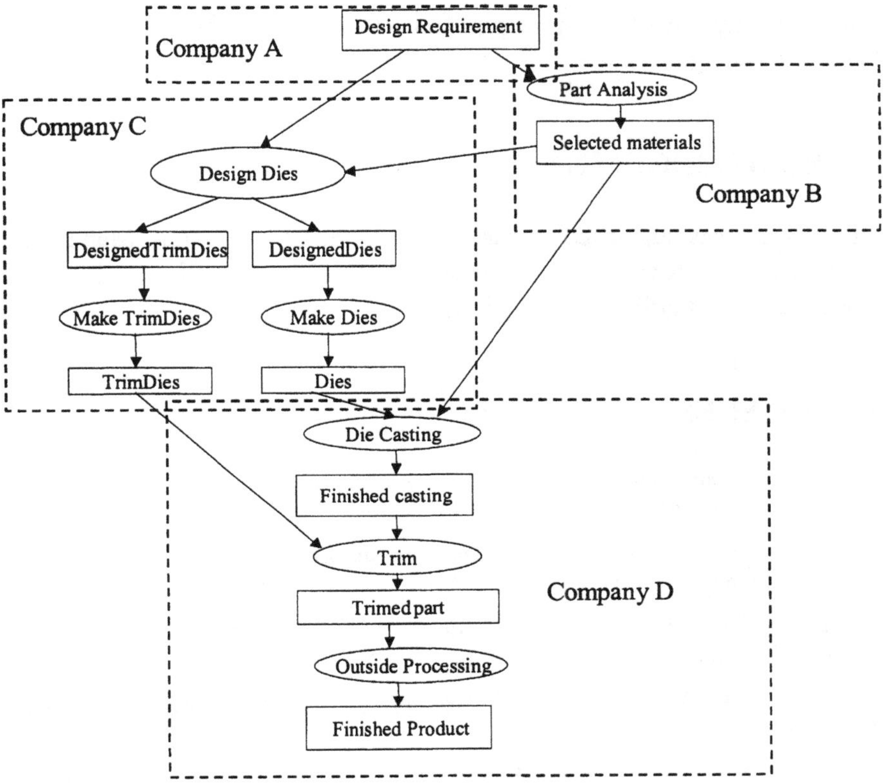

Figure 8: Four Companies in a Die casting Process

9. CONCLUSION

The process grammar approach is well suited to integrated D&M process and to enact the processes and the suppliers among a network of suppliers because it is inherently generative: process grammar provides a natural vehicle for modeling the iterative nature of design, analyzing the effect of one process on other processes, and the exploration of alternative processes. The process diagram captures the process sequence and the detailed tasks, which can be selectively shared to enable collaboration. The sensitivity analysis and percolation provide the effect of particular task to other tasks, thus allowing engineers to select a process and parameters that would meet the manufacturing constraints. The framework can also be used to navigate through the myriad of possible processes to adopt the best manufacturing processes as design changes.

ACKNOWLEDGEMENTS

This work was supported in part by the Division of Design, Manufacture and Industrial Innovation, Directorate of Engineering, National Science Foundation (Grant NSF DMI 0075396).

REFERENCES

Aberdeen Group White Paper, (2000), The e-Business Supply Chain: Meeting Customer Requirements in the Internet Economy.

Artiba, A. and S.E. Elmaghraby, (1997), The planning and scheduling of production systems : methodologies and applications New York : Chapman & Hall, 1997.

Andreoli, J.-M., Pacull, F., and Pareschi, R., (1998), "XPECT: A Framework for Electronic Commerce," *IEEE Internet Computing*, 1, 4, pp. 40—48.

Ashby, M. F., (1995), Materials Selection in Mechanical Design, Butterworth – Heinemann, Oxford, England.

Asimow, W., (1962), "An Overview of the Structure of the Design Process," in Emerging Methods in Environment Design and Planning, Ed. G.T. Moore, MIT Press, Cambridge, MA, pp. 285-307.

Baldwin, R. and Chung, M.J., (1995a), Design Methodology Management: A Formal Approach, IEEE Computer, February 1995, pp. 54-63.

Baldwin, R. and Chung, M.J., (1995b), "Managing Engineering Data for Complex Products," Research in Engineering Design, 1995, Vol. 7, pp. 215-231.

Berners-Lee, T., Cailliau, R., Luotonen, A., Frystyk Nielsen, H., and Secret, A., (1994), "The World-Wide Web," *Communicatiosn of the ACM*, 37, 8.

BYUPLAN Generative process Planning System User manual, (1985), Computer-Aided Manufacturing Laboratory, Brigham Young University, Provo, Utah.

Chan, F., Spiller, M., Newton, R., (1998), "WELD-An Environment for Web-Based Electronic Design", ACM Design Automation Conference, 1998, pp. 146-151.

Chang, T-C, Wysk, R. A. and Wang, H-P, (1991), Computer-Aided Manufacturing, Prentice-Hall.

Charles S., (1986), "The Impact of Applying Artificial Intellegence with Group technology and Computer Aided Process Planning System," Autofact Conference, Detroit, MI.

Chung, M.J., and Kwon, P., (1998), A Web-based Framework for Design and Manufacturing a Mechanical System, 1998 DETC, Atlanta, Georgia, Sep. 1998, CD-ROM.

Chung, M.J., Kwon, P. and Pentland, B. (2001), "Making Process Visible: A Grammatical Approach To Managing Design Processes," ASME Transaction, Journal of Mechanical Design, accepted.

Chung, M.J., and Kim, S., (1991), The Configuration Management for Version Control in an Object-Oriented VHDL Environment, in Proc. 1991 International Conference on Computer Aided Design, pp. 258-261, Nov. 1991.

Cox, J. F. and M. S. Spencer, (1998), The constraints management handbook. St. Lucie Press ; Falls Church, Va. : APICS.

Couretas, J. M. and Ziegler, B. P., (2001), "DEVs Framework for M&S Based Scalable Integrated Enterprises," http://www.eng.usf.edu/nsf/conference/scalable/scalable_enterprises/Ziegler_Uariz.ppt.

Cutkosky, M. R., Tenenbaum, J. M. And Glicksman, J., (1996), "Madefast: Collaborative Engineering over the Internet," Communication of the ACM, 39,9, pp.78-87.

DCLASS Technical Manual, (1985), CAM Software Research Laboratory, Bringham Young University, Grove, Utah.

Dellen, B. Maurer, F., Pews, G., "Knowledge-based techniques to increase the flexibility of workflow management", Data and Knowledge Engineering, North-Holland, 1997.

Dengler, D., (1994), "An Adaptive Deductive Planning System", Technical Report "RR-94-06", Deutsches Forschungszentrum fur Kunstliche Intelligenz GmbH Erwin-Schrodinger Strasse Postfach 2080 67608 Kaiserslautern Germany", 1994.

Eccles, R. G. and Nohria, N., (1992), Networks and Organizations, Boston: Harvard Business School Press.

Ehrig, H., Introduction to the Algebraic Theory of Graph Grammars, 1st Workshop on Graph Grammars and Their Applications to Computer Science and Biology, (1979), pp. 1-69, Springer, LNCS.

Erkes, J.W., Kenny, K.B., Lewis, J.W., Sarachan, B.D., Sobololewski, M.W., and Sum, R. N., 1996, "Implementing Shared Manufacturing Services on the World-wide Web," Communication of the ACM, 39, 2, pp.34-45.

Ghosh, D. and Tanniru, M (1997). Generating and evaluating process redesign options: A network modeling approach Annals of operations research. 72, 293-299.

Goel, A. K., Ali, K. S., Donnellan, M. W., de Silva Garza, A. G., and Callantine, T. J., (1994), "Multistrategy Adaptive Path Planning," IEEE Expert, 9(6), December 1994.

Gu, J., Purdom, W., Franco, J., and Wah, B. (1996), Algorithms for the Satisfiability (SAT) Problem: A Survey - DIMACS Series in Discrete Mathematics and Theoretical Computer Science.

Hines, K., and Borriello, G., "A Geographically Distributed Framework for Embedded System Design and Validation", ACM. 1998

IDEF, (1998), http://www.idef.com

Jacome, M. F. and Director, S. W., (1996), "A Formal Basis for Design Process Planning and Management," IEEE Transactions on Computer-Aided Design of Integrated Circuits and Systems, 15, 10, pp. 1197-1211.

Jagadish, H. V., (2001), "Coopetitive Design and Manufacturing Across Organizations," http://www.eng.usf.edu/nsf/conference/scalable/scalable_enterprises/Jagadish_Umich.ppt

Janusz, B., (1997), Model-based process redesign. Journal of intelligent manufacturing. 8, no. 5, 345.

Kalpakjian, S., 1992, Manufacturing Processes for Engineering Materials, 2nd Edition, Addison-Wesley, Reading, Massachusetts.

Kamath, M., (2001), "ADVANCED MODELING PROJECT (OSU-CIM)", http://www.okstate.edu/cocim/advmod/

Kempenaers, J., Pinte, J., Detand, J., and Kruth, J., "A Collaborative Process Planning and Scheduling System", Computer in Engineering Vol. 1, ASME, 1994.

Keyes, D., Dillon, L. and Chung, M. J., (1999), "Analysis of a Scheduler for a CAD Framework," International Conference of Software Engineering 1999.

Knapp, D.W. and Parker, A.C, (1991), "The ADAM Design Planning Engine," IEEE Transactions on Computer Aided Design of Integrated Circuits and Systems, pp. 263-285.

Kwon, P., Crimp, M., and Chung, M. J., Automating the Design Process and Powder Fabrication of Functionally Gradient Materials, ASME Int. Symposium Annual Winter Conference, Dallas, Dec. 1997.

Lander, S. E., Staley, S. M. and Corkill, D. D., (1995), "Designing Integrated Engineering Environments : Blackboard-Based Integration of Design and Analysis Tools," IJCAI-95 Workshop Intelligent Manufacturing Systems, AAAI.

Lavana, H., Khetawat, A., Brglez, F., and Kozminski, K., (1997), "Executable Workflows: A Paradigm or Collaborative Design on the Internet," Proceedings of the 34th ACM/IEEE Design Automation Conference, June.

Lubell, J. (2000) "An XML Repository Architecture for Step Modules," NIST Technical report.

Malone, T. W., (1987), Modeling coordination in organizations and markets. , 33, 1317–1332.

Malone, T. and Crowston, K., (1994), "The Interdisciplinary Study of Coordination." ACM Computing Surveys 26: pp. 87-119.

Malone, T. W., Crowston, K. G., Lee, J., Pentland, B., Dellarocas, C.,Wyner, G., Quimby, J., Osborn, C. S., Bernstein, A., Herman, G., Klein, M., & O'Donnell, E., (1999), "Tools for inventing organizations: Toward a handbook of organizational processes," Management Science, 45, 3, pp. 425-443.

Mehra, A., Minis, I., Proth, J., "Hierarchical Production Planning for Complex Manufacturing System", Computer in Engineering Vol. 1, ASME, 1994

Melis, E., and Meier, A., (2000), "Proof Planning with Multiple Strategies", Lecture Notes in Computer Science, vol. 1861, pp. 644-650.

Miura, J. and Shirai, Y., "Knowledge-Based Control of Decision Theoretic Planning - Adaptive Planning Model Selection", 14th Workshop New Results in Planning, Scheduling and Design (PuK2000), pp. 107-114, Berlin, August 2000.

Nau, D.S., Gupta, S.K. et al. (1995) AI planning versus Manufacturing-Operation Planning, A Case Study. IJCAI-95.

Newell, A., & Simon, H. A., (1972), Human Problem Solving. Englewood Cliffs, NJ: Prentice-Hall.

Petrie, C., "Agent-Based Engineering, the Web, and Intelligence", IEEE Expert, Dec, 1996. see http://cdr.stanford.edu/NextLink/Expert.html

Petrie, C., Jeon, H., and Cutkosky, M., "Combining Constraint Propagation and Backtracking for Distributed Engineering", ECAI-96 Workshop on Non- Standard Constraint Processing, Budapest, August, 1996, revised for AAAI-97 Workshop on Constraints and Agents, Providence, RI, Jul, 1997. see also "http://cdr.stanford.edu/ProcessLink/papers/non-stan-const/non-stan-const.html".

Rea, H. J., Falconer, R. A., Murray, J. L., and Simmons, J. A.,(1988), "A Structured approach for assembly planning," ASME International Congress and Exposition – 5, Symposium on Assembly Modeling and Assembly Systems; 1998, Anaheim, CA.

Riemer, K., (2001), ebXML Business Process, http://www.gca.org/papers/xmleurope2001/papers/s18-1.html.

Shaw, M. J., (2001), Information-based manufacturing : technology, strategy and industrial applications. Boston : Kluwer Academic.

Schey, J. A., (1987), Introduction to Manufacturing Processes, 2nd edition, McGraw-Hill, New York, NY

Schlenoff, C., Knutilla, A. and Ray, S., (1996), Unified Process Specification Language: Requirements for Modeling Process, NISTIR 5910, NIST.

Schmidt, L. C., and Cagan, J., (1997), "GGREADA: A Graph Grammar-Based Machine Design Algorithm," Research in Engineering Design, 9, pp. 195-213.

Schurmann, B. and Altmeyer, J., "Modeling Design Tasks and Tools - The Link between Product and Flow Model -," Proceedings of 34th ACM/IEEE Design Automation Conference, June 1997.

Simon, H. A. (1973) "The structure of ill-structured problems," Artificial Intelligence, 4: 181-201.

Sriram, R.D., (1997), Intelligent Systems for Engineering, Springer-Verlag, 1997.

Steinhage, A., and Schoner, G., (1997), "The dynamic approach to autonomous robot navigation", Proc. of the IEEE International Symposium on Industrial Electronics, ISIE'97, pp. SS7--SS12.

Sutton P. R., and Director, S. W., (1998), "Framework Encapsulations: A New Approach to CAD Tool Interoperability," Proceedings of the 35th ACM/IEEE Design Automation Conference, June 1998.

Tate, A., "Representing Plans as a Set of Constraints - the <I-N-OVA> Model", Proceedings of the Third International Conference on Artificial Intelligence Planning Systems, Edinburgh, UK, May, 1996. See also ftp.aiai.ed.ac.uk/pub/documents

Tulkoff, J., (1981), "Process Planning in the Computer Age," Machine and Tool Blue Book.

Veeramani, D. and Garikapati, R., (2000), "XML based Markup Languages for Knowledge Management in Manufacturing Companies," CD-ROM DETC 2000/CIE-14603.

vom Ende, A., Helmke, M., Dobermann, M., and Naumann, T., (2000), "A Concept for an Adaptive Process Management in a Distributed Product Development Environment," ProSTEP Science Days 2000, September 2000.

Whiteside, R.A., Pancerella, C. M., and Klevgard, P. A., (1997), "A CORBA-Based Manufacturing Environment", Proceedings of the Hawaii International Conference on Systems Sciences , January 1997.

Williamson, O., (1975), Markets and Hierarchies, New York: Free Press.

Wong, T.N. and Siu, S. L., (1995), "A Knowledge-based Approach to Automated Machining Process Selection and Sequencing," International Journal of Production Research, 34 (1), 117-133.

WFMC, Workflow Management Coalition , (1998), Workflow Security Considerations –White Paper, Document Number WFMC-TC-1019,. Feb 1998.

Wu, B., (2000), Manufacturing and supply systems management : a unified framework of systems design and operation. London ; New York : Springer.

Yin, G. and Zhang, Q., (1996), Recent advances in control and optimization of manufacturing systems. London ; New York : Springer.

Yourdon, E., (1989), Modern Structured Analysis, Englewood Cliffs, NJ.

Chapter 3

FINITE AUTOMATA MODELING AND ANALYSIS OF SUPPLY CHAIN NETWORKS

Houshang Darabi
University of Illinois at Chicago, Department of Mechanical and Industrial Engineering, Chicago, Illinois, USA.

Mohsen A. Jafari
Rutgers University, Department of Industrial and Systems Engineering, Piscataway, New Jersey, USA.

ABSTRACT

This chapter introduces a new approach for modeling and design analysis of supply chain networks. A supply chain defines a network of enterprises that should work in harmony according to a set of business protocols. Traditionally, the business protocol within a supply chain is pre-defined so that changes in one business may potentially require extensive changes and set-ups on the other business partners. What we are proposing in this chapter provides the framework for a new paradigm so that some aspects of business protocol defined across the supply chain do not have to be pre-defined. Instead the inter- and intra-business logic within the supply chain provides re-configurability and scalability across the network. Our modeling and design paradigm is built upon the results obtained from the discrete event control-theoretic subject area. The results from our methodology provide a blueprint for the design of enterprise software that coordinate the various tasks within an enterprise and between the enterprises in a supply chain network.

1. INTRODUCTION

In this new information age, customers have become increasingly demanding, with higher expectations on the quality of products, services and customization. In response to these pressures, supply chain networks have been undergoing tremendous changes. Supply chains will change their information flows from the current predetermined path to dynamic and flexible flows that span all supply chain participants. Therefore, the workflow models reflecting the traditional supply chain environment with a static view are not suitable, as they require that all the components will be known in advance, and assume that the specifications do not change during the model execution. In reality, some tasks may not be known in advance, or some dependencies may have to be selected or resolved at run-time. Run-time changes can be user requested changes, or changes necessitated due to exceptions and/or driven by market forces. In this chapter we introduce a new modeling approach for modeling and design analysis of supply chain networks.

Our formalism, which uses the results obtained from Discrete Event Systems (DES) control theory, enables scalability while providing a framework for optimization. Furthermore, this formalism can generate a descriptive roadmap for the operation of the enterprise, and yet it is analytical so that it can be used as basis for system validation while addressing such issues as redundancy and conflict between business rules. With our formalism we distinguish between the process (business model) and its control (business rules). A major advantage to this view is that the external changes that require modification to the underlying business process (es) will be dealt with by remodeling the process behavior. This is a very natural view of real systems. For example, in a manufacturing environment the changes in production schedule, product demands, and product mix ratios are all related to the control of the process. However, changes in system production capacity, machine failures, failure in communication and sensory network or changes in production processes are inherent to process behavior. The separation of the process behavior from the rules and laws that govern the system is essential to the fulfillment of the scalability feature. In our view scalability requires that the behavioral aspects of the enterprise (including the governing rules and objectives of the organization) should remain invariant with fluctuation or increase in the level of demand for the system resources. On the other hand, this does not necessarily mean that the computing or the functional resources required to perform the transactions within the enterprise should not be revamped by increase in the number of business requests. With our formalism, scalability means merely duplicating the building blocks of our process model without changing the control laws.

A supply chain is defined as a set of coordinated process activities that are carried out by processing entities to achieve a common business objective [16]. An activity, or a task, is a logical step or description of a piece of work that contributes towards the accomplishment of a process [7, 15]. A task can be a manual or automated processing activity. Processing activities can be humans, application programs or database management systems [6, 15].

Finite Automata techniques have been used for modeling and analysis of many practical systems including Communication [14], Chemical [13], database [12], manufacturing [2] and Traffic [10]. As already mentioned, we will address the application of FA techniques for modeling and analysis of supply chain networks. We specifically use Ramadge-Wonham [14] method to develop the supremal behavior of a supply chain. This behavior, which is shown by a state transition chart, includes every possible sequence of supply chain task execution that is acceptable to the supply chain users. In real problems, the behavior of state space has an exponential growth when the numbers of tasks increases. As an Example, for a supply chain with 10 activities (tasks) and 3 task statues, the number of states is more than 3^{10}. Though this might seem as a disadvantage, we believe that the proposed method is practical and the state space problem can be resolved because of the following reasons:

1. The behavior is generated by an efficient algorithm and the user does not have to build the states one by one [17, 18, 19].
2. Several software are available for an automatic implementation of the process [17, 18, 19].
3. After the creation of the state chart, several reduction techniques are available to reduce the number of states [3, 9, 15].

The objective of this chapter is not to discuss the above works, but to show how FA analysis can be applied for supply chain modeling and analysis. Therefore, the above references can be used to efficiently implement the results of this chapter.

In Section 2, we present the supply chains and DES preliminaries. In Section 3, we provide the reader with a modeling procedure for DES modeling of supply chain networks. In Section 4, we use the established model to analyze different properties of supply chains. Section 5 presents a case study for a restaurant. In Section 6, we drew the conclusions and present the future research potentials.

2. PRELIMINARIES

2.1 Supply Chains: Tasks and Dependencies

A supply chain is a collection of activities, the support of which, are required for the completion of any process. Therefore an activity, or a task, is *any* work, which contributes towards the completion of a process. Hence a supply chain can be thought of as a collection of tasks together with the associated relationship between the various tasks.

Consider a restaurant, which consists of various participating entities like the purchasing section (ingredients), the customer section (service) and the manufacturing section (cooking). Each section by itself represents a supply chain because it has a collection of activities in it, done for finishing the tasks assigned to that particular section. There is also a co-ordination between all the individual sections of that restaurant to achieve the overall objective i.e. the functioning of the restaurant. Thus it is evident that the "*individual*" tasks combine to form the whole "supply chain".

A *task*, which forms the basis of any supply chain, might be expressed as a collection of task primitives and states. Primitives {begin, abort, pre-commit, commit} are commands that change the state of the task. A task can be in one of the following states {Initial, Execution, Done} at any time.

There are certain guidelines to be followed in the execution of any task:

1. Use of the "*begin*" primitive alone can initiate a task.
2. Use of either a "*commit*" or an "*abort*" primitive can terminate a task.
3. A task at any stage can exist in only *one* of the following states {Initiation, Execution, Done}.

As we know, the completion of individual tasks alone cannot signify the completion of a supply chain. It is the combination of tasks, which are in proper coordination with one another, which guarantees the successful execution of a supply chain. This coordinating relationship between the tasks is known as "Task Dependency". Task dependency is broadly categorized into three types: *Control flow* dependency, *Value* dependency and *External* dependency.

Control flow dependencies are specified in terms of the task primitives. Let us consider two tasks tw_i and tw_j which are in the states st_i and st_j respectively. Using the control flow dependency, it is possible to specify the condition like the state st_i of a task tw_i, under which a task tw_j can enter state st_j. Table 2.1 shows some samples of the standard control flow dependencies used in supply chains.

Dependency	Description
Begin-on-commit dependency	Task tw_j can not begin until task tw_i commits
Begin dependency	Task tw_j can not begin until task tw_i has begun
Termination dependency	Task tw_j can terminate only after the completion (abort or commit) of task tw_i.

Table 2.1: Samples of Standard Task Dependencies

For example a business rule which states that the purchasing department is allowed to reorder items only if the inventory falls down to a certain level is shown by a *Begin-on-commit* dependency.

Value dependencies are based on general conditions. Specifically, a value dependency states that a task tw_j can enter state st_j only after the outcome of task tw_i satisfies a condition c_i, where the condition can be a logical expression whose value is 1 or 0.

External dependencies are caused by some parameters external to the system, such as time. However, they can be based on *any* external parameter. An external dependency states that a task tw_i can enter state st_i only if a certain condition c_j is satisfied where the parameters in c_j are external to the system. For example, if it is specified that the execution of a task tw_i can start *only* at 10:00am, it is an external dependency.

2.2 Discrete Event System Theory - A Control Theoretic Approach

This section gives a brief overview of the *Finite Automata* approach of DES. The modeling and design analysis of supply chain networks discussed in this chapter are based on this theory. We begin with the basic concepts and terminology associated with the DES theory.

Discrete event systems are *event driven systems*. An event should be thought of as occurring instantaneously and causing transitions from one state to another. In DES, the state changes depend entirely on the occurrence of *particular discrete events* over time. As an example for a DES, let us consider the working of a machine. This is a general example, which will explain the concept of event driven systems. Let us assume that the machine is initially in the OFF state. When a button is pressed, the machine in response to the identified event, takes a specific action. So the state of the system could undergo a transition from the initial OFF state to the ON state.

Any DES has an underlying finite event set Σ associated with it. The event set Σ could be thought of as the "alphabet" of a language and event sequences are thought of as "words" or "strings" in that language. So, a language that is defined over an event set Σ is a set of finite-length strings formed from events in Σ. As an example, let $\Sigma = \{a,b,c\}$ be the set of events. Then we may define a language $L1 = \{ab, ccc, abc\}$ consisting of three strings only; or another language $L2 =$ {All possible strings of length three starting with event a } which contains a finite number of strings.

The key operation involved in building strings, and thus languages, from a set of events Σ is called *concatenation.* The string ab in language $L1$ is the concatenation of the events a and b. Let Σ^* be the set of *all* finite strings of elements of Σ. If $\Sigma = \{a,b,c\}$, then $\Sigma^* = \{a,b,c,aa,ab,ac,ba,bb,cc,\cdots\}$.

To summarize, a language is the set of all possible orders of events that could happen in a system. With this information, the next step is to know a means of representing the language. This is explained using the automata theory as follows. An automaton is a device that is capable of representing a language according to well-defined rules. Figure 2.1 shows the automaton of the language $L1 = \{ab, ccc, abc\}$. In this figure each rectangle represents a "*state*". The labeled arcs represent the *transitions* between these states. $Q = \{1,2,3,4,5,6,7\}$ · represents the set of states of the automaton. The set of transitions here represent the event set $\Sigma = \{a, b, c\}$ for the finite automaton which are responsible for state transitions. The arcs provide a graphical representation of the *transition function* of the automaton. This transition function is denoted as $f : \Sigma \times Q \to Q$.

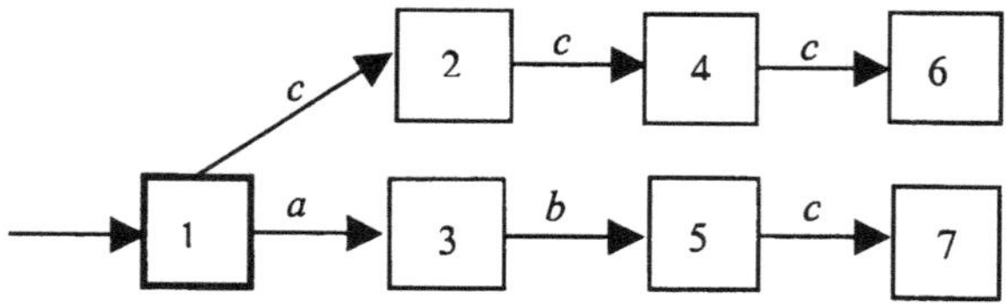

Figure 2.1: State transition diagram for language $L1$

For language $L1$:

$f(1,a) = 3$ $\quad f(1,c) = 2$ $\quad f(2,c) = 4$

$f(3,b) = 5$ $\quad f(4,c) = 6$ $\quad f(5,c) = 7$.

In addition to $\{Q, \Sigma, f\}$ there are two more components required to completely describe a finite automaton. They are

1. An initial state, generally denoted by q_0. Here $q_0 = 1$
2. A subset Q_m of Q that represents the states of Q that are marked. Marked states are also referred to as "final" states. Here $Q_m = \{6, 7\}$.

Formally, a Finite Automaton (FA), G is denoted as $G = (Q, \Sigma, f, q_0, Q_m)$ where Q is the set of *states*, Σ is the finite set of *events* associated with the transitions in G, $f : Q \times \Sigma \to Q$ is the transition function, q_0 is the initial state and $Q_m \subseteq Q$ is the set of marked states. Let the language or behavior generated by $G = \{Q, \Sigma, f, q_0, Q_m\}$ be $L(G) = \{s \in \Sigma^* : f(s, q_0) \text{ is defined}\}$. Here the function for transition from the initial state q_0 by string s can be defined by a sequence of event transitions starting from q_0. Mathematically, $f(s, q_0)$ is defined as follows: If $s = \sigma_1, \sigma_2 .... \sigma_n = \sigma_1 t$, where $\sigma_1, \sigma_2, \cdots, \sigma_n \in \Sigma$, $f(s, q_0) = f(\sigma_1 t, q_0) = f(t, f(\sigma_1, q_0))$ and $f(\sigma_1, q_0)$ is an event transition. Let $f(\sigma_1, q_0) = q_1$ and $t = \sigma_2 w$. Then $f(s, q_0) = f(\sigma_1 t, q_0) = f(t, f(\sigma_1, q_0)) = f(w, f(\sigma_2, q_1))$. In the same way the transition by string w can be broken to individual event transitions. The language $L(G)$ represents all the directed paths that can be followed starting at the initial state. Therefore a string s is in $L(G)$ if and only if it corresponds to an admissible path in the state transition diagram or equivalently, if and only if f is defined at (q_0,s). The automaton G models the "uncontrolled behavior" of the DES. In practice G has to satisfy a set of user-defined specifications or *control specifications*. Therefore the behavior of G, $L(G)$ should be restricted to those strings that satisfy the control specifications. This restriction on the behavior of G is imposed by the *supervisor* (S), which is described in Section 2.3.

Mathematically, the union of n finite automata, $G_i = (Q_i, \Sigma_i, f_i, q_{0_i}, Q_{m_i})$, $i = 1,2,\cdots,n$, is a finite automaton $G = (Q, \Sigma, f, q_0, Q_m)$ where $Q = Q_1 \times Q_2 \times \cdots \times Q_n$, $\Sigma = \Sigma_1 \cup \Sigma_2 \cup \cdots \cup \Sigma_n$, $q_0 = (q_{0_1}, q_{0_2}, \cdots, q_{0_n})$, $Q_m = Q_{m_1} \times Q_{m_2} \times \cdots \times Q_{m_n}$. The transition function f, is defined as, $f((q_1,q_2,\cdots,q_n),e) = (f_1'(q_1,e), f_2'(q_2,e),\ldots, f_n'(q_n,e))$ where

$$f_i'(q_i,e) = \begin{cases} f_i(q_i,e) & \text{if } e \in \Sigma_i \\ q_i & \text{if } e \notin \Sigma - \Sigma_i \end{cases}, \text{ for } (q_1,q_2,\cdots,q_n) \in Q \text{ and } e \in \Sigma.$$

2.3 Supervisory Controller

As already discussed, supervisor **S** is used to restrict the behavior of G to satisfy the control specifications. Roughly speaking, **S** sees (or observes) some, possibly all, of the events that G executes. Then, **S** tells G which events in the current active set of G are allowed next. More precisely, **S** has the capability of disabling some, but not necessarily all, feasible events of G. The decision about which events to disable will be allowed to change whenever **S** observes the execution of a new event by G. In this manner, **S** exerts dynamic feedback control of G. If **S** observes only a subset of events of Σ, then **S** has a partial observation of G. Conditions for proper control of G by **S** under partial observation are given in the Appendix A.4.

Formally, the supervisor **S** consists of two elements **S** $=(S,\Phi)$. $S = (X, \Sigma, g, x_0, X_m)$ is an automaton with state set X, initial state x_0, a marked subset $X_m \subseteq X$, and transition function $g: \Sigma \times X \to X$, while $\Phi: \Sigma \times X \to \{1,0,dc\}$ is a feedback function that selects a control policy $\Phi[x](\sigma)$ for each $x \in X$ and $\sigma \in \Sigma$ such that

$$\Phi[x](\sigma) = \begin{cases} 1 & \sigma \in \Sigma_u, x \in X \\ \{0,1,dc\} & \sigma \in \Sigma_c, x \in X \end{cases}.$$

Where Σ_c is the set of controllable events i.e. these are the events that can be prevented from happening, or disabled, by supervisor **S** and Σ_u is the set of uncontrollable events i.e. the events that cannot be prevented from happening by the supervisor. $\Phi[x](\sigma) = 0$ (1) is used to disable (enable) event σ while the state of S is x. The notation *dc* (don't care) implies that the assignment of 0 or 1 to $\Phi[x](\sigma)$ has no effect on the behavior of G. The finite automata analysis of DES is also called as the RW algorithm, named after W.M. *W*onham and P.J. *R*amadge [14], the developers of key results of this theory. The RW algorithm generates the supervisor for DES G by accepting DES behavior, $L(G)$ and the control specifications. From now on, by the word "*supervisor*", we mean the supervisor which is generated by the RW algorithm.

The control specification of a DES can be modeled by a finite automaton, say H. Automaton H, that can generate the language $L(H)$, is called *specification model.* $L(H)$, is called *desirable behavior.* Let $K = L(\mathbf{S}/G)$ be the coupled language (or controlled language) generated by discrete event system G under the supervision of $\mathbf{S} = (S, \Phi)$. The term (S/G) is interpreted as a language generated by G under the supervision of **S**. As already mentioned, **S** supervises G by enabling and disabling of the events at the right time. $K = L(\mathbf{S}/G)$ is said to be *Supremal* since it only prevents the occurrence of those strings of $L(G)$ that violate the control specifications. Therefore all other strings that are legal are allowed to happen.

In the subsequent sections, we apply the theory of *Supervisory Control* to verify some useful properties of supply chains. We first attempt to build the behavioral and specification models of the supply chain. The *behavioral model* is a finite automaton that includes every possible behavior of the supply chain. The specification FA model shows the desirable behavior of the supply chain. We next attempt to construct a supervisor that coordinates the supply chain tasks. Using the coupled language of the supervisor and the supply chain, we verify properties such as task redundancy, event controllability and scalability of the supply chain.

3. SUPPLY CHAIN MODELING

3.1 Deriving the Behavioral Model

Consider a supply chain, comprising of n tasks $\{tw_i, i = 1,2,\cdots,n\}$. While deriving the behavioral model of this chain, it is assumed that every task, tw_i, is expressed by the FA of Figure 3.1. The initial state IN_i is different from the other states by its bold rectangle. Let L_i be the language generated by the FA of Figure 3.1. Table 3.1 and Table 3.2 describe the states and the events of this FA respectively. In Table 3.2 each event is either controllable or uncontrollable. As already mentioned, a controllable event is one that can be disabled or enabled at any time by the supervisor. An example is an order processing system, which lets only valid customers begin an order-processing task. In this case "begin" event of an order-processing task is considered as "controllable" since it can be enabled/disabled at any time by the system. In the same system, "abort" event of an order is uncontrollable, under the assumption that an order is aborted when a customer cancels the order. This is because the customer's decision regarding finalizing or canceling the order cannot be controlled by the system. The controllability of events given in Table 3.2 (3rd column) is not fixed and might be changed for different problems or tasks. For example event b_i of task tw_i might be controllable and event b_j of task tw_j might be uncontrollable. The problem of "change in controllability of event" will be discussed in Section 4.

For the FA shown in Figure 3.1, we make the following assumptions:

1. Each task must finally be either *committed* or *aborted.*
2. After the commit or abort of any task, it can start again, i.e., the beginning event can happen.

Now let us construct the behavioral model G of the supply chain with the set of tasks $\{tw_i, i = 1,2,\cdots,n\}$. As a result of our previous discussion, we know that G can be constructed by taking the union of finite automata of tasks tw_i, $i = 1,2....,n$ (Figure 3.1).

Although the suggested task automaton in Figure 3.1 has been very useful for us in modeling many practical systems, it should be mentioned that we do not fix this model for all applications. In fact we encourage the designers to develop their own choice of task models depending on their specific application. For Example we have used different task models for the case study in Section 5.

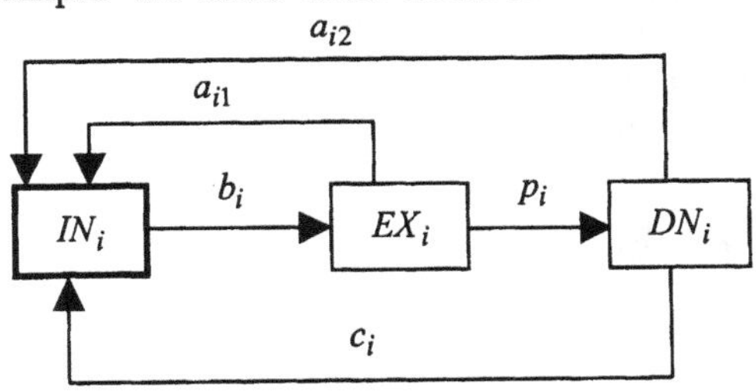

Figure 3.1: Task Automaton

State	Description
IN_i	Initial
EX_i	Execution
DN_i	Done

Table 3.1: Task Automaton states

Event	Description	Controllability
a_{i1}	Abort while execution	Uncontrollable
a_{i2}	Abort when done	Uncontrollable
b_i	Begin	Controllable
c_i	Commit	Uncontrollable
p_i	Pre-commit	Uncontrollable

Table 3.2: Task Automaton events

Example 3.1: An order processing system

Let us consider an order processing system (a supply chain). It is assumed to be comprised of two sequential steps: 1. Order processing 2. Order approval. This supply chain encompasses all the events in the system starting from the placing of an order with the system until the delivery of the product to the customer. It works in the following manner. Whenever an order is placed with the system, the order approval step is handled by either of two inventory managers. Manager 1 is a local manager. He can approve a processed order by signing it or he might reject the order. Manager 2 is a remote manager (not present in the inventory processing system location). But he can approve or reject the processed orders through an existing communication line between the order processing system and his office. In this example, we assume that both managers use the same set of rules for order approval. This assures that there will not be a conflict in their decisions. It is also assumed that to complete the processing operations, the order processing system must receive at least one manager's decision. Three tasks can be considered for this example: 1. Order Processing, (tw_1), 2. Local Manager order approval (tw_2), and 3. Remote Manager order approval (tw_3). Each of these tasks can be represented using the task automaton of Figure 3.1. Figures 3.1a and 3.1b show the finite automata of order processing and local manager approval tasks respectively.

A similar model can be constructed for the remote manger approval also. These three individual models are combined to generate the required supply chain behavioral model. It is noted that each model has its own set of controllable and uncontrollable events. A description of the states and events of this system are given in Table 3.3 and Table 3.4 respectively.

Here we have three tasks each with three states. Hence the behavioral model of this system will have 27 states. A partial representation of this model is given by Figure 3.1c. A complete description of the finite automaton in matrix form is given in the Appendix A.1.

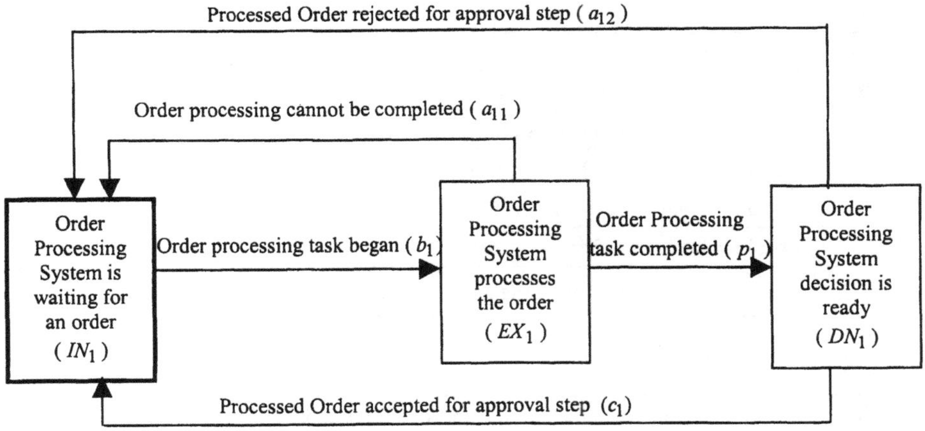

Figure 3.1a: Task automaton for order processing

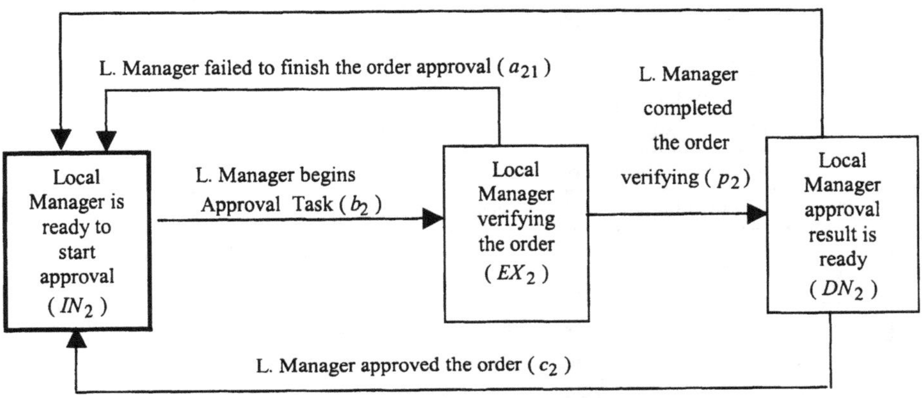

Figure 3.1b: Task automaton for Local Manager Approval

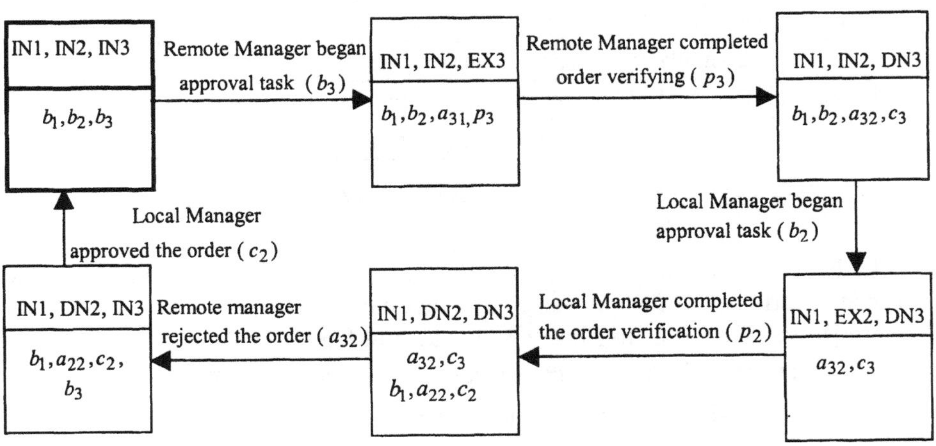

Figure 3.1c: Partial representation of behavioral model for example 3.1

State	Description
IN_1	Order processing unit is waiting for a new order
EX_1	Order processing unit is processing an order
DN_1	Order processing unit has completed the processing operations
IN_2	Local manager is ready to start the order approval operation
EX_2	Local manger is verifying an order
DN_2	Local manager is done with the approval operation
IN_3	Remote manager is ready to start the order approval operation
EX_3	Remote manger is verifying an order
DN_3	Remote manager is done with the approval operation

Table 3.3: States descriptions of the task automata of Example 3.1

Event	Controllability	Description
a_{11}	Uncontrollable	Order processing cannot be completed
a_{12}	Uncontrollable	Processed order rejected for approval step
b_1	Controllable	Order processing task began
c_1	Uncontrollable	Processed order accepted for approval step
p_1	Uncontrollable	Order processing task completed
a_{21}	Uncontrollable	Local manager failed to finish the order approval
a_{22}	Uncontrollable	Local manager rejected the order
b_2	Controllable	Local manager began approval task
c_2	Uncontrollable	Local manager approved the order
p_2	Uncontrollable	Local manager completed the order verifying
a_{31}	Uncontrollable	Remote manager failed to finish the order approval
a_{32}	Uncontrollable	Remote manager rejected the order
b_3	Controllable	Remote manager began approval task
c_3	Uncontrollable	Remote manager approved the order
p_3	Uncontrollable	Remote manager completed the order verifying

Table 3.4: Events descriptions of the task automata of Example 3.1

In Figure 3.1c, each rectangle represents a particular state of the system [upper section] and the various events that have the potential to happen in that state [lower section]. In Figure 3.1c, starting from the initial state (IN1, IN2, IN3), 3 events $\{b_1, b_2, b_3\}$ could cause the existing state of the system to undergo a transition to another state. Suppose event b_3 happens. This takes the system to the state (IN1, IN2, EX3). If however event b_1 had occurred, the system would have undergone a transition to state (EX1, IN2, IN3) (not shown in Figure 3.1c). Continuing from state (IN1, IN2, EX3) where the remote manager has begun the approval of the task, any of the events $\{b_1, b_2, a_{31}, p_3\}$ could happen. If, as in our Example, event p_3 occurs, we can note that the state of task tw_2 changes from (EX3) to (DN3). i.e. the remote manager after processing the order has approved the order. Hence there is another transition from state (IN1, IN2, EX3) to state (IN1, IN2, DN3). If however event a_{31} had occurred, implying that the remote manager failed to finish the order approval, the transition would be from state (IN1, IN2, EX3) back to the initial state according to our assumption that a system could return to the beginning state after the abort of an event. The same logic, when extended, gives the complete set of all the state transitions of the system.

3.2 Deriving the Specification Model

This step succeeds the derivation of the behavioral model in the FA Modeling approach. At this point, we recall the concept of *task dependencies*, which basically enforce supply chain specification control.

In Figures 3.2a to 3.2c, we have presented the finite automata models of the dependencies of Table 2.1. There is no formal rule for naming the states. It is sufficient if different states have different names.

Consider the automaton in Figure 3.2a, which models the "Begin-on-commit" dependency. Starting from the initial state of this FA, the set of strings R = $\{c_i, c_ib_j, c_ib_jc_i, c_ib_jc_ib_j, ...\}$ can be generated. The FA generates a language that contains the alternate occurrences of c_i and b_j (starting from c_i). The intersection of this language with the behavioral model language establishes the set of supply chain behavior that complies with the "Begin-on-commit" dependency of tasks tw_i and tw_j. In fact, any other language generated by the behavioral model, except this intersection, will violate the "Begin-on-commit" dependency.

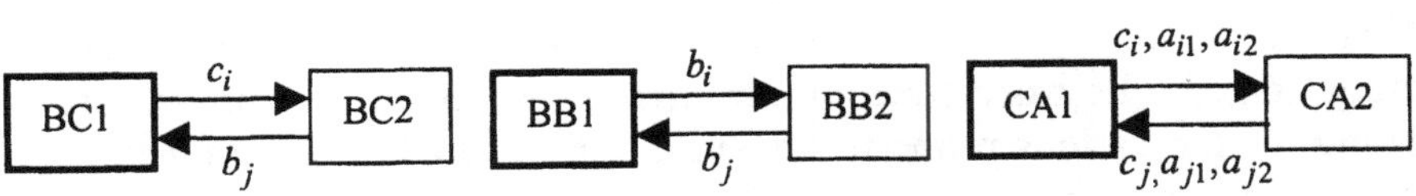

Figure 3.2a: Begin-on-commit dependency

Figure 3.2b: Begin dependency

Figure 3.2c: Termination dependency

Unfortunately, as a limitation of FA modeling, no specific modeling procedure exists that can be used for all type of dependencies. In the case study in Section 5, we introduce more examples of dependency modeling.

Example 3.2 Begin-on-commit Dependency
Consider the Example 3.1. Here we introduce a constraint (or control specification) that both local and remote managers cannot start approval operations until processed order is accepted for the approval step. In other words, events b_2 or b_3 for an order can only happen if event c_1 has already happened for that order. This specification is a begin-on-commit dependency and can be shown by Figure 3.3. The condition to be remembered is that here we assume that any specific order is sent to only one manager and not both.

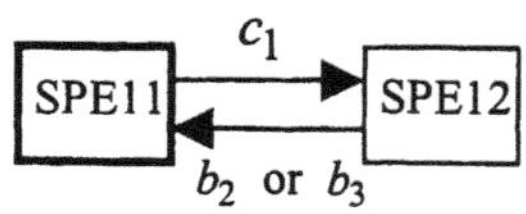

Figure 3.3: Begin-on-commit dependency for Example 3.2

4. SUPPLY CHAIN ANALYSIS

It can be shown [12] that for two FA's G_1 and G_2 if $\Sigma_{G_2} \subseteq \Sigma_{G_1}$ then $L(G_1 \cup G_2) = L(G_1) \cap L(G_2)$. Therefore using the union operator defined in Section 2.2, we can derive the intersection of the supply chain and the control specification languages. The resulting language formed from the intersection of the supply chain language with all control specification languages is the desirable language as defined in Section 2. Although the desirable language satisfies all supply chain task dependency constraints, it still may not be useful. In fact, this language might not be controllable. Let $L(H)$ and $L(G)$ be the supply chain desirable and behavioral languages respectively. Then $L(H)$ is not controllable if $\exists \sigma \in \Sigma_u$ and $s \in L(G)$, where $s \in L(H) \subseteq L(G)$, $s\sigma \in L(G)$ and $s\sigma \notin L(H)$ where Σ_u is the set of uncontrollable task primitives of the supply chain. In Section 4, we introduce a supervisor that does not allow the supply chain to experience any uncontrollable behavior. The language that supply chain generates under the control of this supervisor plays a fundamental role in the analysis of supply chain properties.

This section deals with the analysis of the supply chain modeled in the previous stage. The basic function performed by this analysis procedure is checking for the various properties desired in that supply chain model. The properties include consistency, specification redundancy, event redundancy, event controllability analysis and scalability. The supervisor incorporated with the feedback function plays an important part here.

In this section we assume for a given supply chain G and its control specification H, the behavior and desirable languages are shown by $L(G)$ and $L(H)$, where $L(H) \subseteq L(G)$. We also assume the supervisor **S** is available such that $K = L(S/G)$, or the behavior of G when it is controlled by **S**, is supremal. Let $G = \{Q, \Sigma, f, q_0, Q_m\}$ and $S = (X, \Sigma, g, x_0, X_m)$, as defined in Section 2.3. The reader can use one of the software in [17, 18, 19] to construct **S** (We do not discuss the details of the construction algorithm here).

4.1 Supply Chain Consistency

An important question while designing a supply chain network is whether the current supply chain can satisfy the requirements or if the supply chain is consistent with the management requirements. For example, can the order processing system successfully finish the order approval process? If the management requirements can be modeled as control specifications, then the answer to this question can be found by the Finite Automata theory. If the supremal language of the supply chain under the control of its supervisor is not empty, then it is guaranteed that a part of the supply chain behavior exists that satisfies every management requirement. On the other hand if the language is empty then the control specifications (or management requirements) cannot be satisfied or the supply chain is not consistent. We recall that the supremal language includes the maximum set of feasible behavior of the supply chain. Therefore if this language set is empty, no feasible behavior exists. The following Example illustrates the application of supremal language in determining the consistency of the supply chain.

Consider the order processing system in Example 3.1. Here we will impose the specification provided by Figure 3.3 i.e. the "Begin-on-commit dependency". The supremal language is generated using the finite automata technique provided by RW algorithm. The feedback policy (Φ) and the supervisor state transition description (in Tabular format) are in the Appendix A.2 and A.3. The supremal language has 36 states as shown by the Table A.2 in the Appendix. Figure 4.1 shows a partial representation of the supremal language.

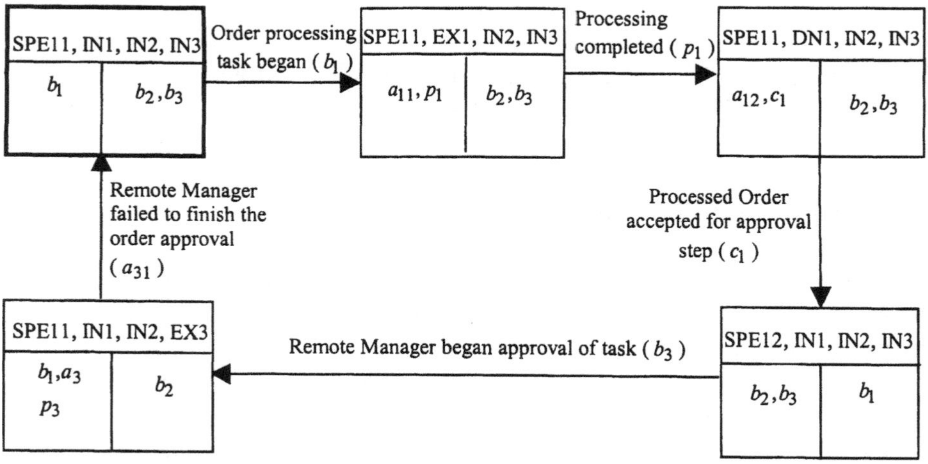

Figure 4.1: Partial Representation of Supremal Language of Example 4.1

From Figure 4.1, we have a system in its initial state (SPE11, IN1, IN2, IN3). Each box in this figure consists of three parts. The upper left part shows currents state of the system. The upper right shows the events that can happen in that state. The lower part represents the events that are disabled at that state. Here the supervisor enables/disables events so as to enforce conformity. The feedback table gives the events that are enabled/disabled at the various states. Here in state 1, the events b_2 and b_3 are disabled by the supervisor. If the event "*Order processing task begin*" $\{b_1\}$ happens, the state of task tw_1 changes from IN1 to EX1. Hence the state of the entire system undergoes a transition from state (SPE11, IN1, IN2, IN3) to state (SPE11, EX1, IN2, IN3). In the new state, the events $\{a_{11}, p_1\}$ could cause the system to attain different states. If in the task tw_1, the order processing task is completed $\{p_1\}$, then the state of this task automaton changes from EX1 to DN1 and the state of the system changes from (SPE11, EX1, IN2, IN3) to (SPE11, DN1, IN2, IN3). At state (SPE11, DN1, IN2, IN3) events a_{12} and c_1 can happen. If the processed order is accepted for approval step $\{c_1\}$, it causes a transition to state (SPE12, IN1, IN2, IN3). According to our constraint, events b_2 and b_3 can happen only after c_1 happens. In state (SPE11, DN1, IN2, IN3) (or before c_1 happens), b_2 and b_3 are disabled. But now in state (SPE12, IN1, IN2, IN3)(or after c_1 occurred), these two events can occur. If the remote manager begins approval on the processed order from the approval step event b_3, then the state of the task becomes EX3 and the state of the system undergoes a transition to state (SPE11, IN1, IN2, EX3). If the remote manager fails to finish the order approval $\{a_{31}\}$, then the system can return to the beginning state 1 and wait for the next order.

To show how supremal language can be used to reveal the inconsistency of a supply chain with respect to its control specification, we add a new constraint (control specification) to the supply chain of Example 4.1. Suppose that the occurrence of p_1 or a_{11} is dependent on the beginning events of one manager (b_2 or b_3). Figure 4.1a models this dependency.

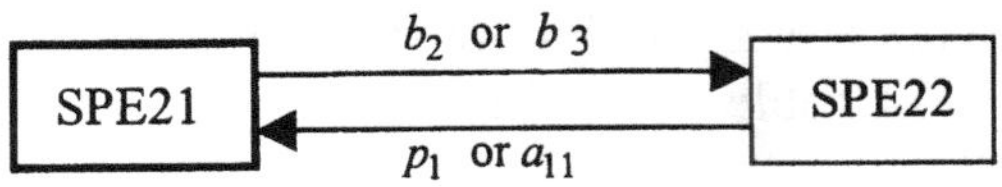

Figure 4.1a: New Constraint for Example 4.1

If we add this constraint to the problem, the resulting supremal language is empty, i.e. the system behavior cannot satisfy this new requirement. We notice that the new constraint is in conflict with the constraint of Figure 3.3 and so both of them cannot be satisfied simultaneously. Although for the given example this conflict can be identified by looking at the problem, in practice, where thousands of tasks and specification models exists, the automata procedure of supremal language construction is the only way to identify these conflicts.

4.2 Redundancy Checking

4.2.1 Control Specification Redundancy

Suppose that the supremal language of a supply chain model G is constructed from the languages generated by m given specification models $Spe_1, Spe_2, \cdots, Spe_m$ and G itself. Specification model Spe_i where $1 \le i \le m$, is called *redundant* if removing of such a specification does not affect the supremal language.

This definition of specification redundancy is based on the removal of a specification from the set of current ones. But the result can also be used when we consider adding a new specification that is redundant to the set of current ones. The following lemma provides necessary and sufficient conditions for the redundancy of a specification.

Lemma Given behavioral model G and desired behavior model H, the specification Spe_i where $1 \le i \le m$, is redundant if and only if $L(\mathbf{S}/G) = L(\mathbf{S}'/G)$ where $\mathbf{S}$ and $\mathbf{S}'$ are the supremal languages of G before and after the removal of Spe_i respectively.

Proof It is directly concluded from the definition of supremal language.

Example 4.2 Redundant specification

We modify the control specification of Example 4.1 by adding a new specification model shown by Figure 4.2. In this case the supremal language does not change. This shows that the specification model of Figure 4.2 is redundant. This specification allows an abort before the completion of order processing task only when this task has already begun. But this behavior is an inherent part of supply chain behavior and therefore will not affect its behavior. If we use the RW algorithm for the new problem, we get the same supremal language as of Table A.2 in the appendix.

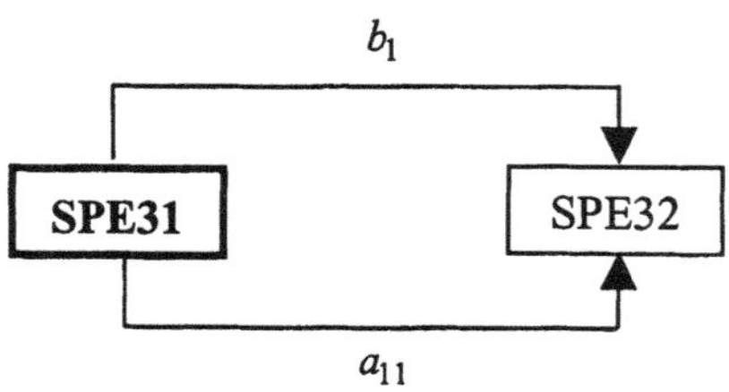

Figure.4.2: A redundant specification for Example 4.1

4.2.2 Event Redundancy (partial observation)

Event redundancy checking of a supply chain model is important and it can be investigated only by the Finite Automata technique. Given a supply chain model $G = \{tw_i, 1 \le i \le n\}$ and the specification model H, event $e \in \{a_{i1}, a_{i2}, b_i, c_i, p_i$ where $1 \le i \le n\}$ is called redundant if the information on the occurrence of this event is not essential in control of G, i.e., the observation of this event is not a necessary condition for having correct control while other events are observed.

Using the above definition any subset of $\{a_{i1}, a_{i2}, b_i, c_i, p_i$ where $1 \le i \le n\}$ is said to be redundant if the observation of the events of that subset is not necessary for the correct control while other events are observable. The following example illustrates the idea.

Example 4.3 Redundant events

The partial observation theory can be used to identify the redundant events of a DES. In Example 4.3, we identify some of the redundant events of Example 4.1. The reader should refer to the Appendix A.4 for the background theory. The reader should not be confused by the word "redundant", which we use for an event. Redundancy of an event does not mean that the occurrence of the event is not important. It only means that the occurrence information is not essential for the supervisor to decide about the enabling/ disabling of the system events.

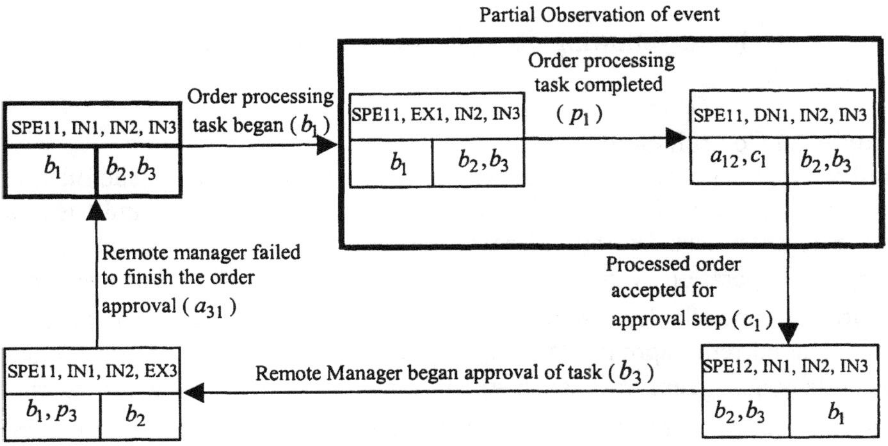

Figure 4.3: Redundant Events

Figure 4.3 gives the partial representation of the coupled language when event p_1 is not observed. The figure is interpreted similar to our previous represenatations. The outer rectangle in the figure conveys the information that when the supervisor sees the system, it cannot observe event p_1. But this does not affect the control enforced by the supevisor because both states SPE11, EX1, IN2, IN3 and SPE11, DN1, IN2, IN3 are disabling the same set of events $\{b_2, b_3\}$. Therefore, for the supervisor it is enough to know that after the observation of event b_1 in state (SPE11, IN1, IN2, IN3), it should disable $\{b_2, b_3\}$ until it observes c_1.

4.3 Event Controllability Analysis

In Example 3.1, we assumed that only beginning events of each task of a workflow are controllable and all other events are uncontrollable. In some cases this assumption might not hold. In this section we introduce the idea of change in controllability. A change in controllability is one of the following:
1. Controllable to uncontrollable (CU)
2. Uncontrollable to controllable (UC)
The following lemma [5] can be used in the event controllability analysis.
Lemma Let event σ be initially a controllable event. The supremal language $K = L(S / G)$, is not affected by $CU(\sigma)$ if and only if event σ has not been disabled in any state of the supervisor **S**.

It is concluded from the properties of $K = L(\mathbf{S} / G)$ that for any event σ that is initially uncontrollable, a $UC(\sigma)$ will not make the language K smaller, however K may not be supremal anymore. This property guarantees that any UC change of events will not cause a violation from system specification. In the following example we illustrate the application of the above lemma.

Example 4.4 Controllability analysis

In Example 3.1 assume that the remote manager is replaced by automatic approval software, which scans the system in some random time and tries to start the approval task for any order waiting for approval. The program starts the approval operation even if there is no order waiting. In this case event b_3 is no longer controllable. Therefore a CU change must be analyzed for this event. The current supremal language, given in Appendix A.2 is no longer acceptable since event b_3 has been disabled in several instances of the coupled language. The reader notices that this change will cause an empty supremal language set. The reason is that when the order processing system starts from its initial state ($IN1, IN2, IN3, SPE11$), the event b_3 can happen and thus is in conflict with the specification defined by Figure 3.3.

4.4 Scalability

Finite Automata Analysis provides a fair structure for handling the scalability of a supply chain model. The Scalability problem deals with the predicting change in supply chain behavior upon a shift in one of its attributes (or resources).

A supply chain is scalable with respect to an attribute (or resource) if its control behavior can be explained as a function of the number of that attribute (or resource). The approach to scalability has two dimensions. 1. *Task* scalability 2. *Specification* Scalability

4.4.1 Task Scalability

Assume that the number of tasks in a supply chain model $G = \{tw_i, 1 \le i \le n\}$ makes a shift h where $-n \le h < \infty$. Using the union operator discussed in Section 2.2, the new supply chain behavioral model will have 3^{n+h} states and $5(n+h)$ events. Assuming the control specifications do not change, the scaled supply chain model and the current specification can be fed to RW algorithm and the new supremal language and its related control policy can be derived. This mechanism will help to handle the task scalability in an automatic manner. The following example illustrates the idea.

Example 4.5 Adding a new task

In Example 4.1 suppose a new task "order request" is introduced. The event information of this task is shown in Table 4.7. Applying the FA procedure for the new shuffled model with four tasks will yield a new supremal language with 108 states (not shown in this chapter). It is recalled that Finite Automata approach accepts the supply chain behavioral and desired behavior models as two separate models. Therefore a change in the number of tasks only affects the behavioral model.

Event	Controllability	Description
a_{41}	Uncontrollable	Order request cannot be completed
a_{42}	Uncontrollable	Order request cannot be satisfied
b_4	Controllable	Order request task began
c_4	Uncontrollable	Order is accepted for processing step
p_4	Uncontrollable	Order request task completed

Table 4.1: Order Request events

4.4.2 Specification Scalability

While the scalability of tasks can be formalized, it is difficult to provide a general formalism for specification scalability. This is due to different specification models in real world problems. Therefore we end this section by providing an example, which addresses a specific type of specification scalability. Developing a formal way of handling the specification scalability remains an open area for research.

Example 4.6 Adding to the number of simultaneous orders

Again considering Example 4.1, the presence of specification of Figure 3.3 will restrict the controlled workflow to accept the maximum number of 3 orders at the same time (one is processed by the order processing system and one by each manager). Suppose we would like to formulate a scalable specification model that is a function of number of orders in the system. Having this generic model, and given the maximum number of orders, one can feed the behavioral model and the specifications model to RW algorithm and automatically generate the control structure. The generic model for the specification of Example 3.1 designed for the maximum of w orders in the systems is given in Figure 4.4. It is noted that the mechanism of this model is in fact acting as a buffer that can accept $w-3$ order waiting for either manger to pick them up for approval operations.

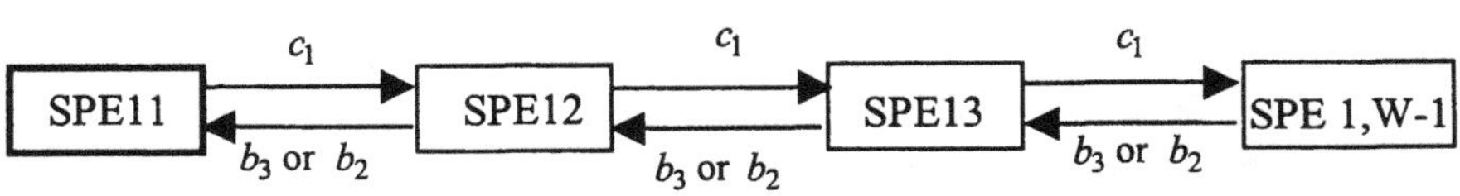

Figure 4.4: A Scalable specification model

Table 4.2 summarizes the analysis techniques discussed so far in this chapter with some sample applications.

Analysis Technique	Application
Consistency Checking and supremal language generation	-Verifying if the current supply chain routines can satisfy all business rules, constraints and requirements. -Creating a supply chain workflow that is used to co-ordinate the supply chain activities and their secure execution (meeting the supply chain constraints).
Control Specification Redundancy	-Verifying if the current business rules are redundant -Analyzing the supply chain reaction to the introduction of a new business rule (or supply chain constraint).
Event Redundancy	-Identifying the necessary information for supply chain coordination decisions. -Identifying the information that is generated but not necessary for the control of the supply chain.
Event Controllability Analysis	-Investigating the reaction of supply chain to the addition of extra control or removal of current control means.
Task Scalability	-Automatic updating of supply chain procedures, when a new task is introduced or a current one is removed.
Specification Scalability	-Developing a parametric formulation of business rules. For example, verifying how an inventory system execution reacts to different number of orders.

Table 4.2: Analysis Techniques with Sample Applications

5. "GOURMET-TO-GO"- A CASE STUDY

In this section, we present a case study, which illustrates the application of finite automata modeling and analysis in the design of a supply chain network of a restaurant. For two reasons, our focus is more on the modeling part than the analysis of the restaurant supply chain:

1. We used a modified version of the modeling approach that we presented in Section 3. Therefore the details of the modeling phase can help the reader understand how the DES method is applied to real world problems.
2. The analysis techniques discussed in Section 4 can be automated and the prescribed rules can be applied despite the problem size. We believe that the examples given in Section 4 provide the reader with the required insight to the analysis phase. At the same time, due to space limitation we are not able to include the analysis results of the case study here. The interested reader can contact the authors to receive a copy of the results.

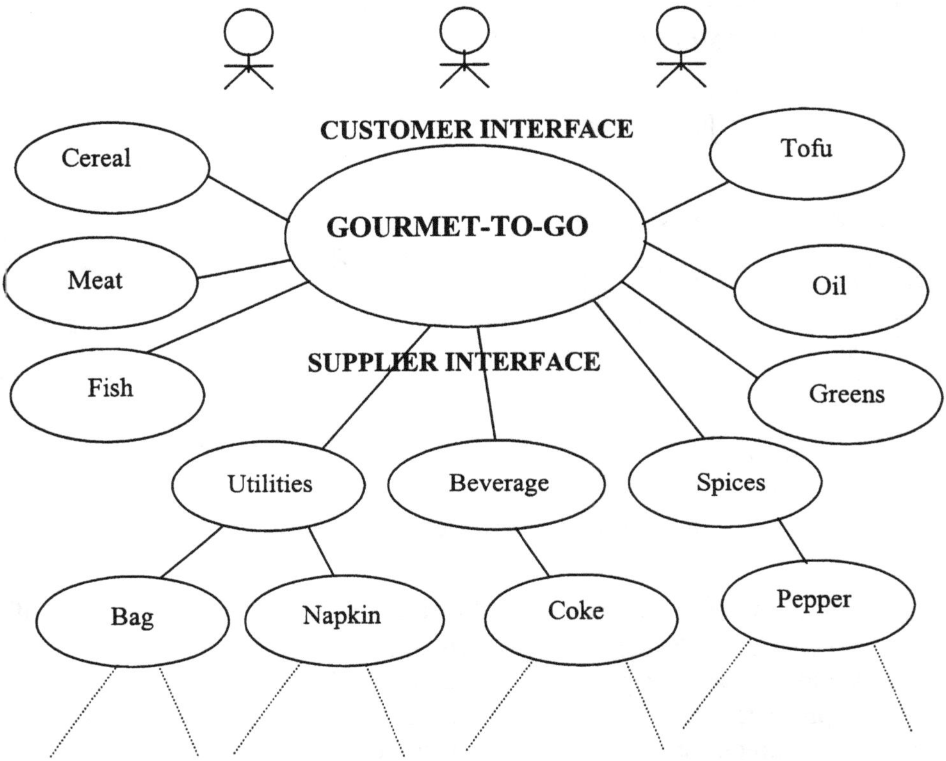

Figure 5.1: Gourmet-To-Go Supply Chain Network

System Description:

Figure 5.1 represents the "GOURMET-TO-GO" supply chain network. In this Figure, each oval represents a particular business type that is connected to the restaurant and also to its own suppliers and service providers. The customers place the order over the Internet, by phone or in person.

As evident from the Figure 5.1, our supply chain is embedded in a network of suppliers and customers. So any change in the demand or supply pattern in this network affects its business. Assuming that each oval represents a business, the connectivity between the various businesses is governed by the business rules of that business and all other businesses that it may have interaction with. This connectivity based on business rules is defined within the business itself. An intermediary agent defines this connection between "Gourmet-To-Go" and any other businesses in the network. Now let us study the analogy between this supply chain and the finite automata theory. Here we recollect the idea of how a supply chain can be modeled using the finite automata technique. This requires us to take a view of a subset of the processes that are going on in "Gourmet-To-Go".

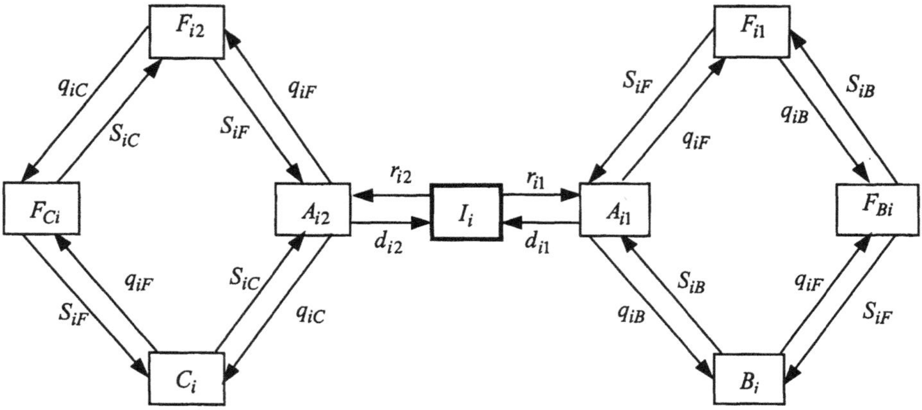

Figure 5.2: Finite Automaton of the customer service task of server i

Let us consider a server i $(i=1,2)$ of the restaurant that takes care of customer service. The server model, which fills an order, is equivalent to a finite automaton and is represented in Figure 5.2. Each rectangle in the above figure represents a state of the server i and the arcs represent the transition of the server from one state to another as a result of the various events

The two sides of Figure 5.2 are mirror images: the left side represents the filling of an order for fries and chicken, and the right side represents the filling of an order for fries and beef. This is an illustration of an enterprise having two products, each product having two components to be assembled. In the initial state (I_i), the server is idle. The arrival of an order for product 1 (r_{i1}) (or beef and fries) transitions the server to the state of preparing the order (A_{i1}). If a component of the assembly (fries, beef) is not available at this level of the production, a requisition for replenishment is issued (q_{iF}, q_{iB}). When a requested component arrives (S_{iF}, S_{iB}) the preparation of an order can continue (A_{i1}). If we consider a complex process, A_{i1} may represent the aggregation of a supply chain of discrete tasks. There is another process going on in "Gourmet-To-Go", which is related to the server process. This is the requisition process in Figure 5.3. We will discuss how a finite automaton for a requisition process (fries) is modeled.

The requisition process for server i is initialized at D_{i0}, the state in which the server has no requisitions outstanding. The event q_{iF} that also occurs in Figure 5.2, transitions the requisition process to the state where a requisition is submitted (D_{i1}). The requisition has to be satisfied by a higher level of the "Gourmet-To-Go" inventory system. There are two inventory policies that have been modeled in Figure 5.3. The state D_{i1} is the state where the requisition has been received and the state D_{i2} represents the state in which a purchase order has been processed. The right side transition (p_1, q_i) shows a policy where, each time a requisition is received (D_{i1}), it naturally follows that a purchase order is released (D_{i2}). This is a "lot-for-lot reorder" policy, which implies that no inventory is being stored at this stage. The transitions in the left side are more complicated. There are two possibilities. The transition set (p_2, e_i) occurs when the state of the inventory is $I \geq Q$; i.e., the inventory level is greater than the requisitioned quantity. These transitions result in state k_i, in which it is deemed that the requisition can be immediately satisfied from the existing inventory of this server. The transition set (p_2, q_i) occurs when the status of the inventory is $I < Q$; i.e., the inventory level is less than the requisitioned quantity. This results in reaching the state D_{i2}; i.e., a purchase order is sent. Finally, the transition S_{iF} is the event that marks the completion of the delivery of the requisitioned material to the customer. This transition also appears in the server process of Figure 5.2, where it returns that process to state A_{il} and the server can now assemble the order. Thus, the server process and the requisition process are linked by common transitions (q_{iF}, S_{iF}).

The requisition process is related to another process, the purchasing process; it is the unfilled requisition that triggers the release of a purchase order. The purchasing finite automaton is shown in Figure 5.4. From the initial state, P_{00}, the requisition of server $i (i = 1,2)$ results in a purchase order being sent (q_i) and the purchasing automaton enters the state where an outstanding order for server i exists. The subsequent delivery of that order is marked by the event r_i. It is noted that the purchase order is related to the requisition process by the common transition q_i.

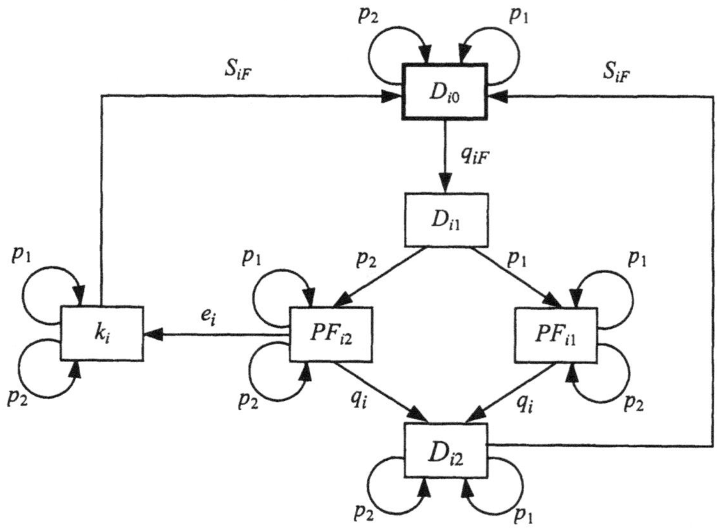

Figure 5.3: Fries requisition finite Automaton

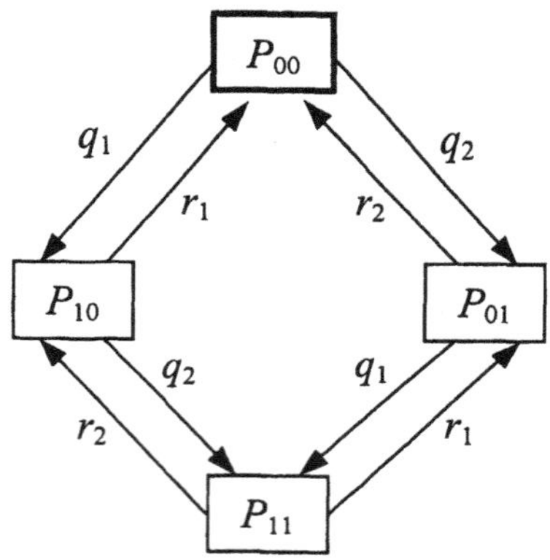

Figure 5.4: Purchase Finite Automaton

The finite automata models described so far show the manner in which related processes in the "behavioral model" are activities and are linked through common events. When all the relevant subsystem processes under consideration are appropriately specified, the collection of models is referred to as the "behavioral model". Summing up, our "behavioral model" consists of the server processes, requisition processes and the purchasing processes. This model can be shown by an automaton with 9*9*6*6*6*4 = 69984 states.

Business Rule Models or Control Specifications:

The business rules governing the business transitions from one state to another should be able to capture the logic of the order in which the transitions should occur during the execution of the restaurant tasks. The following dependencies were considered in the studied case.

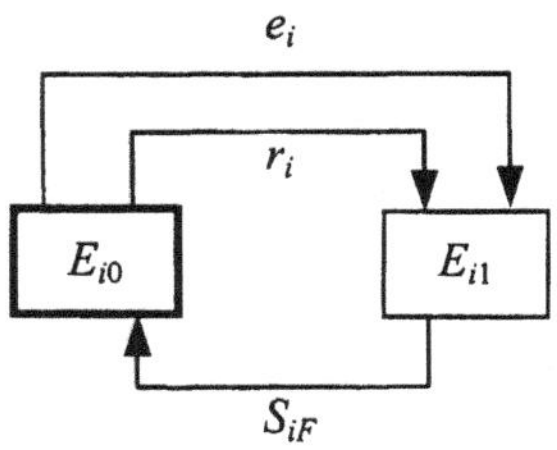

Figure 5.5a: Sequential Rule

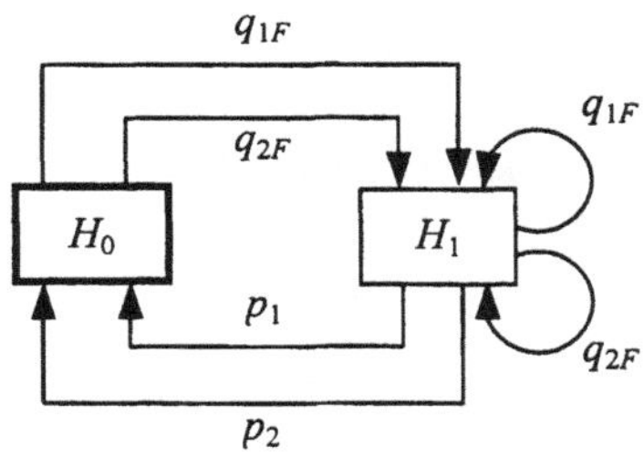

Figure 5.5b: Sequential Rule

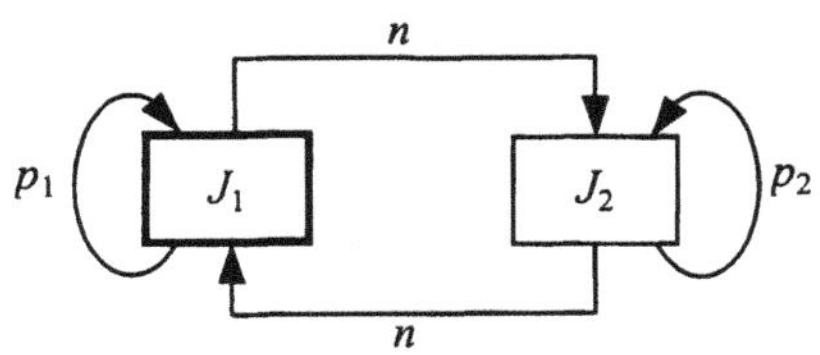

Figure 5.5c: Decision Rule

In Figure 5.5a, the finite automaton is initialized in state E_{i0}. This model tells that in any instance of a system, a transaction for e_i or r_i must occur before the transaction S_{iF}. In other words, regardless of which policy is used in the requisition process, there needs to be sufficient inventory (e_i) or a new order must arrive (r_i) before a re-supply to the server is made (S_{iF}).

Figure 5.5b illustrates another rule: A requisition must be processed before the relevant requisition policy is executed. The "relevant policy" is a variable that may be selected by management. In Figure 5.5c, state J_1 is the state in which requisition-processing policy 1 (Right-hand side of Figure 5.3) is in use. State J_2 is requisition policy 2 (left-hand side of Figure 5.3). The transition "n" is a management decision event. In our formalism, these rules are defined and/or changed separately from the underlying processes of the behavioral model. In a final step the business rules are combined (coupled) to the behavioral model in order to constrain the evolution of the business processes to behave according to these business rules.

Resulting Supremal Language:

When the set of business rules are coupled to the behavioral model by RW algorithm, the resulting diagram (supremal language) shows the possible ways in which the business transactions can evolve within the constraints of the business rules. These are referred to as "business scenarios", a partial set of which is shown in the following Figure 5.6. (The complete supremal language for this case study has 2100 states, which is not shown here). In the first sequence, policy 1 is in use. The sequence terminates in state 8. It is noted that the path generated by the digraph is legal in the sense that the constraints of Figures 5.5a, 5.5b and 5.5c hold. The second sequence begins when r_{22} transitions into state 9. Coincidentally, management is changing one of its business rules; i.e., it switches to policy 2 (stock replenishment). The resulting sequence again follows the business rules.

The supremal language for this case study was generated automatically by the software in [19]. The resulting process is dynamic; i.e., business rule or policy changes can be automatically reflected in a new legal workflow. The properties presented in Section 4 can be investigated for this supply chain. For example, it can be shown that the delivery event $\{d_{11}\}$ in Figure 5.2 may be unobservable. In other words there is no need to receive the information on the occurrence of this event to guarantee the conformance of the restaurant activities with its business rules.

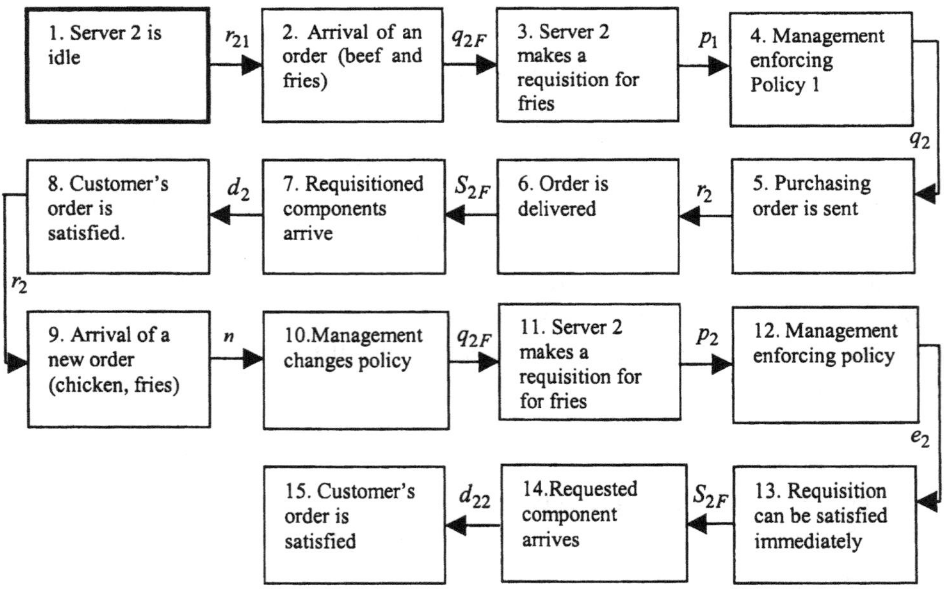

Figure 5.6: Partial representation of supremal language

6. CONCLUSION

A supply chain can be considered as a set of business operations (or tasks) and a set of business rules. One important objective of supply chain management is to assure that the business operations do not violate the business rules. With such a view, we used the theory of supervisory control of discrete event systems to model and analyze the supply chain task executions. In the modeling phase, we introduced two models, namely behavioral and desirable behavior models. These models were used to derive the largest set of supply chain behavior, which satisfy the business rules. This set was the basis for all the analysis techniques introduced in Section 4. Despite the exponential growth of supply chain state space, due to its complexity, the available state reduction techniques and implementation software tools can be efficiently and effectively applied for the real world cases. The following research directions might be taken as a continuation of this work:

1. Developing a standard procedure for modelling the business rules
2. Developing special purpose software for DES-supply chains
3. Applying the formalism to more real problems
4. Developing a hybrid formalism to use the performance optimization and DES tools together. Such formalism will incorporate the well-known operation research tools with the introduced DES view here. The optimization of supply chain behaviour is possible through FA method if one considers feasible behaviour generated by the FA method as a feasible space of an optimization problem. This means the controlled behaviour of a supply chain, in fact, constitutes different feasible alternatives of accomplishing the tasks of that supply chain. By defining an objective function, one might select the best execution policy of tasks among the feasible alternatives. The objective function might express any performance measure, such as time, and inventory.

APPENDIX:

A.1. Behavioral Model for Example 3.1

		a_{11}	a_{12}	b_1	c_1	p_1	a_{21}	a_{22}	b_2	c_2	p_2	a_{31}	a_{32}	b_3	c_3	p_3
IN1, IN2, IN3	1			2					3					4		
EX1, IN2, IN3	2	1*				6			5					7		
IN1, EX2, IN3	3			5			1				8			9		
IN1, IN2, EX3	4			7					9			1				10
EX1, EX2, IN3	5	3				11	2				12			13		
DN1, IN2, IN3	6		1		1				11					14		
EX1, IN2, EX3	7	4				14			13			2				15
IN1, DN2, IN3	8			12				1		1				16		
IN1, EX2, EX3	9			13			4				16	3				17
IN1, IN2, DN3	10			15					17				1		1	
DN1, EX2, IN3	11		3		3		6				18			19		
EX1, DN2, IN3	12	8				18		2		2				20		
EX1, EX2, EX3	13	9				19	7				20	5				21
DN1, IN2, EX3	14		4		4				19			22				6
DN1, IN2, EX3	15	10				22			21				2		2	
EX1, IN2, DN3	16			20				4		4		8				23
IN1, DN2, EX3	17			21			10				23		3		3	
DN1, DN2, IN3	18		8		8			6		6				24		
DN1, EX2, EX3	19		9		9		14				24	11				25
EX1, DN2, EX3	20	16				24		7		7		12				26
EX1, EX2, DN3	21	17				25	15				26		5		5	
DN1, IN2, DN3	22		10		10				25				6		6	
IN1, DN2, DN3	23			26				10		10			8		8	
DN1, DN2, EX3	24		16		16			14		14		18				27
DN1, EX2, DN3	25		17		17		22				27		11		11	
EX1, DN2, DN3	26	23				27		15		15			12		12	
DN1, DN2, DN3	27		23		23			22		22			18		18	

*Event a_{11} makes a transition from state 2 to state 1.

Table A.1: Tabular Format of FA for Example 3.1

A. 2. Supremal Language for Example 4.1

DESCRIPTION	STATE	a_{11}	a_{12}	b_1	c_1	p_1	a_{21}	a_{22}	b_2	c_2	p_2	a_{31}	a_{32}	b_3	c_3	p_3
SPE11,IN1,IN2,IN3	1			36												
SPE12,IN1,DN2,DN3	2							17		17			18		18	
SPE11, DN1,DN2,DN3	3		10		2			20		20			22		22	
SPE12,IN1,EX2,DN3	4						17				2		23		23	
SPE12,IN1,DN2,EX3	5							21		21		18				2
SPE11,EX1,DN2,DN3	6	10				3		24		24			26		26	
SPE11,DN1,EX2,DN3	7		14		4		20				3		27		27	
SPE11,DN1,DN2,EX3	8		15		5			25		25		22				3
SPE12,IN1,EX2,EX3	9						21				5	23				4
SPE11,IN1,DN2,DN3	10			6				28		28			30		30	
SPE11,EX1,EX2,DN3	11	14				7	24				6		31		31	
SPE11,EX1,DN2,EX3	12	15				8		29		29		26				6
SPE11,DN1,EX2,EX3	13		19		9		25				8	27				7
SPE11,IN1,EX2,DN3	14			11			28				10		33		33	
SPE11,IN1,DN2,EX3	15			12				32		32		30				10
SPE11,EX1,EX2,EX3	16	19				13	29				12	31				11
SPE12,IN1,IN2,DN3	17								14				34		34	
SPE12,IN1,DN2,IN3	18							34		34				15		
SPE11,IN1,EX2,EX3	19			16			32				15	33				14
SPE11,DN1,IN2,DN3	20		28		17								35		35	
SPE12,IN1,IN2,EX3	21								19			34				17
SPE11,DN1,DN2,IN3	22		30		18			35		35						
SPE12,IN1,EX2,IN3	23						34				18			19		
SPE11,EX1,IN2,DN3	24	28				20							36		36	
SPE11,DN1,IN2,EX3	25		32		21							35				20
SPE11,EX1,DN2,IN3	26	30				22		36		36						
SPE11,DN1,EX2,IN3	27		33		23		35				22					
SPE11,IN1,IN2,DN3	28			29									1		1	
SPE11,EX1,IN2,EX3	29	32				25						36				24
SPE11,IN1,DN2,IN3	30	26						1		1						
SPE11,EX2,EX3,IN3	31	33				27	36				26					
SPE11,IN1,IN2,EX3	32			29								1				28
SPE11,IN1,EX2,IN3	33			31			1				30					
SPE12,IN1,IN2,IN3	34								33					32		
SPE11,DN1,IN2,IN3	35		1		34											
SPE11,EX1,IN2,IN3	36	1				35										

Table A.2: Tabular Format of FA for Example 4.1

A.3. Feedback Policy for Example 4.1

	a_{11}	a_{12}	b_1	c_1	p_1	a_{21}	a_{22}	b_2	c_2	p_2	a_{31}	a_{32}	b_3	c_3	p_3
1	d c	dc	1	dc	dc	dc	dc	0	dc	dc	dc	dc	0	dc	dc
2	dc	dc	0	dc	dc	dc	1	dc	1	dc	dc	1	dc	1	dc
3	dc	1	dc	1	dc	dc	1	dc	1	dc	dc	1	dc	1	dc
4	dc	dc	0	dc	dc	1	dc	dc	dc	1	dc	1	dc	1	dc
5	dc	dc	0	dc	dc	dc	1	dc	1	dc	1	dc	dc	dc	1
6	1	dc	dc	dc	1	dc	1	dc	1	dc	dc	1	dc	1	dc
7	dc	1	dc	1	dc	1	dc	dc	dc	1	dc	1	dc	1	dc
8	dc	1	dc	1	dc	dc	1	dc	1	dc	1	dc	dc	dc	1
9	dc	dc	0	dc	dc	1	dc	dc	dc	1	1	dc	dc	dc	1
10	dc	dc	1	dc	dc	dc	1	dc	1	dc	dc	1	dc	1	dc
11	1	dc	dc	dc	1	1	dc	dc	dc	1	dc	1	dc	1	dc
12	1	dc	dc	dc	1	dc	1	dc	1	dc	1	dc	dc	dc	1
13	dc	1	dc	1	dc	1	dc	dc	dc	1	1	dc	dc	dc	1
14	dc	dc	1	dc	dc	1	dc	dc	dc	1	dc	1	dc	1	dc
15	dc	dc	1	dc	dc	dc	1	dc	1	dc	1	dc	dc	dc	1
16	1	dc	dc	dc	1	1	dc	dc	dc	1	1	dc	dc	dc	1
17	dc	dc	0	dc	dc	dc	dc	1	dc	dc	dc	1	dc	1	dc
18	dc	dc	0	dc	dc	dc	1	dc	1	dc	dc	dc	1	dc	dc
19	dc	dc	1	dc	dc	1	dc	dc	dc	1	1	dc	dc	dc	1
20	dc	1	dc	1	dc	dc	dc	0	dc	dc	dc	1	dc	1	dc
21	dc	dc	0	dc	dc	dc	dc	1	dc	dc	1	dc	dc	dc	1
22	dc	1	dc	1	dc	dc	1	dc	1	dc	dc	dc	0	dc	dc
23	dc	dc	0	dc	dc	1	dc	dc	dc	1	dc	dc	1	dc	dc
24	1	dc	dc	dc	1	dc	dc	0	dc	dc	dc	1	dc	1	dc
25	dc	1	dc	1	dc	dc	dc	0	dc	dc	1	dc	dc	dc	1
26	1	dc	dc	dc	1	dc	1	dc	1	dc	dc	dc	0	dc	dc
27	dc	1	dc	1	dc	1	dc	dc	dc	1	dc	dc	0	dc	dc
28	dc	dc	1	dc	dc	dc	dc	0	dc	dc	dc	1	dc	1	dc
29	1	dc	dc	dc	1	dc	dc	0	dc	dc	1	dc	dc	dc	1
30	1	dc	1	dc	dc	dc	1	dc	1	dc	dc	dc	0	dc	dc
31	1	dc	dc	dc	1	1	dc	dc	dc	1	dc	dc	0	dc	dc
32	dc	dc	1	dc	dc	dc	dc	0	dc	dc	1	dc	dc	dc	1
33	dc	dc	1	dc	dc	1	dc	dc	dc	1	dc	dc	0	dc	dc
34	dc	dc	0	dc	dc	dc	dc	1	dc	dc	dc	dc	1	dc	dc
35	dc	1	dc	1	dc	dc	dc	0	dc	dc	dc	dc	0	dc	dc
36	1	dc	dc	dc	1	dc	dc	0	dc	dc	dc	dc	0	dc	dc

Table A.3: Feedback policy of Example 4.1

A.4. Partial Observation and Switching Theory

A projection (mask) P is a partition over the event set Σ, which splits it to two disjoint sets:

Σ_o^P : The set of events that are observed, and

$\Sigma - \Sigma_o^P$: The set of events that are not observed such that

$$\begin{cases} P(\sigma) = \sigma \ if \ \sigma \in \Sigma_o^P \\ P(\sigma) = \varepsilon \ if \ \sigma \in \Sigma - \Sigma_o^P \end{cases}$$

where ε is the null string. Given the coupled language $K = L(\mathbf{S}/G)$ and the plant language L, there exists a supervisor with partial observation of events under the projection P, which realizes K if and only if K is observable with respect to language L and projection P. K is observable with respect to language L and projection P if and only if

$(s_1,s_2) \in \ker P \Rightarrow (s_1,s_2) \in act_{K,L}$ where $(s_1,s_2) \in \ker P \Leftrightarrow P(s_1) = P(s_2)$

and

$$(s_1,s_2) \in \mathrm{act}_{K,L} \Leftrightarrow \begin{cases} A_{\overline{K},L}(s_1) \bigcap IA_{\overline{K},L}(s_2) = \phi = IA_{\overline{K},L}(s_1) \bigcap A_{\overline{K},L}(s_2) \text{ and} \\ s_1 \in \overline{K} \bigcap L \text{ and} \\ s_2 \in \overline{K} \bigcap L \Rightarrow (s_1 \in K \Leftrightarrow s_2 \in K) \end{cases}$$

For $s \in \Sigma^*$, the sets $A_{\overline{K} \cap L}(s)$ and $IA_{\overline{K} \cap L}(s)$ are defined as follow:

$$A_{\overline{K} \cap L}(s) = \begin{cases} \{\sigma | s\sigma \in \overline{K}\}, & \text{if } s \in \overline{K} \\ \phi, & \text{otherwise} \end{cases}$$

$$IA_{\overline{K} \cap L}(s) = \begin{cases} \{\sigma | s\sigma \in \overline{K}\}, & \text{if } s \in L - \overline{K} \\ \phi, & \text{otherwise} \end{cases}$$

where $\overline{K}$ is the set of all prefixes of strings in K.

REFERENCES

[1] N. R. Adam, V. Atluri and W. K. Huang. Modeling and analysis of workflows using petri nets. *Journal of Intelligent Information Systems*. 10:131-158(1998).

[2] B.A. Brandin. The real-time Supervisory Control of an Experimental Manufacturing Cell. *IEEE Transactions on Robotics and Automation*, 12 (1): 1-14, February 1996.

[3] C. G. Cassandras and S. Lafortune. *Introduction to Discrete Event Systems*. Kluwer Academic Publishers, 1999.

[4] V. Chandra, B. Oruganti, and R. Kumar, ``UKDES: A Graphical Software Tool for the Design, Analysis & Control of Discrete Event Systems". *IEEE Transactions on Control Systems Technology*.

[5] H. Darabi and M. A. Jafari. Recovery analysis of supervisory control of discrete event systems. In *Proceedings of the IEEE International Conference on Systems, Man and Cybernetics*, San Diego, California, pp. 704-709 (1998).

[6] A.K. Elmagarmid and W. Du. "*Workflow Management: State of the Art vs. State of the market*", In Advances in Workflow Management systems and Interoperability, pages 1-17, August 1997.

[7] D. Georgakopoulos, M. Hornick and A. Sheth. " *An Overview of Workflow Management: From Process Modeling to Workflow Automation Infrastructure*", Distributed and Parallel Databases, pp. 119-153, 1995.

[8] J. E. Hopcroft and J. D. Ulman. *Introduction to Automata Theory, Languages, and Computations*. Addison-Wesley, Reading, MA. (1979).

[9] A. Haji-Valizadeh K. A. Loparo, "Minimizing the cardinality of an events set for supervisors of discrete-event dynamical systems", *IEEE Transactions on Automatic Control*, vol. 41, no. 11, pp. 1579-1593, 1996.

[10] A. Hsu, F. Eskafi, S. Sachs, and P. Varaiya. Protocol Design for an Automated Highway System. Discrete Event Dynamical Systems: Theory and Application, 2(3/4): 183-206, 1993.

[11] R. Kumar, S. Nelvagal, and S.I. Marcus. A discrete event systems approach for protocol conversion. *Discrete Event Dynamical Systems: Theory and Applications*. Wiley, NY, 1990.

[12] S. Lafortune. "Modeling and Analysis of Transaction Execution Database Systems". *IEEE Transactions on Automatic Control*. Volume: 33. 5, May 1998. Page(s): 439-447.

[13] E.P Patsidou E.C. Yamalidou and J.C. Kantor. Modeling discrete-event dynamical systems for chemical process control- a survey of several new techniques. Computers and Chemical Engineering, pages 14 (3): 281-299, 1990.

[14] P. J. Ramadge and W. M. Wonham. Supervisory control of a class of discrete event Processes. *SIAM J. of Control and Optimization*. 25(1): 206-230(1987).

[15] Rusinkiewicz, M., and Sheth, A., "*Specification and Execution of Transactional Workflows, in Modern Database Systems: The Object Model, Interoperability, and Beyond*", Addison-Wesley, 1994.

[16] A. F. Vaz and W. M. Wonham, "On supervisor reduction in discrete-event systems", Int. J. of Control, vol. 44, no. 2, pp.475-491, 1986.

[17] Workflow Reference Model, "*Technical Report*", Workflow Management Coalition, Brussels, Brussels, 1994.

[18] http://www.engr.uky.edu/~kumar/jpub.html#software

[19] http://www.eecs.umich.edu/umdes/

[20] http://www.control.utoronto.ca/people/profs/wonham/wonham.html

Chapter 4

DISTRIBUTED CONTROL ALGORITHMS FOR SCALABLE DECISION-MAKING FROM SENSORS-TO-SUPPLIERS

Vittal Prabhu

Associate Professor of Industrial Engineering
Pennsylvania State University
310 Leonhard Building
The Pennsylvania State University
University Park, PA 16802, USA

Abstract: In this chapter we will review distributed algorithms that have been developed for decision-making and control across enterprises. We will identify the key barriers in scaling such decision making across multiple functions including production scheduling, maintenance scheduling, inventory management, and transportation. After this we will review a class of algorithms based on distributed arrival time control. These algorithms can also be scaled in size because of their non-combinatorial nature and inherent parallel/distributed nature. The mathematical techniques required for modeling and analysis of these control systems are also reviewed. Applications and extensions of the distributed arrival time control algorithms for various functions are presented.

Key Words: Distributed algorithms, sensors-to-suppliers models, control theory

1. INTRODUCTION

Competitive enterprises resort to decentralization and flat/distributed organizations to manage and control complexity (Kim 1985, Weindhal and Scholtissek 1994), which will become increasingly important in the 21st century (Bollinger 1998). The need for research in intelligent autonomous controllers and distributed real-time control techniques has been repeatedly identified as high priority in several studies (NRC 1991, NGM 1997, NRC 1998). By the year 2020, information in manufacturing enterprises will be integrated such that information on all aspects of manufacturing enterprises and the marketplace will be instantly available throughout the enterprise. There is a need to develop decision-making tools that can take advantage of such highly integrated information technologies and to develop models that encompass entire manufacturing enterprises. Several new open architectures and standards have been proposed for efficient interchange of information throughout the manufacturing enterprise ranging from the sensor level to the supply chain level. The technologies necessary for rapidly integrating distributed manufacturing systems such as open architecture controllers are being extensively deployed in the industry (Procter and Albus 1997). This will allow a considerable amount of intelligence to be incorporated at the equipment level for improvements in productivity and quality (Wright 1995, Koren et al. 1996). However, there is little scientific basis for characterizing the behavior of the resulting interactions in such intelligent systems (Antsaklis 1994) or for predicting the integrated effect of their behavior.

The analytical complexity of modeling entire enterprises is usually formidable and some form of approximations and decomposition procedures are required to tackle the problem. One example of such decomposition that has been used is for solving complex scheduling problems in large semiconductor facilities (Ovacik and Uzsoy 1996). Another example is to use continuous variables for approximating discrete variables and thereby reduce computational and analytical complexity of the resulting mathematical models. In control of high volume manufacturing systems, treating the flow of parts as a continuous variable by aggregating over the large number of parts produced has led to tractable distributed scheduling policies (Perkins and Kumar 1989) and flow controllers (Gershwin 1989, Caramanis and Sharifnia 1991). However, because of aggregation, such techniques are unsuitable for real-time event-level control required for the unified control approach.

The manner in which decision-making and control problems are decomposed and solved determines the resulting architecture. A spectrum of architectures ranging from centralized to fully distributed have been developed for manufacturing systems (Dilts et al. 1991). Distributed manufacturing control systems are organized as a large number of autonomous entities that retain a minimal amount of global information, and are characterized by the absence of centralized, supervisory controllers. In these manufacturing systems, control is highly distributed among its autonomous entities such as part entities (workpieces), machine tools, robots, automated guided vehicles, and human operators. This results in a flat, or heterarchical control architecture (Hatvany 1985, Duffie and Prabhu 1994). Fully distributed control of discrete manufacturing systems further reduces complexity by eliminating supervisory controllers (Duffie et al

1988). Such systems pose significantly different set of modeling and analysis issues compared to distributed process control for distributed parameter systems (Prabhu and Duffie 1995, Prabhu and Duffie 1996, Prabhu and Duffie 1999).

2. FEEDBACK CONTROL OF DISCRETE EVENT-TIMING

The difficulty in predicting the behavior of distributed manufacturing control systems and guaranteeing their good performance are the two barriers that prevent wider acceptance of such systems in the industry at present. Recent efforts towards addressing this problem have led to the development of an approach for feedback control of discrete-event timing, which lends itself to highly distributed control architectures. Some of the applications of this approach include real-time production scheduling (Prabhu 2000), maintenance scheduling (Koomsap 2000), inventory control (Masin and Prabhu 2001), and vehicle routing for just-in-time logistics (Phukan 2002). In general this approach is suitable for controlling queuing systems in which the duration of the events being controlled is much smaller than the time it takes to simulate the queuing system.

To illustrate this approach, consider a production scheduling application in which multiple parts need to be processed on one or more machines. To fully distribute control, each part determines its own arrival time at the queue for various machines it needs. This is called the distributed arrival time control for part-driven distributed scheduling in which scheduling decisions are reduced to adjusting the arrival time at which parts start seeking the required machines (Prabhu 1995). Furthermore, by treating arrival time as a continuous variable, the approach lends itself to continuous control algorithms and makes it possible to predict the dynamics of the distributed arrival time control systems using control theoretic techniques. Feedback of predicted performance can be obtained from deterministic distributed simulation based on the current state of the manufacturing system as shown in Figure 1. Shop floor events are dynamically rescheduled using the simulation trace when arrival times with better global performance are discovered in simulations. In contrast to combinatorial approaches, dynamics of continuous controllers for discrete-event timing can be tractably modeled using differential equations. Moreover, the complexity of the resulting model grows linearly because there is one equation for each controlled event, which makes it highly scalable.

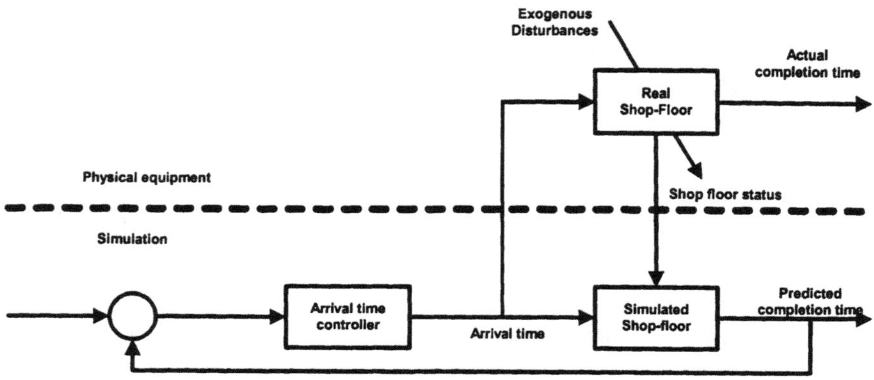

Figure 1. Overview of arrival time control.

Treating arrival times as continuous variables allows this fundamentally discrete-event system to be analyzed as a continuous control system, eliminating combinatorial complexity in making scheduling decisions. However, part processing couples the controller outputs discontinuously because a particular part's position in a queue at a machine or cell can change discontinuously depending on the relative arrival times of other parts in the queue, making the system highly non-linear.

Parts are processed in the order in which they arrive and if a part arrives when another part is being processed at the machine, it incurs queuing time. Therefore, if arrival times are close to each other, then parts interact via queuing as shown in Figure 2.

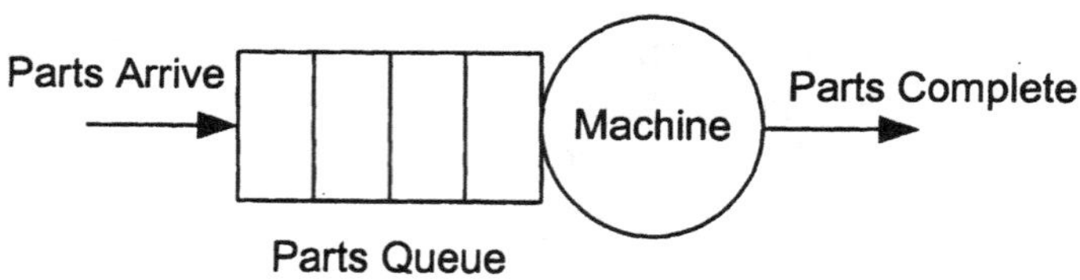

Figure 2. Parts interact via queuing

The time when a part entity arrives is crucial because it will determine the processing sequence; i.e., it will determine how a part interacts with other parts because of its relative position in the queue. It should be noted that the first part in the queue does not incur queuing time thus defining the beginning of a queue. There can be more than one part with zero queuing time leading to "fragmented queues". The machine idles between queue fragments because there are no parts to process. Parts in different queue fragments do not interact, and the rest of this development therefore focuses on a single queue fragment.

The ith part in a given queue fragment incurs a queuing time of q_{qi} after which the machine processes it for a duration of p_{qi} and then completes it at c_{qi}. (The subscript q will be used henceforth to denote that analysis is being performed on a queue fragment.) The completion time for the ith part in the queue can be expressed as

$$c_{qi}(t) = a_{qi}(t) + q_{qi}(t) + p_{qi}(t) \qquad (1)$$

and because of the FCFS queuing discipline,

$$q_{qi}(t) = a_{q1}(t) + p_{q1}(t) + p_{q2}(t) + \ldots + p_{q(i-1)}(t) \qquad (2)$$

Equation 2 is valid for any given queue fragment. From Equation 1 and 2, completion time can be expressed as

$$c_{qi}(t) = a_{q1}(t) + p_{q1}(t) + p_{q2}(t) + \ldots + p_{q(i-1)}(t) + p_{qi}(t) \qquad (3)$$

It is important to note that, except for the first part, completion time of a part is not affected by its own arrival time as long as the ordering of arrival times of parts in the queue is not changed. However, when the ordering of arrival time changes the processing sequence, the corresponding change in completion time is discontinuous. This makes completion time a highly nonlinear function of arrival time, with dead-zones and discontinuities, unless arrival times are so far apart that each part is in its own queue fragment and incurs zero queuing time.

Depending on the arrival times, the dynamics of a system of n parts can be divided into three regions: the decoupled region, the dead-zone region, and the discontinuity region. In the decoupled region, M_C, there is no interaction among controllers, all queuing times are zero, and the dynamic behavior of each part can be analyzed independently. In the dead-zone region, M_Z, two or more parts interact and at least one queuing time is greater than zero. This region is called the dead-zone region because the partial derivative of the completion time with respect to the arrival time is zero for any part which incurs queuing time. In the discontinuous region, M_S, two or more arrival times are equal and this region can be expressed as

$$M_S = \{\mathbf{a} \mid a_i = a_j \text{ for } i \neq j\} \qquad (4)$$

which is an (n-1)-dimensional hyperplane in $\mathbf{R}^n$. Each hyperplane $\{a_i = a_j\}$ divides $\mathbf{R}^n$ into two half-spaces in such a way that on one side of the hyperplane, $a_i < a_j$ and Part i is processed before Part j. On the other side of the hyperplane, $a_i > a_j$ and Part i is processed after Part j. The set M_S consists of

$$\binom{n}{2} = \frac{n(n-1)}{2} \tag{5}$$

such hyperplanes which form the boundary of n! mutually exclusive regions corresponding to n! possible processing sequences. The dead-zone region M_Z can be expressed as the following intersection of half-spaces:

$$M_Z = \{\mathbf{a} \mid (a_i < a_j + p_j) \cap (a_j < a_i + p_i), a_j \neq a_i \text{ for } i \neq j\} \tag{6}$$

The decoupled region M_C is the compleme nt of the union of M_Z and M_S which can be expressed as

$$M_C = (M_Z \cup M_S)^c \tag{7}$$

Figures 3 and 4 illustrate the various regions in the arrival time space and completion space, respectively, for two parts and a single machine. Figure 5 illustrates the nonlinear mapping from the arrival-time space to the completion-time space. In Figure 3, the discontinuity region M_S is the line passing through the origin where $a_1 = a_2$. FCFS is not defined when arrival times are equal and mapping of this line into completion time space is not defined. Above this line, $a_1 < a_2$ and the processing sequence will be <1, 2>. Below this line $a_2 < a_1$ and the processing sequence will be <2, 1>. The shaded region in Figure 3, excluding M_S, is the dead-zone region M_Z. In the decoupled region M_C, parts do not interact because their arrival times are far apart and in this region the mapping from arrival-time space to completion-time space is linear.

Because only one part can be processed by a machine at a time, the completion time of parts will differ by at least the processing time of the second part processed. This gives rise to an infeasible region for the completion-time vector and is shown as the shaded region in Figure 4. From Equation 3, in the dead-zone region of sequence <1, 2> the partial derivative of c_2 with respect to a_2 is zero because c_2 is independent of a_2. This is illustrated in Figure 5 where all points in a segment parallel to the a_2 axis in the dead zone are mapped to one point on the boundary of the infeasible region in the completion-time space. Likewise in the dead-zone region of sequence <2, 1> the partial derivative of c_1 with respect to a_1 is zero. Consequently each segment parallel to the a_1 axis in the dead-zone is mapped to one point on the boundary of the infeasible region.

Figures 6 and 7 further illustrate the nature of nonlinearities in the two part, one machine system. In Figure 6, as a_2 is increased from zero towards a_1 which is held constant, c_2 linearly increases from p_2 to a_2+p_2. At $a_1 = a_2$ there is discontinuity at which c_2 jumps to $a_1+p_1+p_2$. Increasing a_2 then has no effect on c_2 while a_2 remains less than a_1+p_1. In this region, as shown in Figure 7, any change in a_2 causes an equal change in q_2, no change in c_2, and gives rise to the dead-zone type nonlinear relationship between a_2 and c_2. In the decoupled region where a_2 is greater than a_1+p_1, q_2 remains zero because arrival times are far apart and any change in a_2 causes an equal change in c_2 which gives rise to a linear relation between a_2 and c_2.

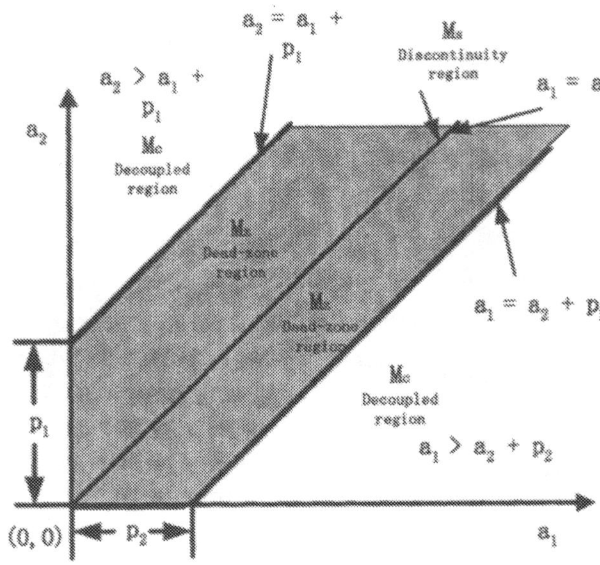

Figure 3. Illustration of various arrival time regions

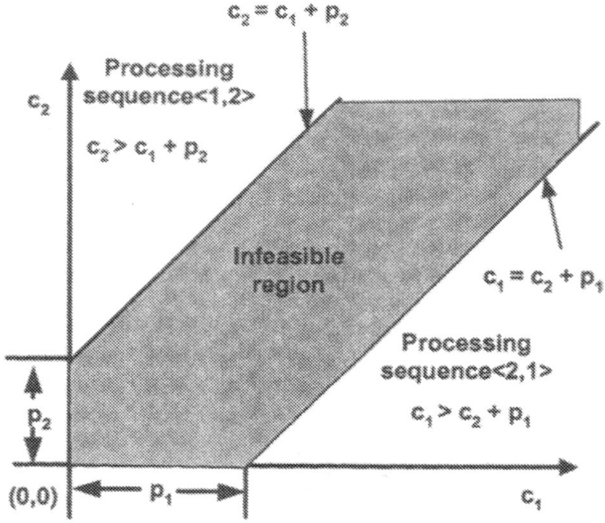

Figure 4. Illustration of completion time regions

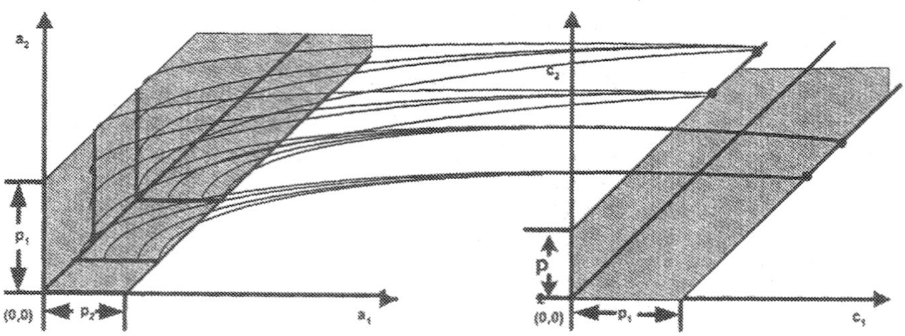

Figure 5. Illustration of nonlinear mapping from arrival to completion times.

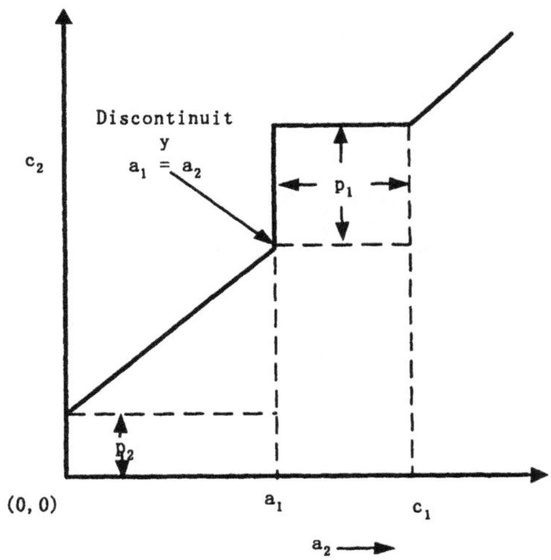

Figure 6. Effect of arrival time on completion time in two part, one machine case.

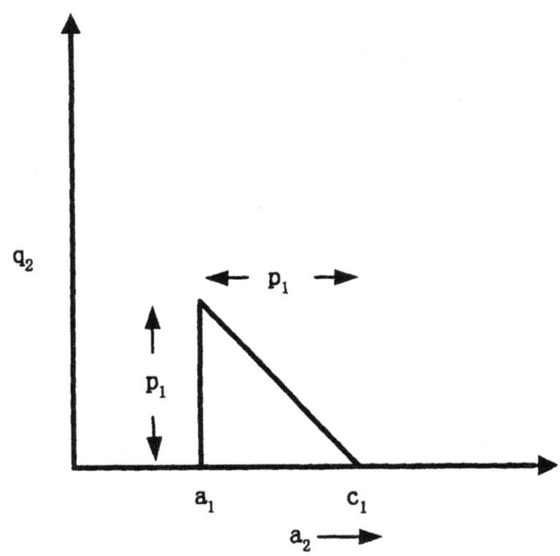

Figure 7. Effect of arrival time on queuing time in two part, one machine case.

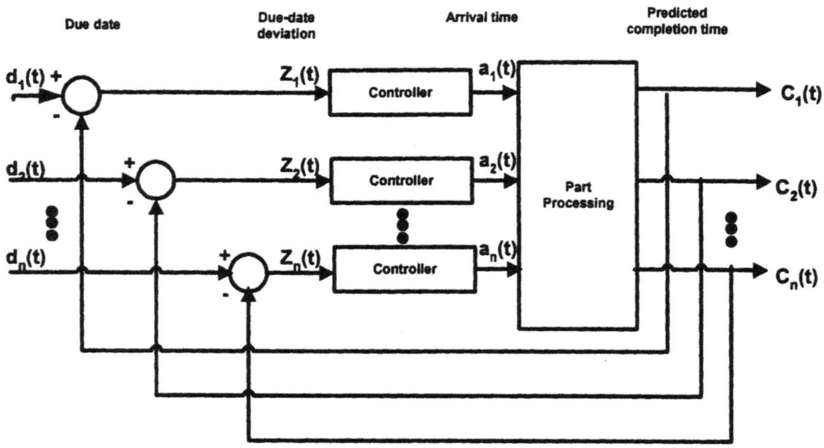

Figure 8. A generic local arrival-time controller

Controlling arrival time for multiple parts gives rise to a multivariable control system with highly nonlinear coupling between arrival and completion times that can include dead-zone and discontinuity-type nonlinearities. Explicit linearization or decoupling is ruled out because it would require global information and rigid relationships. Instead, system nonlinearities can be ignored, aggregated as a disturbance, and compensated using feedback in a local arrival time controller for each part as shown in Figure 8 where z_{qi} is the due date deviation and d_{qi} is the due date. The disturbance aggregates all of the relevant global information into one variable and couples the controllers. As arrival times change dynamically, the disturbance and the coupling between controllers will also change dynamically. The arrival time controller in each part therefore uses a local control law and local feedback for adjusting its arrival time. No assumptions made by arrival time controllers about the number of parts, processing sequence, processing times, due dates of other parts, machine(s) used, or queuing discipline. This results in high levels of autonomy and control distribution retaining minimal global information which are design principles of heterarchical systems.

The choice of the local control law used in each part will have a significant impact on system stability, dynamics, and analyzability. The fact that there is no "plant" in the traditional sense can be used advantageously in selecting a controller because it is well known that for a constant disturbance input, error will asymptotically become zero when integral control is used for a plant with no dynamics (Bollinger and Duffie 1988). Therefore the following integral control law was chosen:

$$a_j(t) = k_j \int_0^t (d_j(\tau) - c_j(\tau))d\tau + a_j(0) \quad (8)$$

where k_j is the controller gain for the jth part in the system, and $a_j(0)$ is an arbitrary initial condition. This simple algorithm only requires local feedback for computing the arrival time. When there is sufficient capacity to meet processing demands and when there is no queuing, disturbances due to queuing will be zero, and error between due-date and completion time (the due-date deviation of the part) will asymptotically become zero. When there is insufficient capacity, this control law adjusts arrival time to make the time average of the part due-date deviation zero with equal penalty for earliness and tardiness.

Even though the arrival times are continuous variables, the system of differential equations that arise from Equation 8 for a n-part system is discontinuous because infinitesimal changes in arrival times can change part processing sequence. The next section reviews discontinuous differential equation theory and this theory is applied to analyze dynamics of arrival time control.

3. MODELING EVENT TIMING CONTROL USING DISCONTINUOUS DIFFERENTIAL EQUATIONS

Classical differential equations are of the form

$$\dot{\mathbf{x}}(t) = f(\mathbf{x}(t)), \text{ where } \mathbf{x}(t) \in \mathbf{R}^n \quad (9)$$

where $\dot{x}(t)$ is the derivative of $\mathbf{x}(t)$ with respect to time t and the function f is continuous in t and $\mathbf{x}$. Solving the above equation means finding a differentiable function

$$\mathbf{x}(t) = \mathbf{x}(0) + \int_0^t f(\mathbf{x}(s))\, ds \quad (10)$$

To date, much of the work in this field focused on equations where f is continuous (Coddington and Levinson 1955). This can be attributed to the fact that continuous functions are reasonably well-behaved and hence amenable to a variety of mathematical manipulations, and also to the fact that there are a large number of applications which can be modeled using such equations (Voxman and Goetschel 1981).

At the turn of the 20th century a new and general approach to integration was developed by Lebesgue which brought about a substantial cohesiveness and simplicity for many areas of analysis (Voxman and Goetschel 1981). Later, in the early twentieth century, Caratheodory used Lebesgue integration theory to study differential equations of the form of Equation 9 where the function f is continuous in **x** but possibly discontinuous in t (Hajek 1979). Solution of this equation is an absolutely continuous function **x**(t) such that the above equation holds at all times except when f is discontinuous instantaneously. Differential equations with discontinuous functions of time have been used to study dynamics of bang-bang optimal control (Hajek 1991).

In 1960 Filippov generalized the concept of solution of differential equations in order to solve Equation 9 where the function f is possibly discontinuous in **x** and t and the solution encounters a line (or surface) of discontinuity infinitely often in a finite time (Filippov 1960). Many of the results in the classical theory of differential equations have been shown to be valid for these discontinuous differential equations (Filippov 1988). These developments have been greatly stimulated by applications such as sliding mode control (Slotine and Sastri 1983, Utkin 1992). Sliding mode control uses a switching control law where the gain of the feedback control system is switched as the state crosses a discontinuity surface. The sliding mode occurs when the system state repeatedly crosses and immediately re-crosses a switching surface because all motion in the neighborhood of the switching surface is directed towards it. Once a trajectory reaches such a switching surface, it is constrained to remain on it and this motion is called sliding mode. Sliding mode control is a mature technique for robust control of uncertain plants, and it has been used in applications including machine drives, flight control, robotics, and automotive control (Zinober 1994).

In contrast to sliding mode control where the primary issue is the design of the discontinuity surfaces, in distributed arrival time control, discontinuity is inherent to the plant because of the manner in which the part processing sequence changes with arrival times as explained in Section 2. Filippov's approach can be used here to solve the case with an arbitrary number of parts and a single machine. Briefly, solution to such an equation is an absolutely continuous function whose derivative vector is a weighted average of all its possible values in the neighborhood. The solution is found by taking the intersection of the discontinuity surface with the closure of the convex hull of all possible derivative vectors in the neighborhood. The key issue in the dynamics of this discontinuous system is that the velocity of the trajectory may not be uniquely determined by the state of the system on the discontinuity surface but belongs to a set of feasible values which depend "loosely" on the state.

Next, the mathematical principles and definitions required to understand Filippov's approach are described. The technique is then used to derive arrival time trajectories in distributed arrival time control problem for *n* parts in a queue fragment and one machine when all of the *n* parts are in the decoupled region, or all are in the dead-zone, or all are in the discontinuity region. The closed-form expression for the final value of arrival times is then derived by showing that the convex hull lies in a (n-1)-dimensional hyper-plane whose intersection with the discontinuity surface results in a zero velocity. These results are then built upon to derive arrival time trajectories when a queue fragment consists of some parts in the dead-zone and some parts in the discontinuity region.

3.1 Definitions

The three main mathematical concepts that are needed to understand Filippov's approach to the solution of discontinuous differential equations are closure of a convex hull, measure zero, and absolutely continuous functions. Detailed expositions on these topics can be found in many texts (Marsden and Hoffman 1993, Voxman and Goetschel 1981). In this sub-section, these concepts are reviewed starting from elementary concepts and definitions.

3.1.1 Closure of Convex Hull

Convex sets are necessary in the study of discontinuous differential equations because the derivative vector on the discontinuity surface belongs to the closure of the convex hull of all the derivative vectors in the neighborhood on the discontinuity and this fact is crucial for proving many theoretical results (Filippov 1960, Filippov 1988).

Definition 1: Line Segment A line segment in $\mathbf{R}^n$ is the set of points along the line between two points represented by vectors $\mathbf{x}_1$ and $\mathbf{x}_2 \in \mathbf{R}^n$. This can be expressed as

$$\{\mathbf{x} : \mathbf{R}^n : \mathbf{x} = \lambda \mathbf{x}_1 + (1-\lambda)\mathbf{x}_2, \lambda \leq 1\} \tag{11}$$

Figure 8 illustrates the line segment between $\mathbf{x}_1$ and $\mathbf{x}_2$ in $\mathbf{R}^2$.

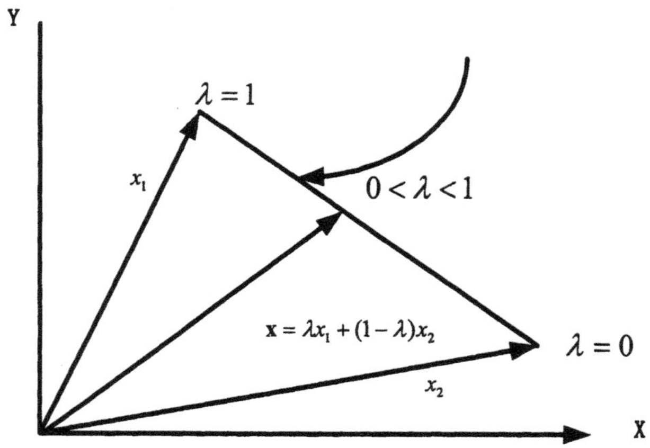

Figure 8. Illustration of a line segment.

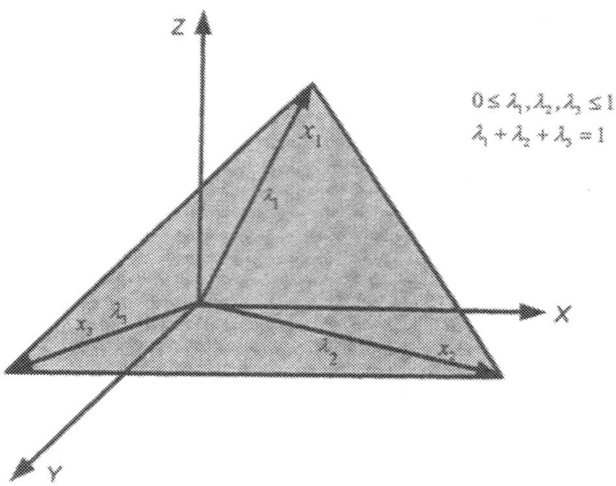

Figure 9. Illustration of a convex combination.

Definition 2: Convex Combination A convex combination of p points in $\mathbf{R}^n$ is the set of points that can be expressed as nonnegative combinations with the total sum of the weights equal to 1:

$$\{\mathbf{x} \in \mathbf{R}^n : \mathbf{x} = \sum_{j=1}^{p} \lambda_j \mathbf{x}_j, \lambda_j \geq 0, \sum_{j=1}^{p} \lambda_j = 1\} \tag{12}$$

Figure 9 illustrates the convex combination of $\mathbf{x}_1$, $\mathbf{x}_2$, and $\mathbf{x}_3$ in $\mathbf{R}^3$.

Definition 3: Convex Set A set of points S in $\mathbf{R}^n$ is said to be convex if the entire line segment joining any two points in the set also belongs to the set S. Figure 10 illustrates the notion of a convex set. It should be noted that set of points in Figure 10 (b) are not convex because there are line segments joining two points that do not belong to the sets.

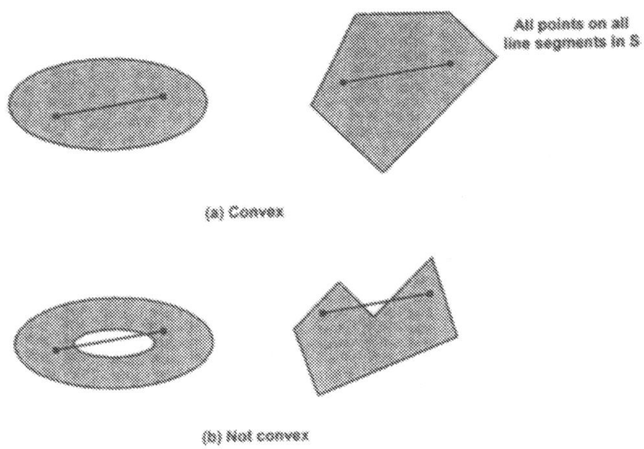

Figure 10. Illustration of convex sets

Definition 4: Convex Hull The convex hull of a set S, Conv (S), is the intersection of all convex sets containing S. Equivalently, Conv(S) can be defined as the smallest convex set containing S. Examples of convex hull are illustrated in Figure 11.

Definition 5: Closure of a Set A point **x** is said to be in the closure of a set S if

$$S \cap B(\mathbf{x}, \delta) \neq \Phi \text{ for every } \delta > 0 \tag{13}$$

where $(\mathbf{x}, \delta)$ is the set of points contained in a ball of radius δ centered at **x**, and ∅ is the empty set. The set of all points in the closure of S is denoted by closure(S).

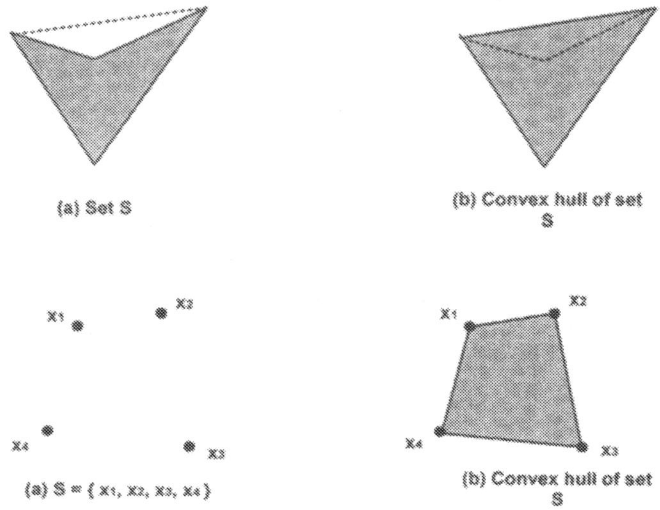

Figure 11. Illustration of convex hull

Definition 6: Closed Set If S = closure(S) then S is said to be closed and is denoted by $\overline{S}$. For example, the set of points in a circle with center (0,0) and radius 1 given by

$$S = (\mathbf{x}_1, \mathbf{x}_2) : \mathbf{x}_1^2 + \mathbf{x}_2^2 \leq 1 \tag{14}$$

is a closed set. It should be noted that in Equation 14 if the inequality $\leq$ is replaced by the strict inequality $<$ then the resulting set S would not be closed.

Definition 7: Closure of a Convex Hull The closure of a convex hull for a set S is the set of all points which belong to the closure of its convex hull and can be denoted as $\overline{Conv}$ (S).

To illustrate the relevance of convex sets to discontinuous differential equations, consider a function $v(t) \in S$ for all t in the interval [a, b] and S is bounded. (A set is said to be bounded if some L exists such that $|x| < L < \infty$ for all $x \in S$). From the mean value theorem (Filippov 1988)

$$v_{mean} = \frac{1}{b-a} \int_a^b v(t)dt \tag{15}$$

Partitioning the domain of integration into small non-overlapping intervals Δ_i, where $a < t_1 < (a+\Delta_1)$, $t_1 < t_2 < (t_1+\Delta_2)$,..., $t_i < t_{(i+1)} < (t_i+\Delta_{i+1})$, ..., $t_{(n-1)} < t_n < (t_{(n-1)} + \Delta_n) < b$ such that

$$b - a = \sum_{i-1}^{n} \Delta_i \tag{16}$$

and in the limit as $\Delta_i \to 0$, the integral in Equation 15 is given by

$$v_{mean} = \sum_{i=1}^{n} \frac{\Delta_i}{b-a} v(t_i) \tag{17}$$

Noting that weight on $v(t_i)$ is

$$\frac{\Delta_i}{b-a} = \alpha_i \geq 0 \text{ and } \sum_{i=1}^{n} \alpha_i = 1 \tag{18}$$

it can be seen that v_{mean} is a convex combination of $v(t_i)$, i.e., $v_{mean} \in \overline{Conv}\, S$. This fact is used in Filippov's approach to compute the derivative vector on discontinuity surfaces (Filippov 1988).

3.1.2 Measure Zero

Intuitively, the term measure of an object refers to the "content" of that object. Some examples of measure include length in $\mathbf{R}^1$, area in $\mathbf{R}^2$, and volume in $\mathbf{R}^3$. An object is said to be of measure zero if it has zero content. The concept of measure can be used in integration of discontinuous functions. In Figure 12, the point t_m where function f(t) is discontinuous has zero width. The integral of f(t) over the interval t_1 to t_2 is shown by the shaded area. The value of f(t) at the discontinuity, $f(t_m)$, has no effect on the integral because the area under the discontinuity is zero. Likewise, even if the discontinuity set in an interval consists of an infinite number of points, it will have no effect on the integral of the function as long as the discontinuity set has a measure zero. Filippov's approach to the solution of discontinuous differential equations requires that the set of points where the function is discontinuous form a set of measure zero. The concept of measure zero will be made more precise by the following definitions.

Definition 8: Countable Set A set S is said to be countable if there is a one-to-one correspondence between set S and set **J** where **J** is the set consisting of all positive integers.

Definition 9: Cover of a Set A cover of a set S is a collection of sets $\{U_1, U_2, ...\}$ whose union contains S.

Definition 10: Interval An interval in $\mathbf{R}^n$ means the set of points $\mathbf{x} = (x_1, x_2, x_3, ..., x_n)$ such that $a_i \le x_i \le b_i$, $(i = 1, 2, 3, ..., n)$.

Definition 11: Volume If U is an interval in $\mathbf{R}^n$ then its volume is defined as

$$\text{volume}(U) = \prod_{i=1}^{n} (b_i \; a_i) \tag{19}$$

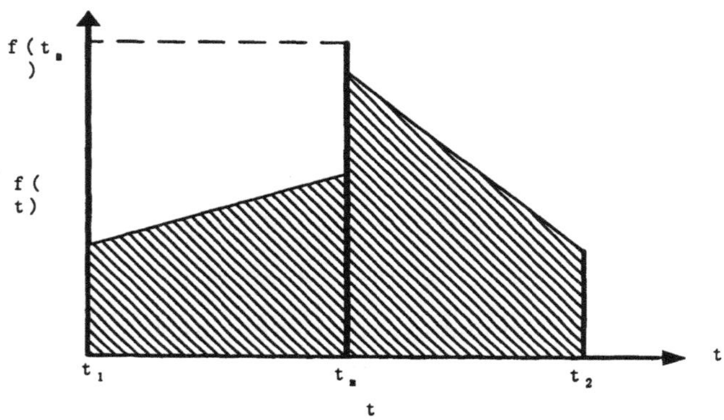

Figure 12. Integration over a set of measure zero

Definition 12: Measure Zero A set S in $\mathbf{R}^n$ is said to have measure zero if for every $\epsilon > 0$ there is a covering C consisting of a countable number of intervals, $C = \{U_1, U_2, ...\}$, such that the total volume

$$\sum_{i=1}^{\infty} \text{volume}(U_i) << \epsilon \tag{20}$$

For instance, a point has zero measure in $\mathbf{R}^1$ and a line segment has zero measure in $\mathbf{R}^2$ even though it may have a positive measure in $\mathbf{R}^1$; in general, measure is defined for a particular space (Marsden and Haoffman 1993). Likewise a 2-dimensional plane will have zero measure in $\mathbf{R}^3$ and a (n-1)-dimensional hyperplane will have zero measure in $\mathbf{R}^n$.

3.1.3 Piecewise Continuous Function

Discontinuous differential equations often have piecewise continuous functions on the right-hand side.

<u>Definition 13: Piecewise Continuous Function</u> (Filippov 1988) A function f(**x**), $\mathbf{x} \in \mathbf{R}^n$, is said to be piecewise continuous in a finite domain $G \in \mathbf{R}^n$ if G consists of a finite number of domains G_i (i = 1...m) in which function f is continuous up to the boundary of each G_i and a set M of measure zero which consists of the boundary points such that

$$G = \left(\bigcup_{i=1}^{m} G_i \right) \cup M \tag{21}$$

The function f is said to be continuous up to the boundary of a domain G_i if, when its argument approaches each point of the boundary, the function tends to a finite limit, possibly to different limits for different boundary points.

For example, Figure 13 shows the absolute value function which is continuous everywhere. Figure 14 shows its derivative which is piecewise continuous in the domain G_1: $x > 0$, and in the domain G_2: $x < 0$;. the function is discontinuous on their boundary M: $x = 0$, which has measure zero.

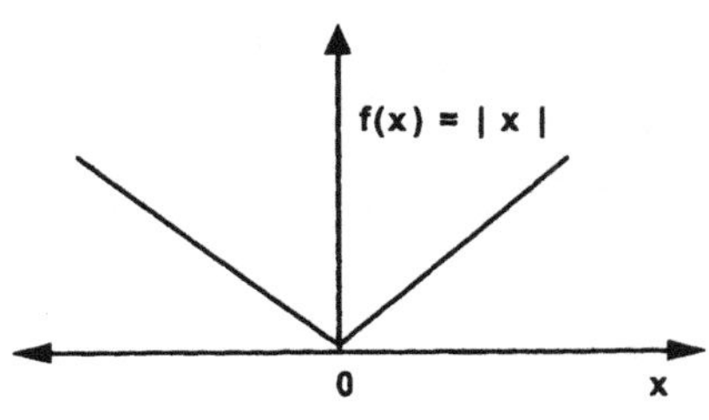

Figure 13. Absolute value function

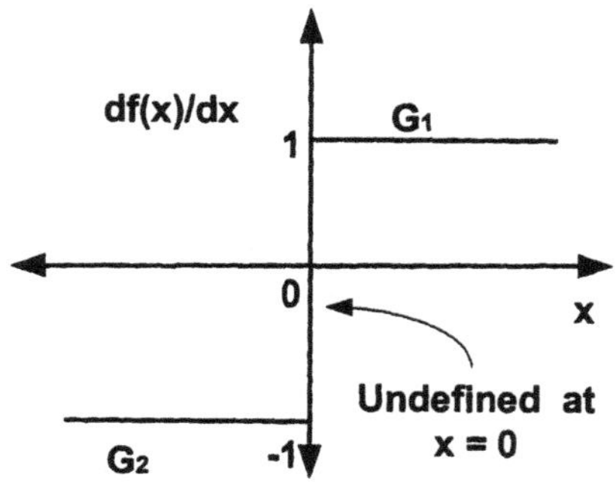

Figure 14. Derivative of the absolute value function

3.1.4 Absolutely Continuous Function

Discontinuous differential equations that can be analyzed using Filippov's approach have solutions that are absolutely continuous functions.

Definition 14: Absolutely Continuous Function A function f which maps the interval [a, b] into $\mathbf{R}^n$ is called absolutely continuous if for every $\varepsilon > 0$ there exists a corresponding $\delta > 0$ such that

$$\sum_{i=1}^{p} \left| f(b_i) - f(a_i) \right| \leq \varepsilon \tag{22}$$

for each finite collection of disjoint subintervals $[a_1, b_2]$, ..., $[a_p, b_p]$ of [a, b]

$$\sum_{i=1}^{p} \left(b_i - a_i \right) \leq \delta \tag{23}$$

An absolutely continuous function has bounded variation. It has a finite derivative except at most on a set of measure zero. For brevity, the term "almost everywhere" can be used in place of "except at most on a set of measure zero". Furthermore, a continuous function is the integral of its derivative if and only if it is absolutely continuous. The absolute value function in Figure 13 is an example of an absolutely continuous function.

3.2 Solution of Discontinuous Differential Equations

Filippov's approach for solving differential equations with discontinuities is reviewed in this sub-section (Filippov 1960, Filippov 1988). To illustrate a typical "pathology" encountered in differential equations discontinuous in **x**, consider the following example due to Filippov (Filippov 1988):

$$\dot{x} = 1 - 2\,\mathrm{sign}(x) \tag{24}$$

Therefore

$$\dot{x} = 3 \quad \text{for } x < 0$$

$$\dot{x} = -1 \quad \text{for } x > 0 \tag{25}$$

Which gives the solutions

$$x(t) = 3t + c_1 \quad \text{for } x < 0$$

$$x(t) = -t + c_2 \quad \text{for } x > 0 \tag{26}$$

As shown in Figure 15, all solutions regardless of the initial conditions reach the line x = 0 as t increases. The solution slides on this discontinuity line and is prevented from leaving this line because as soon as it leaves, the direction field (the velocity vectors) forces it back to the line on either side. The solution on this line is $x(t) = 0$ and $\dot{x}(t) = 0$ but this solution does not satisfy Equation 24 in the usual sense and thus the "pathology".

In reality, discritization due to the computational methods used will determine the actual trajectory followed. This trajectory will lie within some distance δ from the mathematical ideal as shown in Figure 16(b). Intuitively, in the limit as $\delta \to 0$, the actual trajectory will cross the discontinuity infinitely often and tend towards the ideal trajectory shown in Figure 16(a).

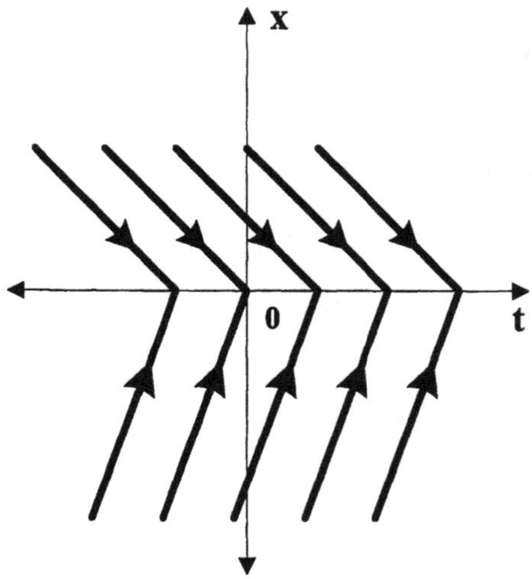

Figure 15. An Example of discontinuous differential equation

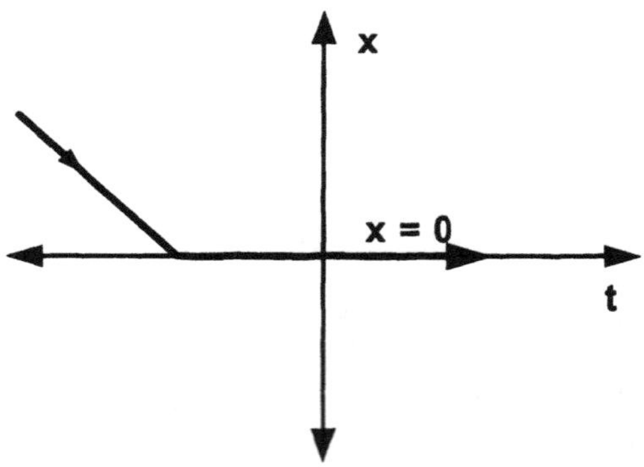

Figure 16(a). Ideal trajectory

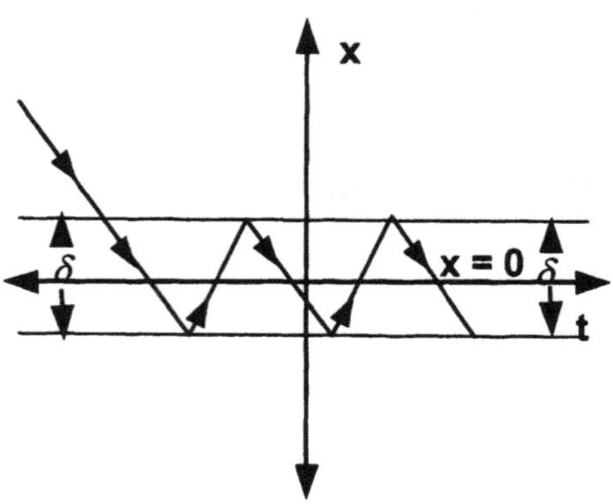

Figure 16(b). Computational effects near discontinuity

The main issue in the solution of discontinuous differential equations is to find the velocity, $\dot{\mathbf{x}}(t)$, for solutions which remain on a discontinuity surface or on an intersection of discontinuity surfaces. Consider a function in the form of Equation 9 where f is a piecewise continuous function in the domain G ∈ $\mathbf{R}^n$ and M is a set of points, with measure zero, where f is discontinuous. The solution specified by Filippov is an absolutely continuous function $\mathbf{x}(t)$ defined on an interval I:[0, T] for which

$$\dot{\mathbf{x}}(t) \in F(\mathbf{x}(t) \tag{27}$$

almost everywhere on I. If the point $\mathbf{x}(t) \notin M$ then $F(\mathbf{x}(t)) = f(\mathbf{x}(t))$ which is a single point in $\mathbf{R}^n$ as in classical solutions to continuous differential equations. More importantly, if the point $\mathbf{x}(t) \in M$ then $F(\mathbf{x}(t))$ is given by

$$F((\mathbf{x}(t)) = \bigcap_{\delta > 0} \bigcap_{M} \overline{\mathrm{Conv}}\ f(\ B(\mathbf{x}(t), \delta) - M\ ) \tag{28}$$

The set $F(\mathbf{x}(t))$ depends on the behavior of the system in the δ-neighborhood of the discontinuity and is the smallest convex closed set containing all the limit values of the function $f(\mathbf{x}^*(t))$ as $\mathbf{x}^*(t) \to \mathbf{x}(t)$, with t constant and $\mathbf{x}^*(t) \notin M$. This accomplished in Equation 28 by taking the intersection over all sets with $\delta > 0$. In Equation 28, "minus the set M of measure zero" allows possible "misbehavior" of f on M to be ignored in forming the set $F(\mathbf{x}(t))$. It should be emphasized that the equation may not have a solution, may have more than one solution, or has exactly one solution on M depending on whether the set $F(\mathbf{x}(t))$ is empty, more than one, or exactly one point, respectively. If there is no solution on M then a trajectory reaching M will not remain in M; if there is more than one solution then a trajectory reaching M will remain on M but its direction is not unique; and if there is one solution then a trajectory reaching M will remain on M and move in a unique direction. The intersection with set M constrains the solution to lie in M. If the point $\mathbf{x}(t) \in$ M lies on the intersection of two or more domains G_i (i = 1, ..., k) then all points $f_i(\mathbf{x}(t))$ will be contained in $F(\mathbf{x}(t))$ which could be a line segment, convex polygon, or a polyhedron with vertices $f_i(\mathbf{x}(t))$, with t constant, where

$$f_i\mathbf{x}(t)) = \lim_{\substack{\mathbf{x}^*(t) \in G_i \\ \mathbf{x}^*(t) \to \mathbf{x}(t)}} f_i(\mathbf{x}^*(t)) \tag{29}$$

Figure 17(a) illustrates the above concepts applied to the example of Equation 24 when, initially $x(0) = 2$ and $\dot{x}(t) = -1$. The solution trajectory reaches the discontinuity set M, the line $x = 0$, at point $(t, x) = (2, 0)$ which lies on the boundary of the two domains G_1: $x > 0$ and G_2: $x < 0$. From Equation 25, with t constant, the limit of $\dot{x}$ as $x^* \rightarrow (2, 0^+)$ is shown as the vector $\dot{x}^+$ starting at (2, 0) with end point at (2, -1) and the limit of $\dot{x}(t)$ as $x^* \rightarrow (2, 0^-)$ is shown in Figure 17(b) as the vector $\dot{x}^-$ starting at (2, 0) and end point at (2, 3). The closure of the convex hull of $\dot{x}^+$ and $\dot{x}^-$ is the line segment joining their end-points (2, 3) and (2, -1), respectively, shown as a thick dashed line. In this case the closure of the convex hull intersects M at (2, 0) which means the set F(t, x(t)) is a single point (2, 0). Therefore in this case $\dot{x}(t)$ has a unique solution on M which starts at (2, 0) and ends at (2, 0). Since these two points coincide $\dot{x}(t) = 0$ on M which gives $x(t) = 0$ as expected. In this example both velocity vectors $\dot{x}^+$ and $\dot{x}^-$ have opposite directions and are directed towards M which causes the solution to slide on the discontinuity line. Consider a modification of the example such that $\dot{x}^+$ and $\dot{x}^-$ are directed in the same direction. Then one of them would be directed away from M causing the convex hull to lie on one side of the discontinuity line and not intersect it as shown in Figure 17(c). In such a case there would be no solution for $\dot{x}^+$ on the discontinuity line and any solution would cross the discontinuity line instantaneously.

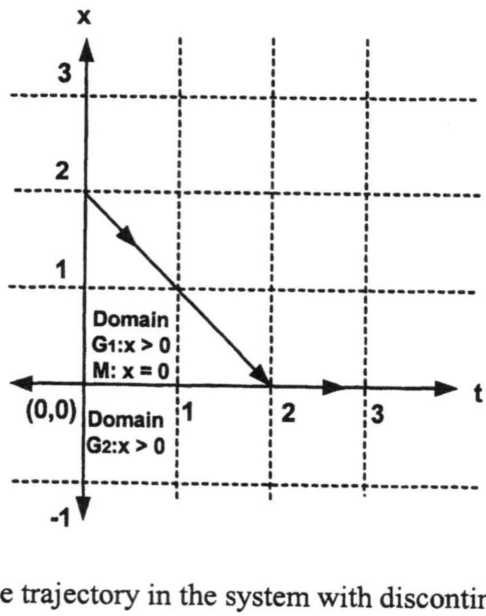

Figure 17(a). The trajectory in the system with discontinuities

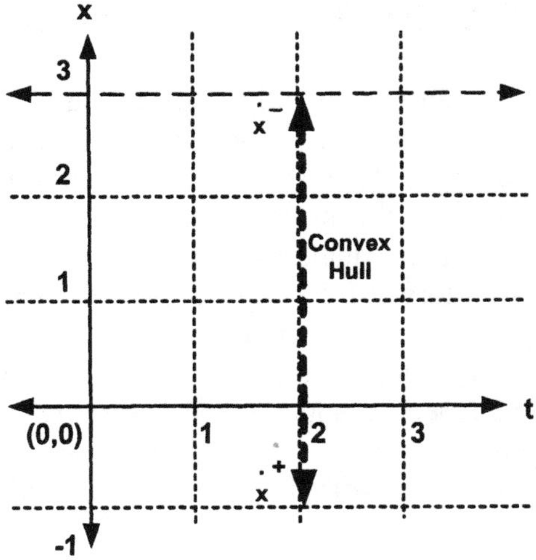

Figure 17(b). Convex hull intersects the discontinuity region

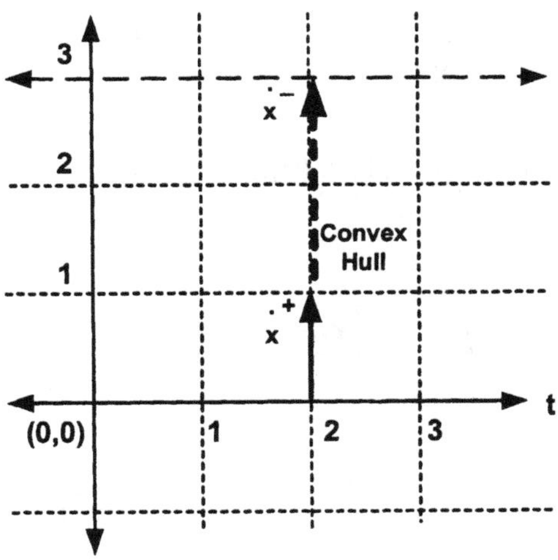

Figure 17(c). Convex hull does not intersect the discontinuity region

3.3 Distributed Arrival Time Control Solution

As discussed in Section 2, as the arrival time vector changes dynamically; it can be in the decoupled, the dead-zone, or the discontinuity regions. The key is to solve the differential equations in order to understand the fundamental dynamics of the system and determine the steady-state values. Solving the differential equations in the decoupled region and the dead-zone region is straightforward. However, "pathological" conditions are possible even in two-parts, single-machine case, where steady-state arrival times converge to an equal value, the discontinuity region. Moreover, for n > 2 there are n! possible sequences and the average arrival times cannot be readily derived. Filippov's approach can be used to solve such dynamical equations for arrival time control with the following integral control law

$$a_j(t) = k_j \int_0^t z_j(\tau)d\tau + a_j(0) \tag{30}$$

Differentiating both sides of Equation 30 gives

$$\dot{a}_j(t) = k_j z_j(t) \tag{31}$$

In the decoupled region, M_C, there is no interaction among controllers and all queuing times are zero. From Section 2, the dynamical model in the decoupled can be expressed as

$$z_1(t) = d_1 - k_1 \int_0^t z_1(\tau)d\tau - a_1(0) - p_1 \tag{32}$$

$$z_2(t) = d_2 - k_2 \int_0^t z_2(\tau)d\tau - a_2(0) - p_2 \tag{33}$$

$$z_j(t) = d_j - k_j \int_0^t z_j(\tau)d\tau - a_j(0) - p_j \tag{34}$$

where $z_j(t)$ is the due date deviation for the jth part, d_j is the due date, $a_j(0)$ is the initial arrival time, and p_j is the processing time. Using Equations 30 and 31, Equations 32 to 34 can be rewritten as

$$\dot{a}_1(t) = -k_1 a_1(t) - k_1 p_1 + k_1 d_1 \tag{35}$$

$$\dot{a}_2(t) = -k_2 a_2(t) - k_2 p_2 + k_2 d_2 \tag{36}$$

$$\dot{a}_j(t) = -k_j a_j(t) - k_j p_j + k_j d_j \tag{37}$$

Equations 35 to 37 can be expressed in the form of Equation 9 as

$$\dot{a} = f(a) \quad a \in M_c \tag{38}$$

where $\mathbf{a} = [a_1\ a_2 \cdots a_n]^T$ and $f(\mathbf{a})$ is given by

$$f(\mathbf{a}) = \begin{bmatrix} k_1 & 0 & 0 & \cdots & 0 \\ 0 & k_2 & 0 & \cdots & 0 \\ & & & \vdots & \\ 0 & 0 & 0 & \cdots & k_n \end{bmatrix} \begin{bmatrix} a_1 \\ a_2 \\ \vdots \\ a_n \end{bmatrix} - \begin{bmatrix} k_1 & 0 & 0 & \cdots & 0 \\ 0 & k_2 & 0 & \cdots & 0 \\ & & & \vdots & \\ 0 & 0 & 0 & \cdots & k_n \end{bmatrix} \begin{bmatrix} p_1 \\ p_2 \\ \vdots \\ p_n \end{bmatrix}$$
$$+ \begin{bmatrix} k_1 & 0 & 0 & \cdots & 0 \\ 0 & k_2 & 0 & \cdots & 0 \\ & & & \vdots & \\ 0 & 0 & 0 & \cdots & k_n \end{bmatrix} \begin{bmatrix} d_1 \\ d_2 \\ \vdots \\ d_n \end{bmatrix}$$

In the dead-zone region, M_z, two or more controllers interact and at least one queuing time is greater than zero. From Section 2, the dynamic model of a queue fragment in the dead-zone region can be expressed as

$$z_{q1}(t) = d_{q1} - k_{q1}\int_0^t z_{q1}(\tau)d\tau - a_{q1}(0) - (p_{q1}) \tag{39}$$

$$z_{q2}(t) = d_{q2} - k_{q1}\int_0^t z_{q1}(\tau)d\tau - a_{q1}(0) - (p_{q1} + p_{q2}) \tag{40}$$

$$z_{qi}(t) = d_{qi} - k_{qi}\int_0^t z_{q1}(\tau)d\tau - a_{q1}(0) - (p_{q1} + p_{q2} + \ldots + p_{qi}) \tag{41}$$

Using Equations 30 and 31, Equation 39 to 41 can be rewritten as

$$\dot{a}_{q1}(t) = -k_{q1}a_{q1}(t) - k_{q1}p_{q1} + k_{q1}d_{q1} \tag{42}$$

$$\dot{a}_{q2}(t) = -k_{q2}a_{q1}(t) - k_{q2}(p_{q1} + p_{q2}) + k_{q2}d_{q2} \tag{43}$$

$$\dot{a}_{qi}(t) = -k_{qi}a_{q1}(t) - k_{qi}(p_{q1} + p_{q2} + \cdots + p_{qi}) + k_{qi}d_{qi} \tag{44}$$

Equation 42 to 43 can be expressed in the form of Equation 9 as

$$\dot{\mathbf{a}} = f(\mathbf{a}) \quad \mathbf{a} \in M_Z \tag{45}$$

where $\mathbf{a} = [a_{q1}\ a_{q2} \cdots a_{qn}]^T$ and $f(\mathbf{a})$ is given by

$$
f(\mathbf{a}) = \begin{bmatrix} k_{q1} & 0 & 0 & \cdots & 0 \\ k_{q2} & 0 & 0 & \cdots & 0 \\ & & & \vdots & \\ k_{qn} & 0 & 0 & \cdots & 0 \end{bmatrix} \begin{bmatrix} a_{q1} \\ a_{q2} \\ \vdots \\ a_{qn} \end{bmatrix}
- \begin{bmatrix} k_{q1} & 0 & 0 & \cdots & 0 \\ k_{q2} & k_{q2} & 0 & \cdots & 0 \\ & & & \vdots & \\ k_{qn} & k_{qn} & k_{qn} & \cdots & k_{qn} \end{bmatrix} \begin{bmatrix} p_{q1} \\ p_{q2} \\ \vdots \\ p_{qn} \end{bmatrix}
+ \begin{bmatrix} k_{q1} & 0 & 0 & \cdots & 0 \\ 0 & k_{q2} & 0 & \cdots & 0 \\ & & & \vdots & \\ 0 & 0 & 0 & \cdots & k_{qn} \end{bmatrix} \begin{bmatrix} d_{q1} \\ d_{q2} \\ \vdots \\ d_{qn} \end{bmatrix}
$$

As the arrival time vector changes dynamically the processing sequence can change. The function f($\mathbf{a}$) is continuous in each of the n! regions corresponding to the n! possible processing sequences, except in the discontinuity region M_S which forms their boundary. Because M_S is the union of a finite number of hyperplanes, it has a measure zero in $\mathbf{R}^n$, and according Definition 13, f(a) is a piece-wise continuous function in $\mathbf{R}^n$. Therefore using Filippov's approach, solution to the following differential equation

$$\dot{\mathbf{a}} = f(\mathbf{a}) \quad \mathbf{a} \in \mathbf{R}^n \tag{46}$$

is specified as a solution to

$$\dot{\mathbf{a}} = F(\mathbf{a}(t)) \tag{47}$$

In the decoupled and dead-zone regions $\mathbf{a}(t) \notin M_S$, $F(\mathbf{a}(t))$ is a single point, and $\mathbf{a}(t)$ can be found by solving the corresponding equations using classical techniques.

3.3.1 Solution in Decoupled Region

In the decoupled region, $\mathbf{a}(t)$ is found by solving Equation 38 which gives

$$a_1(t) = (d_1 - p_1)(1 - e^{-k_1 t}) + a_1(0)\, e^{-k_1 t} \tag{48}$$

$$a_2(t) = (d_2 - p_2)(1 - e^{-k_2 t}) + a_2(0) e^{-k_2 t} \tag{49}$$

$$a_j(t) = (d_j - p_j)(1 - e^{-k_j t}) + a_j(0)\, e^{-k_j t} \tag{50}$$

3.3.2 Solution in Dead-Zone Region

In the dead-zone region, $\mathbf{a}(t)$ is found by solving Equation 45 which gives

$$a_{q1}(t) = (d_{q1} - p_{q1})(1 - e^{-k_{q1} t}) + a_{q1}(0) e^{-k_{q1} t} \tag{51}$$

$$a_{q2}(t) = \frac{k_{q2}}{k_{q1}}(d_{q1} - p_{q1})(1 - e^{-k_{q1} t}) - \frac{k_{q2}}{k_{q1}} a_{q1}(0)(1 - e^{-k_{q1} t}) + k_{q2}(d_{q2} - d_{q1} - p_{q2})t + a_{q2}(0) \tag{52}$$

$$a_{qi}(t) = \frac{k_{qi}}{k_{q1}}(d_{q1} - p_{q1})(1 - e^{-k_{qi} t}) - \frac{k_{qi}}{k_{q1}} a_{q1}(0)(1 - e^{-k_{qi} t}) + k_{qi}(d_{q2} - d_{q1} - p_{q2} - \dots - p_{qi})t + a_{qi}(0) \tag{53}$$

3.3.3 Solution in Discontinuity Region

In the discontinuity region **a**(t)$\in$ M_s, dynamics are not defined and the set F(**a**(t)) will depend on the behavior of the systems in the δ-neighborhood of **a**(t). If n arrival times are equal at **a**(t) then n! sequences are possible in its δ-neighborhood, the point **a**(t) lies at the boundary of n! domains, and the convex hull of f(**a***(t)) in the corresponding dead-zones needs to be considered as prescribed by Filippov's approach in Equation 28. It should be pointed out that in this case, M_s is an intersection of n hyperplanes which is a line in n-dimensional space. From Equations 31 and 45 it is evident that f(**a***(t)) = **Kz**(t) where **K** is the diagonal gain matrix with diagonal(**K**) = [k_1 k_2 k_3...k_n] and **z**(t) is the due date deviation vector [z_1(t) z_2(t) ... z_n(t)]^T for the corresponding processing sequence. Since **z**(t) depends on the due-date vector **d** = [d_1 d_2 ... d_n], the location of **d**, the geometry of the convex hull of **Kz**(t), and the manner in which the convex hull intersects M_s will determine the dynamics of the system in the discontinuity region.

Interpreting Filippov's approach from an engineering point of offers the following important insights:

1. The velocity of the arrival time trajectory is not unique; i.e., velocity is not given by an equation. Instead the velocity belongs to a set of possible values given by the set F(**a**(t)).

2. The velocity at any point in the discontinuity region depends "loosely" on the point. Specifically, the velocity in this discontinuous region depends on the δ-neighborhood of the point instead of depending only on the point as in the continuous region.

3. The velocity of an arrival time trajectory in the discontinuity region is a weighted average of all feasible velocities in the neighboring dead-zone region. Weights depend on the residence time in a particular processing sequence region. The convex hull of velocities in the neighboring dead-zone region will contain their weighted average. Filippov's constrains the average velocity to lie in the discontinuity region by taking the intersection of the convex hull and the discontinuity set.

4. Another noteworthy feature of Filippov's approach is that by allowing the discontinuity region to be ignored in forming the velocity set, it eliminates the need for tie-breaking rules when arrival times are equal. As discussed in Section 3.1.2, this is possible because the discontinuity region has a measure zero and does not affect any integrals which contain it. Consequently, dynamics in the discontinuity region need not be defined for determining the dynamical behavior in the infinitesimal neighborhood of this region.

3.3.4 Convex Hull Geometry

In this subsection the terminology to describe the geometry of the convex hull of due-date deviation vectors will be defined first. After this it will shown that the convex hull lies in a (n-1)-dimensional plane. From Equations 31 and 45 it can be seen that the velocity of the arrival time vector is a linear transform of the due-date deviation vector. Therefore the geometry of the convex hull of velocity vectors will be similar to the geometry of the due-date deviation vectors. In the discontinuity region the convex hull of due date deviation vectors forms an n!-sided polygon which can be viewed as the base of a pyramid in the arrival-time space whose apex is **a**(t) and edges are the n! **Kz**(t) vectors; this will be henceforth be denoted as the z_a-pyramid. Likewise in the completion-time space, convex hull of completion time vectors form a n!-sided polygon which can be viewed as the base of two pyramids. The first pyramid with apex **d** and edges **z**(t) vectors will henceforth be denoted as the z_d-pyramid. The second pyramid with apex **a**(t) and edges consisting of n! disturbance vectors **r** = $[(q_1+p_1)\ (q_2+p_2)\ \ldots\ (q_n+p_n)]$ will henceforth be denoted as the r-pyramid.

In the discontinuity region M_S, **a**(t) may not have a solution, may have more than one solution, or may have exactly one solution depending on whether the set F(**a**(t)) is empty, more than one, or exactly one point, respectively. If there is no solution on M_S then a trajectory reaching M_S will not remain in M_S; if there is more than one solution then a trajectory reaching M_S will remain on M_S but its direction is not unique; and if there is one solution then a trajectory reaching M_S will remain on M_S and move in a unique direction. In distributed arrival time control, the convex hull of **Kz**(t) lies in an (n-1)-dimensional plane and whenever it intersects M_S the intersection is a unique point and hence $\dot{\mathbf{a}}$ is unique, and **a**(t) is unique. This will be shown by showing that the base of the r-pyramid lies in a (n-1)-dimensional plane; i.e., its sides lie in a (n-1)-dimensional space. Sides of the base are the segments joining completion time vectors obtained by interchanging a pair of neighboring parts in a sequence. The n! completion time vectors for sequences <1,2,3,...,n>, <2,1,3, ...,n>, ..., <n,...,3,2,1> which form the edges of the r-pyramid, are given by

$$\mathbf{r}_{<1,2,3,\ldots,n>} = \begin{bmatrix} p_1 \\ p_1 + p_2 \\ \vdots \\ p_1 + p_2 + \cdots + p_n \end{bmatrix} \quad (54)$$

$$\mathbf{r}_{<2,1,3,\ldots,n>} = \begin{bmatrix} p_1 + p_2 \\ p_2 \\ \vdots \\ p_1 + p_2 + \cdots + p_n \end{bmatrix} \tag{55}$$

$$\mathbf{r}_{<n,n-1,\ldots,2,1>} = \begin{bmatrix} p_1 + p_2 + \cdots + p_n \\ p_2 + \cdots + p_n \\ \vdots \\ p_n \end{bmatrix} \tag{56}$$

The side connecting $\mathbf{r}_{<1,2,3,\ldots,n>}$ and $\mathbf{r}_{<2,1,3,\ldots,n>}$ can be obtained by interchanging the position of Part #1 and Part #2 can be expressed as a vector difference

$$\mathbf{r}_{<1,2,3,\ldots,n>} - \mathbf{r}_{<2,1,3,\ldots,n>} = \begin{bmatrix} p_1 \\ p_1 + p_2 \\ \vdots \\ p_1 + p_2 + \cdots + p_n \end{bmatrix} - \begin{bmatrix} p_1 + p_2 \\ p_2 \\ \vdots \\ p_1 + p_2 + \cdots + p_n \end{bmatrix} \tag{57}$$

$$\mathbf{r}_{<1,2,3,\ldots,n>} - \mathbf{r}_{<2,1,3,\ldots,n>} = \begin{bmatrix} -p_2 \\ p_1 \\ \vdots \\ 0 \end{bmatrix} \tag{58}$$

For some sequence S with neighboring parts Part #u and Part #v interchanged to give the sequence S' such that Part #v is processed before Part #u in S' then the column corresponding to $\mathbf{r}_S - \mathbf{r}_{S'}$ will have $+p_v$ in the uth row and $-p_u$ in the vth row because interchanging causes the completion time of Part #u to increase and Part #v to decrease by p_u and p_v, respectively. The completion time of other parts is not affected because Part #u and Part #v are neighbors. By repeating the neighbor interchanging process, all n! sides can be obtained and the space spanned by these vectors can be expressed in an nxn! matrix with the following properties:

1. It has n rows representing the number of parts and n! columns representing the number of possible sequences.
2. Corresponding to the pair interchanged there are two non-zero entries in any column; one corresponding to the part that moved ahead in the sequence is less than zero and one that moved behind is greater than zero.
3. The sum of row elements is zero because the base is a closed polygon.

The rank of this matrix determines the dimensions of the space spanned by the base of the r-pyramid. The first nxn portion of the nxn! B matrix is of the form

$$\begin{bmatrix} -p_2 & 0 & 0 & \cdots & 0 & 0 \\ p_1 & p_3 & p_4 & \cdots & p_n & 0 \\ 0 & -p_2 & 0 & \cdots & 0 & p_4 \\ 0 & 0 & -p_2 & \cdots & 0 & -p_3 \\ \cdots & \cdots & \cdots & \vdots & \cdots & \cdots \\ 0 & 0 & 0 & \cdots & -p_2 & 0 \end{bmatrix} \quad (59)$$

By inspecting Equation 59, the rank of this matrix is at least (n-1) because the first (n-1) columns are linearly independent and the rank can be at most n because there are only n rows. Gausian elimination can be used to determine the rank as follows:

Step 1 $$R2 = R1 * (p1/p2) + R2 \quad (60)$$

Step 2 $$R3 = R2 * (p2/p3) + R3 \quad (61)$$

Step 3 $R4 = R3 * (p3/p4) + R4$ (62)

Step 4 $R5 = R4 * (p4/p5) + R5$ (63)

Step 5 $R6 = R5 * (p5/p6) + R6$ (64)

•••

Step n-2 $Rn\text{-}1 = Rn\text{-}2 *(pn\text{-}2/pn\text{-}1) + Rn\text{-}1$ (65)

Step n1 $Rn = Rn\text{-}1 *(pn\text{-}1/pn) + Rn$ (66)

gives the following matrix

$$\begin{bmatrix} -p_2 & 0 & 0 & \cdots & 0 & 0 \\ 0 & p_3 & p_4 & \cdots & p_{n\text{-}1} & 0 \\ 0 & 0 & \dfrac{p_4 p_2}{p_3} & \cdots & \dfrac{p_n p_2}{p_3} & p_4 \\ 0 & 0 & 0 & \cdots & \dfrac{p_n p_2}{p_4} & 0 \\ \cdots & \cdots & \cdots & \cdots & \cdots & \cdots \\ 0 & 0 & 0 & \cdots & 0 & 0 \end{bmatrix}$$

Since only (n-1) pivots are non zero, its rank is (n-1) and base of the r-pyramid an (n-1)-dimensional plane. The normal to this plane can be found by making use of the fact that columns of a matrix are orthogonal to its left null space. The left null space of the matrix in Equation 59 is a one dimensional line $\mathbf{p} = [p_1\ p_2\ p_3\ \dots\ p_n]^T$. The inner product of **p** and any column of B with $+p_v$ in the uth row and $-p_u$ in the vth row will be $p_v p_u - p_v p_u = 0$. Therefore **p** is orthogonal to all n! columns of B even though only n were considered above. Furthermore, sum of the dimensions of the left null space and column space is equal to the number of rows in a matrix. In this case the left null space has one dimension and therefore the columns of the B matrix span (n-1)-dimensions, verifying that the B matrix has a rank of (n-1).

The completion times **c**(t) lie on the base of r-pyramid; therefore, the equation for the plane containing the base can be expressed as

$$p^T(\mathbf{x}(t) - \mathbf{c}(t)) = 0 \tag{67}$$

where $\mathbf{x}(t) \in \mathbf{R}^n$. The base of the r-pyramid is also the base of the z_d-pyramid; therefore, Equation 67 also represents the base of the z_d-pyramid. The z_a-pyramid is a "magnified" z_d-pyramid with **K** being the magnification factor. Therefore, the base of the z_a-pyramid will be contained in a plane which can be expressed as

$$(K^{-1}p)T(y(t) - Kz(t)) = 0 \tag{68}$$

where $\mathbf{y}(t) \in \mathbf{R}^n$.

When the convex hull of the z_a-pyramid base intersects the discontinuity region, the a vector starts at **a**(t) and ends at the plane given by Equation 65 and can be expressed as

$$(\mathbf{K}^{-1}\mathbf{p})^T \dot{\mathbf{a}}(t) = \mathbf{p}^T \mathbf{z}(t) \tag{69}$$

Since the end point of $\dot{\mathbf{a}}(t)$ lies at the discontinuity region M_s given by the line [1 1 ... 1], all n components of $\dot{\mathbf{a}}(t)$ will be equal. Using this and Equation 66, each of the n components of $\dot{\mathbf{a}}(t)$, which are equal, can be expressed as a scalar (scalars are denoted in plain type: a(t), a (t), etc., whereas vectors are denoted in bold type: $\mathbf{a}(t)$, $\dot{\mathbf{a}}(t)$ etc.)

$$\mathbf{a}(t) = \frac{\mathbf{p}^{T}\mathbf{z}(t)}{(\mathbf{k}^{-1}\mathbf{p})^{T}[1\,1 \ldots 1]} \tag{70}$$

Substituting $\mathbf{z}(t) = \mathbf{d} - \mathbf{a}(t) - \mathbf{r}$ in Equation 70 gives

$$\dot{\mathbf{a}}(t) = \tilde{\mathbf{k}}(\alpha - \mathbf{a}(t)) \tag{71}$$

where

$$\tilde{k} = \frac{p_1 + p_2 + \ldots + p_n}{\frac{p_1}{k_1} + \frac{p_2}{k_2} + \frac{p_n}{k_n}}$$

and

$$\alpha = \frac{\mathbf{p}^{T}(\mathbf{d} - \mathbf{r})}{p_1 + p_2 + \ldots + p_n}$$

Solving Equation 71 gives

$$a(t) = \alpha(1 - e^{-\tilde{k}t}) + a(0)e^{-\tilde{k}t} \tag{72}$$

for all components of $\mathbf{a}(t)$. It needs to be emphasized this solution holds for all n parts as long as their arrival times are equal and in the discontinuity region. Interestingly, dynamics in the discontinuity region can be modeled as a scalar regardless of n, but the exponential convergence rate depends on processing times and gains k.

3.3.5 Steady-State Arrival Time

From Equation 69, $\dot{\mathbf{a}}(t)$ will be zero when

$$\mathbf{p}^T\mathbf{z}(t) = 0 \tag{73}$$

Under this condition the system will reach an equilibrium, and $\mathbf{a}(t)$ will converge to its steady state value denoted by $\mathbf{a}_{ss}$. The left side of Equation 73 is the projection of $\mathbf{z}(t)$ on the base of the normal to the r-pyramid and can become zero under the following two conditions.

Feasible Due Dates

The first condition is $\mathbf{z}(t) = 0$, i.e., when all due-date deviations are zero. Under this condition the due dates and completion times are equal and this can happen only when due dates are feasible. From Section 2, $\mathbf{a}$ in the decoupled region is mapped to $\mathbf{c}$ in the feasible region. In other words when the due dates are feasible the equilibrium point $\mathbf{a}_{ss}$ will be in the decoupled region. Since all queuing times are zero in the feasible region, $\mathbf{a}_{ss}$ can be expressed as

$$\mathbf{a}_{ss} = \mathbf{a}(t) = \mathbf{d} - \mathbf{p},\ \mathbf{a}_{ss} \in \mathbf{M}_c\ ,\ \mathbf{d}\ \in \text{Feasible} \tag{74}$$

For a given $\mathbf{d}$ and $\mathbf{p}$, $\mathbf{a}_{ss}$ is unique and independent of $\mathbf{K}$. The steady state sequence is also unique and is determined by $\mathbf{d}$. Furthermore, from Equations 48 to 50, in the decoupled region $\mathbf{a}(t)$ exponentially converges and the decoupling allows the dynamic equations to be treated as scalar equations in the vicinity of the equilibrium point.

Infeasible Due Dates

The second condition when left side of Equation 73 becomes zero is when the $\mathbf{z}(t)$ vectors lie on the base of the r-pyramid. Under this condition the z_a-pyramid will lie in a hyperplane, the starting point and ending point of $\dot{\mathbf{a}}(t)$ will coincide, the system will reach an equilibrium, and $\mathbf{a}(t)$ will converge to its steady state value $\mathbf{a}_{ss}$ in the discontinuity region. From Equation 71 the steady-state condition in the discontinuity region can be expressed as

$$\mathbf{p}^T\mathbf{a}_{ss} = \mathbf{p}^T(\mathbf{d} - \mathbf{r}) \tag{75}$$

The left side of Equation 75 is the projection of the $\mathbf{a}_{ss}$ vector on the normal to the plane of the r-pyramid base and the right side is the projection of the $(\mathbf{d}\text{-}\mathbf{r})$ vector. Making use of the fact that all arrival times are equal in the discontinuity region $\mathbf{a}_{ss}$ can be expressed as a scalar equation where each of its n components will be given by

$$\mathbf{a}_{ss} = \frac{\mathbf{p}^T \mathbf{d} - \mathbf{p}^T \mathbf{r}}{\mathbf{p}^T [11...1]}, \ \mathbf{a}_{ss} \in M_s, \ \mathbf{d} \in \text{Infeasible} \quad (76)$$

For a given set of due dates and processing times, **d**, **p** and **r** are unique and, $\mathbf{a}_{ss}$ will be unique and independent of **K**. It should be pointed out that even though the system converges to an unique arrival time in steady state, the sequence is not unique because the arrival time trajectory can be in any of the n! regions in its δ-neighborhood corresponding to n! possible sequences.

3.3.6 Two Part, One Machine Case

In this subsection, the geometry of Filippov's approach is used to solve the dynamical equations of distributed arrival time control is illustrated for the two-part, one-machine case. This particular case is chosen because it is easier to illustrate in two dimensions even though the theory is valid for arbitrary number of parts.

Figures 18 and 19 illustrate the geometry in arrival-time space and completion-time space, respectively, for the two-part, one-machine case when the due dates are in the feasible region. For a given **a**(t) in M_S, limiting values of f(**a***(t)) as **a***(t)->**a**(t) are the due-date deviation vectors for the two processing sequences. For sequence <1,2> the due-date deviation vector $\mathbf{z}_{<1,2>}(t)$ is

$$\mathbf{z}_{<1,2>}(t) = \mathbf{d} - (\mathbf{a}^*(t) + \mathbf{r}_{<1,2>}) \quad (77)$$

where the disturbance vector $\mathbf{r}_{<1,2>}$ is the sum of queuing and processing times

$$\mathbf{r}_{<1,2>} = [p_1 \ \ p_1+p_2]T \quad (78)$$

Likewise, for sequence <2,1> the due-date deviation vector $\mathbf{z}_{<2,1>}(t)$ is

$$\mathbf{z}_{<2,1>}(t) = \mathbf{d} - (\mathbf{a}^*(t) + \mathbf{r}_{<2,1>}) \quad (79)$$

where the disturbance vector $\mathbf{r}_{<2,1>}$ is the sum of queuing and processing times

$$\mathbf{r}_{<2,1>} = [p1+p2 \ \ p2]T \quad (80)$$

The vector addition in Equation 78 and 79 is illustrated in Figure 19. The size of the z_d-pyramid is determined by the due-date deviation vectors which can change with **a**(t). The size of the z_a-pyramid is determined by the due-date deviation vectors and the gain matrix **K**, and size of the r-pyramid is determined by the processing times which are assumed to remain constant. In this particular example diag(**K**) = [1 1]; therefore, the z_a-pyramid and z_d-pyramid have the same size.

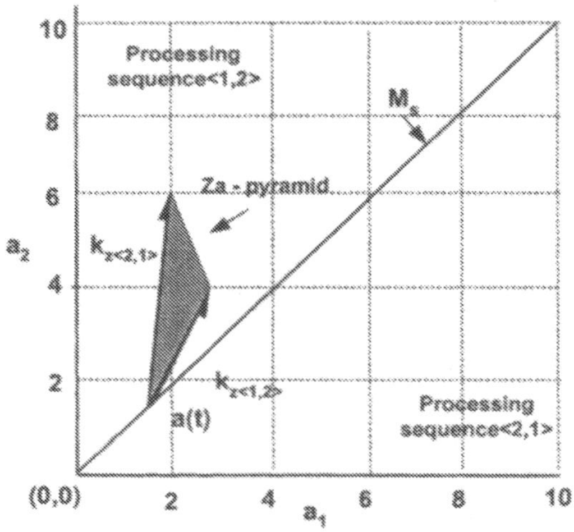

Figure 18. Geometry in arrival-time space for feasible due dates

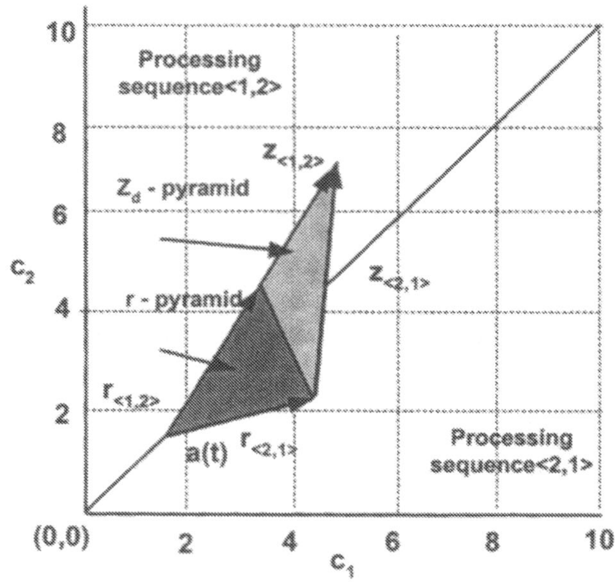

Figure 19. Geometry in completion-time space for feasible due dates

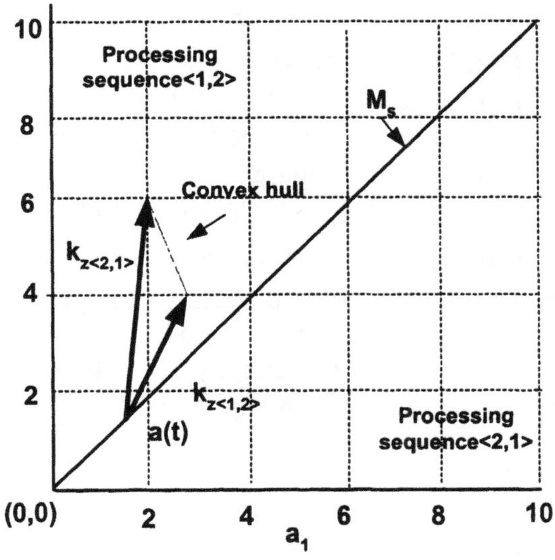

Figure 20. Geometry in completion-time space for feasible due dates

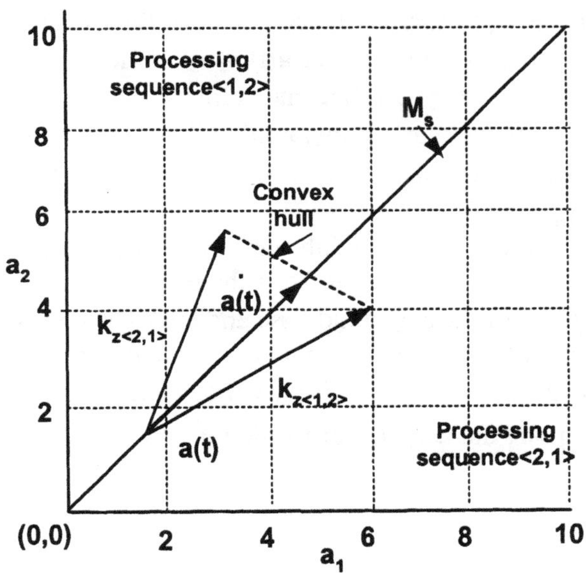

Figure 21. Effect of the "magnification" caused by the gain matrix

The convex hull of $\mathbf{Kz}_{<1,2>}(t)$ and $\mathbf{Kz}_{<2,1>}(t)$ is a line segment, shown in Figure 20 as a dashed line, that lies on one side of M_S and does not intersect

M_S. Therefore, a trajectory in the <2,1> region will have $\dot{\mathbf{a}}(t) = \mathbf{Kz}_{<2,1>}(t)$ and once it reaches M_S it will instantly cross over to the <1,2> region and $\dot{\mathbf{a}}(t)$ will instantly change to $\mathbf{Kz}_{<1,2>}(t)$.

When the gain matrix is diag(**K**) = [3 1] the convex hull is "magnified" along the a_1 axis because $k_1 = 3$ but remains the same along the a_2 axis because $k_2 = 1$ as illustrated in Figure 21. In this case the convex hull intersects the discontinuity region and **a**(t) continues on M_S. The velocity $\dot{\mathbf{a}}(t)$ vector at this point starts at **a**(t) and ends at the point where the convex hull intersects the discontinuity line. Comparing to Figure 20 it can be seen that by changing the gains direction of the **Kz** vectors has changed and caused the solution to lie on the discontinuity line. Therefore, the dynamics of arrival time control can be significantly affected by the choice of gains.

Infeasible Due-Dates

Figures 22 and 23 illustrate the geometry in arrival-time space and completion-time space, respectively, when due dates are in the infeasible region. In this case, both $\mathbf{Kz}_{<1,2>}(t)$ and $\mathbf{Kz}_{<2,1>}(t)$ point towards M_S and their convex hull shown in Figure 24 as a dashed line, intersects M_S.

As **a**(t) moves in the direction of $\dot{\mathbf{a}}(t)$ the r-pyramid will move towards **d** and the magnitudes of $\mathbf{z}_{<1,2>}(t)$ and $\mathbf{z}_{<2,1>}(t)$ will decrease, the sizes of the z_d-pyramid and the z_a-pyramid will decrease, and the magnitude of $\dot{a}(t)$ will decrease. At the steady-state value of $\mathbf{a}(t) = \mathbf{a}_{ss}$, **d** lies on the base of the r-pyramid, $\mathbf{z}_{<1,2>}(t)$ and $\mathbf{z}_{<2,1>}(t)$ become collinear and lie on the base of the r-pyramid, the z_d-pyramid and z_a-pyramid collapse into a line segment, the magnitude of $\dot{\mathbf{a}}(t)$ becomes zero, and the solution remains at $\mathbf{a}_{ss}$ as illustrated in Figures 25 and 26.

Figure 27 illustrates the equilibrium dynamics when due dates are infeasible. The left side of Equation 75 is the projection of $\mathbf{a}_{ss}$ vector on the normal to the plane of the r-pyramid base and the right side is the projection of the (**d**-**r**) vector. It is interesting to note that the projection of both $\mathbf{r}_{<1,2>}$ and $\mathbf{r}_{<2,1>}$ on **p** is the same, and even though their due-date deviations $\mathbf{z}_{<1,2>}$ and $\mathbf{z}_{<2,1>}$ are not zero, the projection $\mathbf{p}^T\mathbf{z}(t)$ is zero.

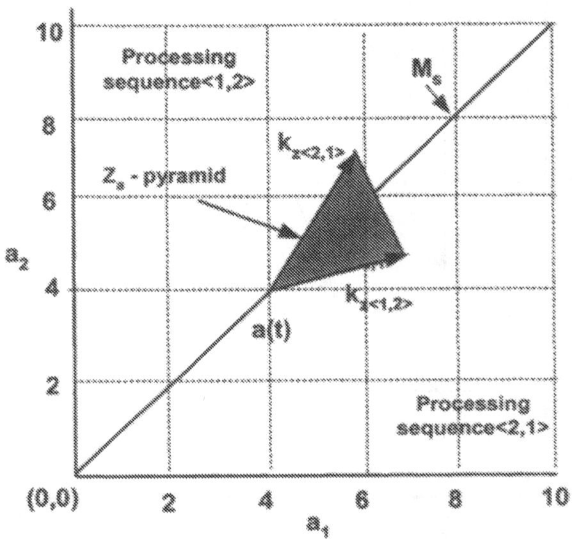

Figure 22. Geometry in arrival-time space for infeasible due dates

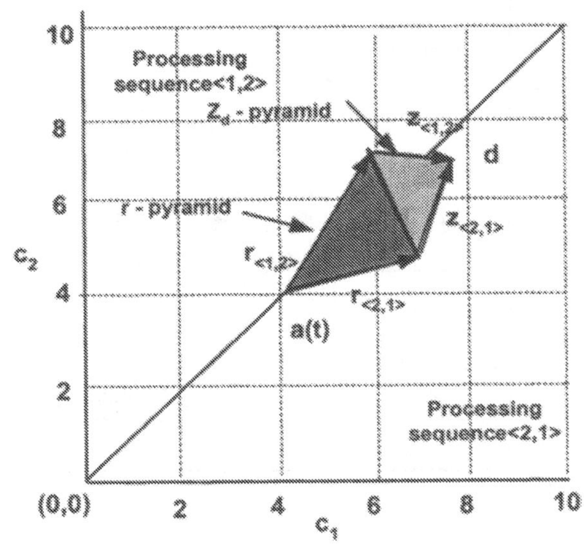

Figure 23. Geometry in completion-time space for infeasible due dates

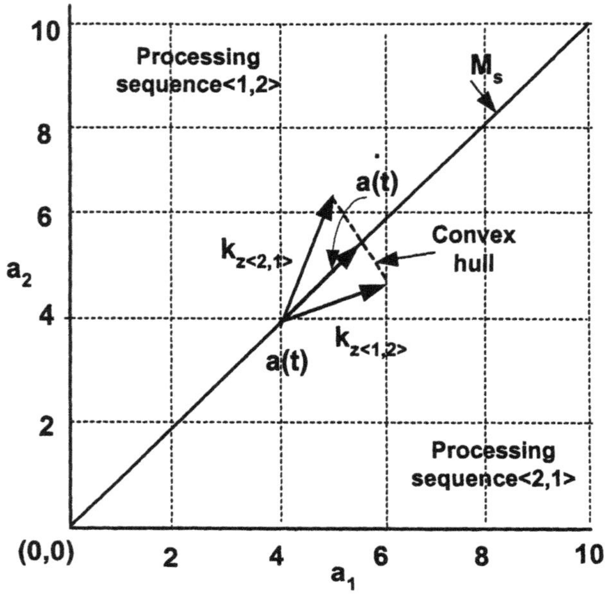

Figure 24. Convex hull of z vectors with infeasible due dates

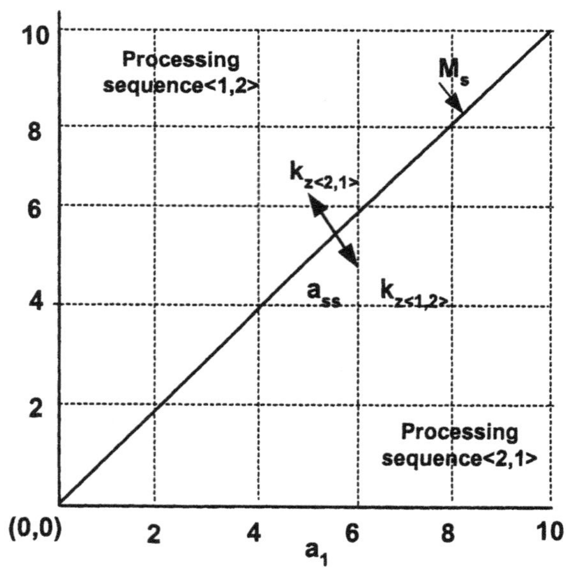

Figure 25. Steady state geometry in arrival-time space with infeasible due dates

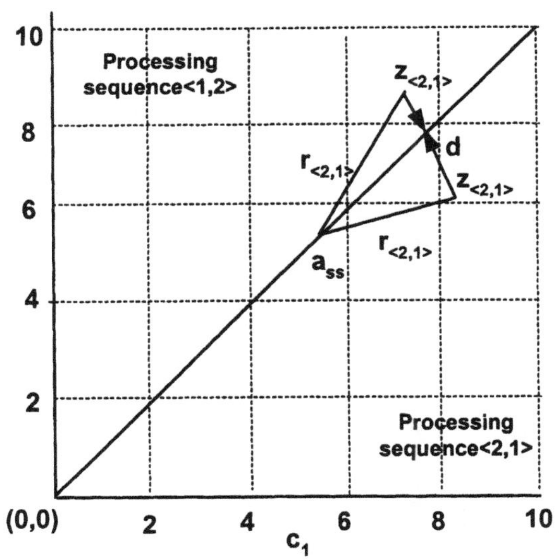

Figure 26. Steady state geometry in completion-time space with infeasible due dates

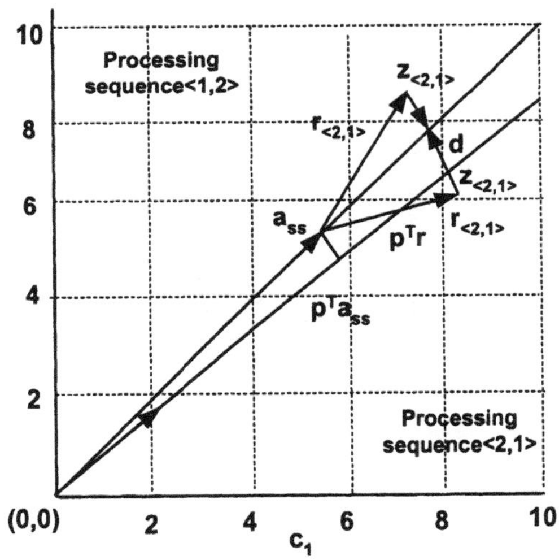

Figure 27. Illustration of steady state conditions

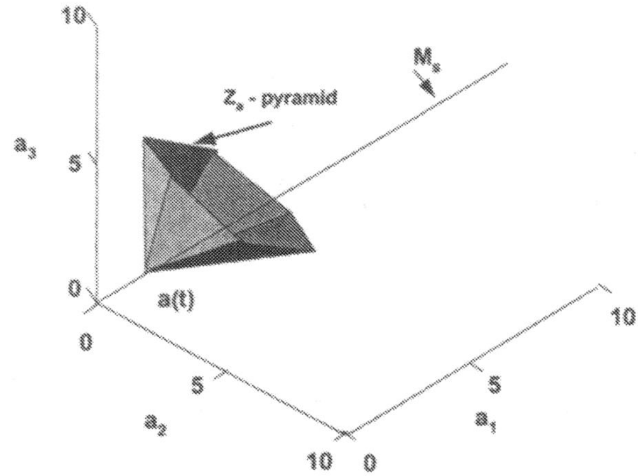

Figure 28. Geometry in three-dimensional arrival-time space for infeasible due dates

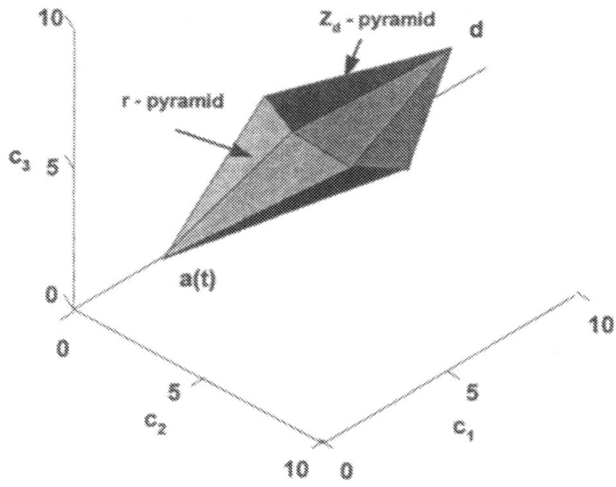

Figure 29. Geometry in three-dimensional completion-time space for infeasible due dates.

3.3.7 Three-Part, One-Machine Case

Figures 28 and 29 illustrate the geometry of arrival-time space and completion time space for the three-part, one-machine case with infeasible due dates. In this case, the pyramids have a six-sided base because six processing sequences are possible when $\mathbf{a}(t)$ is in M_s. The infeasible region is a hexagonal prism in which all the $\mathbf{z}$ vectors ($\mathbf{z}_{<1,2,3>}$, $\mathbf{z}_{<1,3,2>}$, $\mathbf{z}_{<3,1,2>}$, $\mathbf{z}_{<3,2,1>}$, $\mathbf{z}_{<2,3,1>}$, $\mathbf{z}_{<2,1,3>}$) become coplanar and lie on the base of the r-pyramid, the z_d-pyramid and the z_a-pyramid collapse into a plane, the magnitude of $\dot{a}(t)$ becomes zero, and the solution remains at $\mathbf{a}_{ss}$.

3.4 Extensions and Generalizations of Event Timing Control

These results have been extended to multiple machine systems and to more general job shops. Simulation experiments have shown the ability of arrival time control to reduce mean and variance of due-date deviation, tardiness, and completion times. The schedules found have nearly optimal variance (less than 1%) in due-date deviation for single machine systems and better than 20% for all heuristics for more general systems. Some of the recent advances in this area of research can be found in the following publications and reports:

- Multiple machine systems with simple configurations (Duffie et al 2002) and general systems with parallel, dissimilar machines, and sequence dependent setup times in multiple machine cells (Hong 2001, Hong and Prabhu 2002b).
- Analysis of sliding mode dynamics and synthesis of adaptive control algorithms for improving just-in-time scheduling performance (Cho and Prabhu 2002b).
- Scheduling performance of arrival time control for single machine systems (Prabhu 2000), and variance reduction using a vector space approach (Cho and Prabhu 2002a)
- Batch sequencing (Hong et al 2001), scheduling with set-up and batch sizing by integrating arrival time control and reinforcement learning (Hong and Prabhu 2001a, Hong and Prabhu 2002a).
- Complexity of nonlinear dynamics of distributed arrival time control and distributed learning control characterized using theory of chaos (Hong 2001, Hong and Prabhu 2001b).
- Metrics to characterize complexity of scale and design of distributed control systems using software engineering techniques (Kalava 1999, Phukan et al 2002).

- Global stability of arrival time control in machining cells (Prabhu 1998) and stable fault adaptation in general job shops (Prabhu 2002) using nonlinear control theory.

Global stability of such distributed control systems has been proven and is guaranteed if every machining stage has an average capacity strictly greater than zero in order to ensure that the processing times and hence the queuing times for all parts are finite. Robustness of global stability of the system can be guaranteed under most practical faults and uncertainties such as processing delays, queuing delays, due-date changes, communication delays, transportation delays, set-up delays, tool changing delays, machine failure delays, tool breakage delays, tool delivery delays, etc., provided these are bounded.

4. UNIFIED MODELING AND CONTROL FROM SENSORS-TO-SUPPLIERS

There is a need to develop scientific basis for real-time decision-making and controls that can take advantage of information distributed throughout the supply network and the market place. One approach to achieve this is develop unified models that scale across a variety of controls that span distributed process control, distributed production control, and distributed enterprise control. Figure 30 illustrates this unified control approach towards distributed decision-making from sensors-to-suppliers.

A significant barrier to developing mathematical models for unified control is the segregation of the control levels. Process control typically focuses on dealing with variation in processing conditions and inaccuracies in the models by using real-time feedback of physical measurements. Dynamics of such process controls can be modeled using differential equations. In contrast, control at the production and enterprises levels typically focus on dealing with the computational intractability of discrete systems by using efficient heuristics and approximations. Usually these higher-level control systems cannot be modeled using differential equation, which becomes a barrier to model unified controls. However, using event-timing control, described in Sections 2 and 3, offers prospects of overcoming this barrier by using a system of differential equations to develop unified control model. Figure 31 illustrates such a unified control of CNC machine level and production scheduling level.

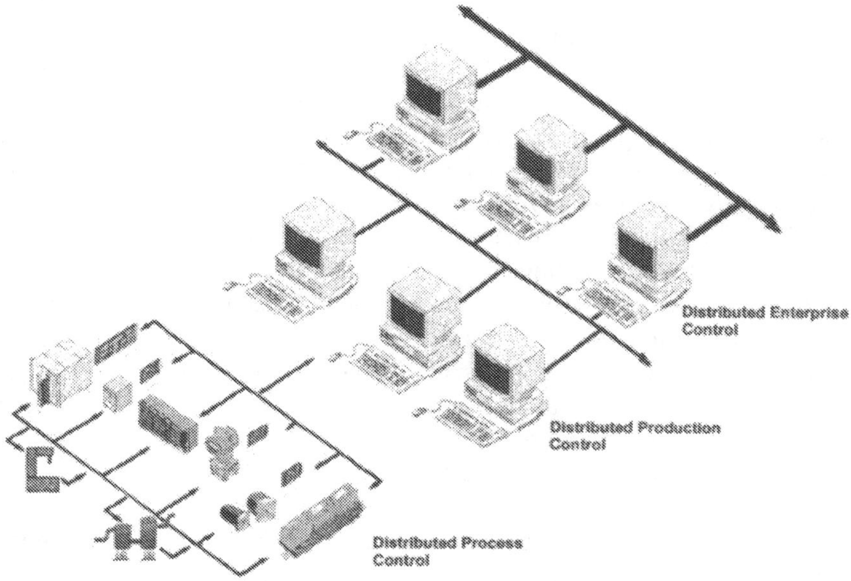

Figure 30. Unified control approach for distributed decision making from sensors-to-suppliers.

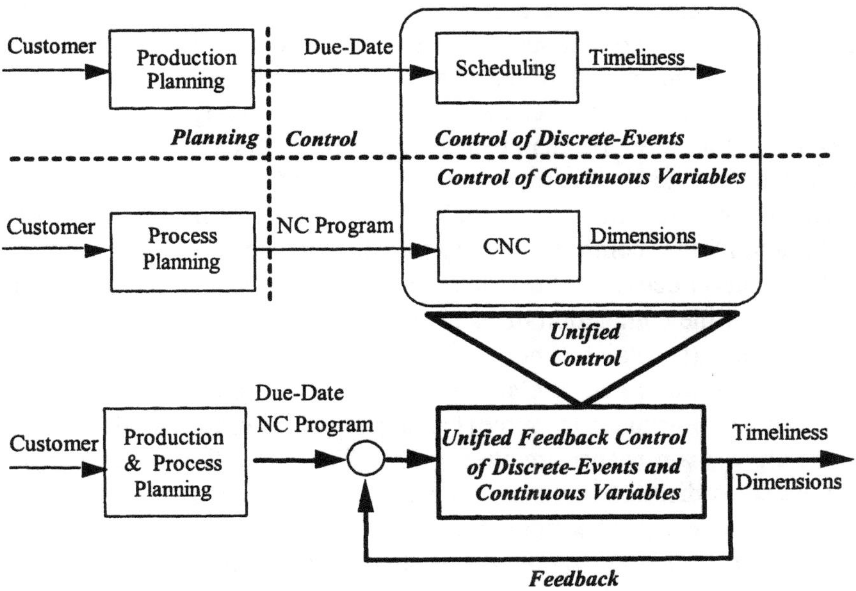

Figure 31. Unified control of process and production levels.

Consider unified control of (a) distributed arrival time control (DATC) for production scheduling; (b) distributed maintenance time control (DMTC) for maintenance scheduling; and (c) distributed machine capacity control (DMCC) for adapting to machinery condition and production demand. Feedback for DACT and DMTC is obtained from distributed simulation, as described in Section 2. Figure 32 illustrates the block diagram of such a unified control system in a job shop with N parts and M machines. (Here subscripts denote a particular part or machine.) A DATC algorithm is embedded in each part entity which tries to meet its due-date, $d_i(t)$, by continuously adjusting its arrival time, $a_i(t)$. Simultaneously, each machine uses a DMTC algorithm to meet its desired maintenance completion time, $r_i(t)$, by continuously adjusting its maintenance request time, $m_i(t)$. Human operators are treated as resources needed for maintenance and modeled in the distributed simulation. Likewise, each machine also uses a DMCC algorithm to meet its production demand, $y_i(t)$, by continuously adjusting its capacity, $c_i(t)$. These distributed controllers are coupled nonlinearly, as shown in dashed lines in Figure 32.

The time interval between a part's arrival time at the queue of a machine and the completion of its processing at the machine is the sum of the queuing time, any down time and processing time at that machine. Machine downtime for maintenance can be viewed as processing time due to "phantom" parts in queue. For the ith part, at the jth machine, this sum is denoted by $z_{ij}(t)$ and the sum of this over all the machines is the amount of time the part remains in the system. Therefore, part processing couples the DATC outputs discontinuously because a particular part's position in a queue at a machine or cell can change discontinuously depending on the relative arrival times of other parts in the queue, making the system highly non-linear. Furthermore, maintenance downtime when a part is in such a queue will couple the corresponding DATC and DMTC outputs, causing additional nonlinearities. Since each part may visit all machines, all of the DATC and DMTC controller outputs may be coupled.

The interval between a machine's maintenance request time for a human operator and the completion of its repair is the sum of the queuing time and repair time, $w_i(t)$. Maintenance couples the DMTC outputs discontinuously because a particular machine's position in a queue for a human operator (or some other maintenance resource) can change discontinuously depending on the relative maintenance request times. These nonlinear couplings are aggregated as disturbances and compensated using feedback control, which reduces complexity. Therefore, feedback control is used in this distributed decision making system to compensate for uncertainties such as those caused by lack of complete information, represented here as disturbance.

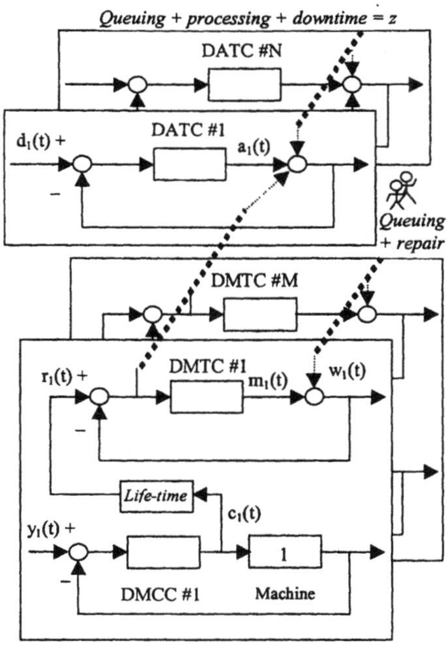

Figure 32. Block diagram of unified control for production scheduling, maintenance scheduling, and condition-based capacity control.

Overall, behavior and performance of the system will depend on the control algorithms. As a first step, an integral control algorithm with constant gain, k, is proposed for all the controllers. The resulting dynamical model for the proposed unified control system can be expressed as

$$\begin{bmatrix} \dot{a}_1(t) \\ \vdots \\ \dot{a}_N(t) \\ \dot{m}_1(t) \\ \vdots \\ \dot{m}_M(t) \\ \dot{c}_1(t \\ \vdots \\ \dot{c}_M(t) \end{bmatrix} = \begin{bmatrix} k\left(d_1(t) - \sum_{j=1}^{M} z_{1j}(t) - a_1(t)\right) \\ \vdots \\ k\left(d_N(t) - \sum_{j=1}^{M} z_{Nj}(t) - a(t)\right) \\ k\left(r_1(t) - w_1(t) - m_1(t)\right) \\ \vdots \\ k\left(r_M(t) - w_M(t) - m_M(t)\right) \\ k\left(y_1(t) - c_1(t)\right) \\ \vdots \\ k\left(y_M(t) - c_M(t)\right) \end{bmatrix} \tag{81}$$

DMCC dynamically controls machine capacity and adapts to changing process condition, production demand, and maintenance operator constraints. For example, when a maintenance operator is not available, it may be better for a machine to operate at a lower than nominal capacity for a longer period of time, compared to operating at the nominal capacity for a much shorter time. Similarly, when there is a higher production demand, it may be better to operate at higher than nominal capacity to improve overall performance. Therefore, the actual processing time for the part can vary depending on a machine's current capacity, coupling the corresponding DMCC and DATC outputs (not shown in Figure 32 for simplicity). The production demand for a machine is a dynamic variable and can be viewed as the average remaining work in the queue of that machine. In order to ensure stability and to avoid saturation of DMCC, its command, $y_i(t)$, is limited to maximum allowable capacity under current machine condition. Machine condition feedback is obtained in real-time from physical measurements and prognostics models (Ray and Tangirala 1997, Katter and Tu 1996). This implies that the DMCC will try to increase its capacity to meet the production demand while satisfying constraints imposed by real-time machine condition. It should be emphasized that this unified control approach allows intelligent machines to fully utilize their local intelligence and sensing capabilities to evaluate their individual failure and utilization characteristics, integrates real-time machinery conditions with maintenance and production scheduling.

Here the machine capacity dynamics has been simplified as a unity gain element for the sake of brevity in explaining the model. Also, it is assumed that capacity is used in conjunction with an appropriate prognostics model to predict life-time and then generate the desired maintenance completion time command for DMTC. The fidelity of this model can be readily increased by using more detailed dynamical models for machine capacity such as adaptively controlled machines (Lauderbaugh and Ulsoy 1989). The result would be high-fidelity analytical models, unifying (a) dynamical models for discrete event timing control and (b) dynamical models of continuous physical processes and their controls.

Recent advances in the area of unified distributed decision-making and control has included the following topics:

- Unified control of process dynamics and predictive maintenance scheduling (Koomsap 2000, Koomsap et al 2001).
- Unified control of machine capacity and scheduling (Cho and Prabhu 2001).
- Unified control of production scheduling and inventory levels (Prabhu and Masin 2001).
- Unified control of enterprise and supply chain inventory levels Masin and Prabhu 2001).

- Unified control of capacitated vehicle routing and delivery scheduling (Chigullapalli 2001, Phukan 2002).

The key here is using simulation feedback combined with continuous variables for controlling discrete events allows scalable distributed decision-making that compensates for absence of global information (Prabhu et al. 1999). This also allows the use of highly scalable parallel and distributed computing architectures to accelerate these distributed decision-making algorithms. Computational speed could be limited mainly by the speed of the simulation technique rather than the usual sequential/centralized architecture. It is possible to use conventional discrete-event simulation techniques for feedback in event control (Hong et al 2001). However the fault-tolerance and generative nature of simulation in this case offers opportunity exploit the underlying distributed architectures to use a continuous computing clock for time-scaled simulation (Sharma 2001) and loosely synchronous (Mahalati 2002) distributed simulation techniques. Moreover, these distributed simulation techniques can also use highly scalable parallel and distributed computing architectures such as network of workstations, cluster of workstations, symmetric multiprocessors, and reconfigurable computers. Reconfigurable computing can also be used for process control and as an effective computing platform for integrating physical control and event-level control within a machine (Koomsap and Prabhu 1999, Surapanani 2001, Patel 2001, Suri 2002).

5. CONCLUSIONS

This chapter described a continuous variable feedback control approach for discrete event timing control. This approach was used for an arrival time control algorithm for production scheduling, which can be modeled using differential equations. The event timing control technique can be used to unify distributed process control, distributed production control, and distributed enterprise control. It lends itself to distributed control architectures, distributed computing architectures, continuous computing clock, and continuous control variables. The synergy arising out such a concurrent paradigm shift promises significant benefits for engineering scalable enterprise systems in terms of reduced complexity and increased reconfigurability, flexibility, fault-tolerance and responsiveness as outlined below.

Impact	***Distributed Control Architecture***	***Distributed Computing Architecture***	***Continuous Computing Clock***	***Continuous Control Variables***
Reduced Analytical Complexity	Moderate	Moderate	Very High	Very High
Reduced Software Complexity	High	Very High	Very High	Very High
Reduced Computational Complexity	Moderate	High	Very High	Very High
Increased Reconfigurability	Very High	High	Moderate	Moderate
Increased Flexibility	Very High	Moderate	High	Moderate
Increased Fault-tolerance	Very High	Very High	Moderate	Moderate
Increased Responsiveness	High	Very High	Moderate	Moderate

Because the actual implementation of the unified control system used simulation based feedback, the granularity of control can be scaled from senor and event level to supplier level. The models will be tractable because each decision will require solution of one differential equation. Moreover, this will allow us to "bridge islands of mathematical models". A host of control theoretic techniques for analysis, design, and optimization can be brought to bear on these problems. The control algorithms can be expected to be linear in time complexity and inherently suitable for parallel and distributed computing, allowing them to exploit Moore's Law and raw bandwidth for scalable decision making.

Acknowledgements This work was partially supported by National Science Foundation Grants DMI-9908267 and DMI-0075572, and Ben Franklin Technology Partnership through Penn State's Center for Manufacturing Enterprise Integration. The author is grateful to Professor Neil Duffie for advice and suggestions during the early stages of this work.

6. REFERENCES

Antsaklis, P., 1994, "Report on the Task Force on Intelligent Control," IEEE Control Systems Magazine, Vol. 14, pp. 4-5, 1994.

Bollinger, J.G, 1998, "Visoneering Manufacturing for the 21st Century," Proceedings of XXVI NAMRC, Atlanta, GA, pp. 3-8, 1998.

Bollinger, J.G. and Duffie, N.A., 1988, Computer Control of Machines and Processes,. Addison Wesley.

Caramanis, M. and Sharifnia, A., 1991, "Near Optimal Manufacturing Flow Controller Design," International Journal of Flexible Manufacturing Systems, Vol. 3, pp. 321-336, 1991.

Chigullapalli, S., 2001, "Condition-based real-time procurement control in semiconductor supply chains," M.S. Thesis, Penn State University, December 2001.

Cho, S. and Prabhu, V.V., 2001, "Distributed Control of Machine Capacity and Production Scheduling," Proceedings of XXIX North American Manufacturing Research Conference, Gainesville, FL May 2001, pp. 311-316

Cho, S. and Prabhu, V.V., 2002a, "A Vector Space Model for Variance Reduction in Single Machine Scheduling," IIE Transactions, Vol. 34, No. 11, pp 933–952, 2002.

Cho, S. and Prabhu, V.V., 2002b, "Sliding Mode Dynamics in Continuous Feedback Control for Distributed Discrete-Event Scheduling," Automatica, Vol. 38, No. 9, pp 1499-1515, 2002.

Coddington, E.A. and Levinson, N., 1955, "Theory of Ordinary Differential Equations," McGraw-Hill, 1955.

Dilts, D. M. et al., 1991, "The Evolution of Control Architectures for Automated Manufacturing Systems," Journal of Manufacturing Systems, Vol. 10, No. 1, pp. 79-93, 1991.

Duffie N.A., Chitturi, R. and Mou, J., 1988, "Fault-tolerant Heterarchical Control of Heterogeneous Manufacturing System Entities", Journal of Manufacturing Systems, Vol. 7, No. 4, pp. 315-328, 1988.

Duffie, N.A. and Prabhu, V.V., 1994, "Real-Time Distributed Scheduling of Heterarchical Manufacturing Systems," Journal of Manufacturing Systems, Vol. 13, No. 2, pp. 94-107, 1994.

Duffie, N.A., Prabhu, V. V., and Kaltjob, P. O., 2002, "A Closed-Loop Approach To Real-Time Distributed Scheduling Of Heterarchical Manufacturing Systems," to appear in, Journal of Manufacturing Systems.

Filippov, A.F., 1960, "Differential Equations with Discontinuous Righthand Sides," American Mathematical Society Translations, Vol. 42, pp. 199-231, 1960.

Filippov, A.F., 1988, "Differential Equations with Discontinuous Righthand Sides," Kluwer, 1988.

Gershwin, S.G., 1989,"Hierarchical flow control: a framework for scheduling and planning discrete events in manufacturing systems," IEEE Proceedings, Vol. 77, No. 1, pp. 195-208, 1989.

Hajek, O., 1979, "Discontinuous Differential Equations I", Journal of Differential Equations, Vol. 32, 1979, pp. 149-170

Hajek, O., 1991, "Control Theory in the Plane", Lecture Notes in Control and Information Sciences, Vol. 153, Springer-Verlag, 1991

Hatvany, J., 1985, "Intelligence and Cooperation in Heterarchic Manufacturing Systems", Annals of CIRP, 14(1), pp. 1-4, 1985.

Hong, J., 2001, "A Heterarchical Shop Floor Control Using Distributed Arrival Time Control and Reinforcement Learning," Ph.D. Thesis, Penn State University, May 2001.

Hong, J., and Prabhu, V. V., 2001a, "Distributed Learning and Control for Manufacturing Systems Scheduling," Engineering of Intelligent Systems (Eds. Monostori, L., Vancza, J., and Ali, M., (Eds.) LNAI 2070, Springer-Verlag, 2001, pp. 582-591.

Hong, J., and Prabhu, V. V., 2001b, "Complexity of Nonlinear Dynamics in Distributed Manufacturing Control System," PSU IME Working Paper 2001.

Hong, J., and Prabhu, V. V., 2002a, "Distributed Reinforcement Learning Control for Batch Sequencing and Sizing in Just-In-Time Manufacturing Systems," to appear in, Applied Intelligence, 2002.

Hong, J., and Prabhu, V. V., 2002b, "Modeling and performance of distributed algorithm for scheduling dissimilar machines with setup," PSU IME Working Paper, 2002

Hong, J., Prabhu, V. V., and Wysk, R.A., 2001, "Real-Time Batch Sequencing using Arrival Time Control Algorithm," International Journal of Production Research, Vol. 39, No. 17, pp. 3863-3880.

Kalava, M., 1999, "Complexity Metrics for Manufacturing Enterprise Software," M.S. Thesis, Penn State University, December 1999.

Katter, J. G. Jr. and Tu, J. F., 1996, "Bearing Condition Monitoring for Preventive Maintenance in a Production Environment," 39, pp. 936-942, (1996)

Kim, S. H., 1985, "Mathematical foundations of manufacturing science: theory and implications," Ph.D. Thesis, MIT, May, 1985.

Koomsap, P. and Prabhu, V. V., 1999, "Field Programmable Gate Arrays for High Performance Manufacturing Automation," International Journal of Intelligent Control and Systems, Vol. 3, No. 4, pp. 701-723.

Koomsap, P., 2000, "Intelligent Control of Laser-Based Manufacturing Processes and Systems," Ph.D. Thesis, Penn State University, May 2000

Koomsap, P., Prabhu, V. V., Schriempf, J.T., and Reutzel, E.W., 2001, "Simulation-Based Design of Laser-Based Free Forming Process Control," Journal of Laser Applications, Vol. 13, No. 2, pp. 47-59, April 2001.

Koren, Y. et al., 1996, "Real-Time Open Control Architectures and Systems Performance," Annals of the CIRP, Vol. 45/1, pp.377-380, 1996.

Lauderbaugh, L. K., and Ulsoy, A. G. 1989, "Model reference adaptive force control in milling", Journal of Engineering for Industry, 111, pp. 13-21, 1989

Mahalati, S., 2002, "Loosely Synchronous Distributed Simulation Using SMP, VIA Cluster and Distributed Clusters," M.S. Thesis, Penn State University, August 2002.

Marsden, J.E. and Hoffman, M.J., 1993, "Elementary Classical Analysis - Second Edition", Freeman and Company, 1993.

Masin, M. and Prabhu, V.V., 2001, "AWIP: A Simulation-based Feedback Control Algorithm for Scalable Design of Self-regulating Production Control Systems", PSU IME Working Paper, 2001

NGM, 1997, "Next-Generation Manufacturing – A Framework for Action," NGM Project Office, Bethlehem, PA, 1997.

NRC, 1991, "The Competitive Edge – Research Priorities for U. S. Manufacturing," National Academy Press, 1991.

NRC, 1998, "Visionary manufacturing challenges for 2020," ," National Academy Press, 1998.

Ovacik, I. M. and Uzsoy, R., 1996, "Decomposition methods for scheduling semiconductor testing facilities," International Journal of Flexible Manufacturing Systems v 8 n 4 Oct 1996, pp. 357-388, 1996.

Patel, J., "Reconfigurable supercomputers for scalable enterprise automation," M.S. Thesis, Penn State University, December 2001.

Perkins, J. R. and Kumar, P.R., 1989, "Stable, Distributed, Real-Time Scheduling of Flexible Manufacturing / Assembly / Disassembly Systems," IEEE Transactions on Automatic Control, Vol. 34, No. 2, pp. 139-148, February 1989.

Phukan, A., 2002, "Algorithms And Information Technology Strategies For Transportation Decision Making In Supply Chains," M.S. Thesis, Penn State University, December 2002.

Phukan, A., Kalava, M., and Prabhu, V.V., 2002, "Complexity metrics for manufacturing control architectures based on software and information flow," PSU IME Working Paper, 2002.

Prabhu, V.V. and Duffie, 1995, "Modeling and Analysis of Non-Linear Dynamics in Autonomous Heterarchical Manufacturing Systems Control," Annals of the CIRP, Vol. 44, No. 1, pp. 425-428, 1995.

Prabhu, V.V. and Duffie, N.A., 1996, "Modeling and Analysis of Heterarchical Manufacturing Systems Using Discontinuous Differential Equations," Annals of the CIRP, Vol. 45, No. 1, pp. 445-448.

Prabhu, V.V. and Duffie, N.A., 1999, "Nonlinear Dynamics in Distributed Arrival Time Control of Heterarchical Manufacturing Systems," IEEE Transactions on Control Systems Technology, Vol. 7, No. 6, November 1999, pp. 724-730.

Prabhu, V.V. and Masin M., 2001, "Scalable Algorithms and Distributed Agent Architectures for Adaptive Enterprises" Presented at Scalable Enterprise Systems Workshop Phoniex, AZ, May 19, 2001, Available athttp://www.ie.psu.edu/cmei/ses_ws/prabhu_files

Prabhu, V.V., 1995, "Real-Time Distributed Arrival Time Control of Heterarchical Manufacturing Systems," Ph.D. Thesis, University of Wisconsin-Madison, 1995.

Prabhu, V.V., 1998, "Lyapunov Stability of Distributed Control in Multiple Machine Heterarchical Manufacturing Cells," Proceedings of XXVI North American Manufacturing Research Conference, Atlanta, GA, May 1998, pp. 311-316.

Prabhu, V.V., 2000, "Performance of Real-Time Distributed Arrival Time Control in Heterarchical Manufacturing Systems," IIE Transactions, Vol. 32, pp. 323-331, 2000.

Prabhu, V.V., 2002, "Stability and Fault Adaptation in Distributed Control of Heterarchical Manufacturing Job Shops," to appear in, IEEE Transactions on Robotics and Automation, 2002.

Prabhu, V.V., Cho, S., and P. Koomsap, 1999, "High Performance Computing and Communication for Real-Time Feedback in Distributed Manufacturing Control Systems," CIRP Journal of Manufacturing Systems, Vol. 29, No. 1, pp. 79-83, 1999.

Procter, F. M. and Albus, J.S., 1997, "Open Architecture Controllers", IEEE Spectrum, June, 1997.

Ray, A. and Tangirala, A., 1997, "A nonlinear stochastic model of fatigue crack dynamics," Probabilistic Engineering Mechanics, Vol. 12, No. 1, pp. 33-40, 1997.

Sharma, A., 2001, "Quality of service models for computing and communication performance in manufacturing control systems", Aug 2001.

Slotine, J.J. and Sastry, S.S., 1983, "Tracking control of non-linear systems using sliding surfaces with applications to robot manipulators", International Journal of Control, Vol. 38, No. 2, 1983, pp. 465-492.

Surapaneni, S., 2001, "System-on-a-chip compiler for industrial automation," M.S. Thesis, Penn State University, December 2001.

Suri, S., 2002, "Reconfigurable Supercomputing Platform for Embedded Control and Decision Making in Manufacturing Enterprises using Distributed Simulations," M.S. Thesis, Penn State University, August 2002.

Utkin, V.I., 1992, "Sliding Modes in Control Optimization", Springer-Verlag, 1992.

Voxman, W.L. and Goetschel, R.H., 1981, "Advanced Calulus", Marcel Dekker, 1981.

Wiendhal, H. P. and Scholtissek, P., 1994, "Management and control of complexity in manufacturing," Annals of the CIRP, Vol. 43, No. 2, 1994, pp. 533-540, 1994.

Wright, P.K., 1995, "Principles of Open-Architecture Manufacturing," Journal of Manufacturing Systems, Vol. 14, No. 3, pp. 187-202, 1995.

Zinober, A.S.I., 1994, Editor "Variable Structure and Lyapunov Control", Lecture Notes in Control and Information Sciences, Vol. 193., Springer-Verlag, 1994.

Chapter 5

COLLABORATIVE MULTIAGENT BASED INFORMATION INFRASTRUCTURE FOR TRANSPORTATION PROBLEM SOLVING

Goutam Satapathy
Intelligent Automation Incorporated
7519 Standish Place
Rockville, MD 20855

Soundar R. T. Kumara
Professor of Industrial and Manufacturing Engineering
310 Leonhard Building
The Pennsylvania State University
University Park, PA 16802, USA

Abstract: In this chapter, we present a multiagent computational approach to solve a procurement problem that falls under the supply chain domain. In this approach we model each transportation company as consisting of a master and several slaves. A master represents the task negotiator and task allocator for a company. A slave represents a vehicle to whom the tasks are allocated by its master. We consider a transportation problem containing several stochastic parameters for real-time decision making. Each master and slave is modeled using BDI (Belief-Desire-Intention) logical model of an agent. The emphasis in this paper is on the information infrastructure for information exchange between the agents. We define the interactions between the agents and model the interactions using Knowledge Query and Manipulation Language (KQML) based syntax and semantics.

Key words: Transportation, Belief-Desire Intention logical model, Multi Agent Systems, KQML

1. INTRODUCTION

Mathematical programming techniques are used to solve transportation problems long before distributed problem solving (DPS) or multiagent system (MAS) are considered for Parallel Distributed Problems (PDP), dial-a-ride problems (DARP) with various load, and time (Golden and Assad, 1988). In real world, due to the presence of uncertainty in travel times, and availability of products/goods for transportation, researchers have also modeled these problems as stochastic programming problems (Laporte et al., 1992; Dror et al., 1989). However, these problems are solved as monolithic problems off-line and their results are then implemented (usually after simulation) in real world. The requirements for today's transportation problem solving, which make MAS[1] an appropriate approach[2] are as follows:

1. A customer assigned transportation problem (e.g. PDP or DARP) needs to be solved in real time. That means, the vehicles that will eventually be assigned with some transportation tasks, have to execute the newly assigned tasks along with the committed tasks and the tasks that are currently being executed.

2. The dynamics of the world must be taken into account during the task assignment and vehicle routing or planning. The task assignments and the plan of the vehicle must take various probable worlds into consideration, for making faster decisions at critical times.

3. Furthermore, if a vehicle fails to execute a committed task then the failed task must be (attempted to be) completed by another vehicle.

4. Since the transportation domain is geographically and functionally distributed (Burmeister, *et al.*, 1997), the entities in the transportation domain such as vehicles (and the driver) should be given more autonomy in planning and execution.

5. The system should allow a customer to pose a problem without any concern to how the tasks are distributed among the transportation

[1] The difference between multiagent system and distributed problem solving is discussed in (Satapathy et al., 1998a; Satapathy et al., 1998b).

[2] We do not argue if the solution to transportation problems solved using multiagent approach is less or more accurate than a mathematical programming choice. We believe that there are certain advantages in solving a transportation problem using multiagent approach which are of practical importance (Satapathy and Kumara, 1999c) such as number of successful recoveries from a plan failure.

companies and vehicles. That means, the final solution of a task distribution depends on a protocol based negotiation among the companies[3]. In a real-world, companies may prefer the solution that is arrived through negotiation than through a centralized task allocation.

The above five requirements make us view the vehicle route generating and the task negotiating systems[4] of the companies as finite state machines which change their state based on the observation of the world, and make decisions or execute decisions accordingly. The states of a vehicle route generating system are defined based on the roads networks, road conditions, the vehicle's mechanical properties and tasks assigned to other vehicles. The states of a task negotiating system are defined based on the plans of the company's vehicles, and the strategic behavior of other companies and their vehicles. As the BDI (Belief-Desire-Intention) model of an agent is similar to a state machine (Benerecetti et al., 1997), we model these software systems as agents.

Research in the application of MAS or DPS to traffic and transportation domain is more than ten years old (Dean and Greenwald, 1992). The work in MAS by Fischer (Fischer et al., 1996) is one of the best known research efforts in cooperative transportation scheduling, which is solved using Contract Net Protocol for task distribution among companies and task assignment among vehicles. Agents were developed using InterRap architecture and tested on AgenDa testbed. Uncertainty in scheduling is not considered in their approach. COSY/DASEDIS (Burmeister and Sundermeyer, 1992; Burmeister et al., 1997) is an agent based system in traffic simulation and guidance. Several agent applications are reported in air traffic control (Ljunberg and Lucas, 1992) and negotiation among airlines for landing in a airport (Sastry et al., 1995) using game theoretic approaches. Wellman presents a market-oriented approach (WALRAS algorithm) to solve an atemporal and steady-state multi-commodity problem, distributedly (Wellman, 1993).

In this chapter, we discuss the modeling of the transportation domain to handle real-time scheduling under uncertainty. The emphasis is on the description of the multiagent system and the interactions among the agents based on KQML specification. We model each transportation company using master-slaves problem solving paradigm. A master represents a task negotiating system of a transportation company, and a slave represents one-vehicle route generating system.

[3] The customer may be aware of the protocol.

[4] We assume that these systems are some kind of human assisted software.

Both the master and slaves are modeled based of the Bill model of an agent and implemented through the object oriented agent model described in (Satapathy and Kumara, 1999b). In section 2, we describe the transportation problem and the master-slaves type structure of a transportation company. In section 3, we present the details of the KQML based interactions among the master and slaves of the transportation companies for solving the transportation problem discussed in section 2. We describe the implementation of each agent (i.e. master and slave), and the multiagent model in section 4.

2. THE TRANSPORTATION PROBLEM

An integrated supply chain information infrastructure facilitates a customer order to be propagated and distributed to various players of the supply chain domain, namely, manufacturers, suppliers, and transporters. A general customer order may be specified as *delivering a set of products at different locations within a time window* - also called as supply chain problem. This problem is solved collaboratively by the transporters, manufacturers, and suppliers. A supply chain problem is composed of several transportation problems, such as, "delivering unfinished products to the manufacturers who will produce finished products", "delivering finished products at the customer specified locations". In this chapter, we consider the following transportation problem:

We are given (1) a set of suppliers (manufacturers) declaring the availability of unfinished products within a time window and (2) a set of manufacturers declaring the capability of producing finished products from the unfinished products, given their arrival time within a time window. The transportation problem comprises of the transportation companies collaboratively deciding the source and destination of each unfinished product, and the companies and their vehicles that must be assigned with each order[5].

It must be noted that a supplier is a *prospective* source of an unfinished product until the transportation problem is solved. Similarly, a manufacturer is also a prospective destination of an unfinished product. The above transportation problem is stochastic in nature due to following reasons:

1. Uncertainty associated with the availability of unfinished products at the suppliers, and the manufacturers.

[5] A transportation order consists of picking a product from a source within certain time window and delivering it at a destination within certain time window

2. Uncertainty associated with travel times.

In order to solve the stochastic transportation problem, we model the transportation domain as a multiagent system where each transportation company is viewed as a cluster of agents. In each cluster, agents are organized in a two level hierarchy consisting of (1) a *master* agent which acts as the task allocator for the vehicles of that cluster and task negotiator with the masters of other clusters and (2) several *slave agents,* each of which represent a vehicle which generates its own vehicle routes. Now, we shall formalize the transportation problem to be solved by the masters and the tasks to be assigned to the slaves i.e. vehicles. (Henceforth, the term vehicle and slave will be used interchangeably.)

Denote **J** as the set of products to be transported from a set of prospective sources to a set of prospective destinations. Denote O_k^j to be an order to pick the product j from a prospective source and deliver it at a prospective destination. We denote $k = 1, \ldots|\mathbf{K}_j|$ such that $\mathbf{K}_j$ is the number of orders involving the product j. A set of orders is a task denoted by T, if

$$((O_{k1}^j \in K_j \in T) \wedge (O_{k2}^j \in K_j \in T) \wedge (k_1 \neq K_2)) \Rightarrow \perp, \forall j \in J \qquad (1)$$

The above logical constraint states that a task cannot contain two orders that involve transportation of the same product j. The symbol stands for a universally false statement.

Denote T as set of tasks such that each task is defined based on the equation (1). Denote J T to be the set of products in J whose transportation orders are contained in the task T. Denote P as the set of transportation companies. Denote task Tr: as the task Tn ET allocated (assigned) to pth company. Tasks are said to be assigned without any conflict, if

$$((T_{n1}^p \in \mathbf{T}) \wedge (\mathbf{T}_{n2}^p \in \mathbf{T}) \wedge (n_1 \neq n_2)) \Rightarrow \perp, \ \forall p \in \mathbf{P} \qquad (2)$$

$$((\mathrm{O}_{k1}^j \in \mathrm{K}_j \in T_{n1}^{p1}) \wedge \mathrm{O}_{k2}^j \in \mathrm{K}_j \in T_{n2}^{p2}) \wedge (T_{n1}^{p1} \in \mathbf{T}) \wedge (T_{n2}^{p2} \in \mathbf{T})) \Rightarrow \perp, \ \forall p_1, p_2 \in \mathbf{P} \qquad (3)$$

The first constraint (Eqn. (2)) represents one task assignment to one company. The second constraint (Eqn. (3)) represents that no two companies are assigned with tasks such that both of their tasks contain an order involving the transportation of the same product j. The second constraint can also be expressed as: $\mathbf{J}_{T_{n1}^{p1}} \cap \mathbf{J}_{T_{n2}^{p2}} = \emptyset$. The objective of the transportation companies is to select and distribute the tasks among themselves satisfying the above constraints, such that the total bid (price for transportation) offered for the transportation of all the products is minimized and the chance of completion of the tasks is maximized.

3. AGENT INTERACTIONS

Figure 5-1 shows two transportation organizations, each of which consisting of one master and several slaves along with four different types of fundamental interactions between them. The interaction links are explained in the following:

- **Link a:** A virtual user representing a customer uses this link to announce the stochastic transportation problem to the masters (managers -physical or software entities) of the transportation companies. Upon the announcement of the problem, the masters negotiate and distribute transportation tasks (using link **b**, **c** and **d**) among themselves. This leads to task allocation to the masters and task commitment by the masters. During the execution of the allocated and committed tasks, the tasks might fail. Therefore, this link also consists of information passage of a master breaking a commitment to a task and another master committing to that task.

- **Link b:** Each master determines the transportation tasks that his company can undertake by using this link for presenting the subtasks to the vehicles as test-assignments. The purpose of the subtasks presentation is to determine the cost and probability of completion of tasks. A master test-assigns subtasks to the selected vehicles by avoiding conflicts so that the vehicles consider all the subtasks together[6]. This interaction link also encompasses the actual assignment of the orders to the vehicle after the negotiation phase (link **d**) among the masters.

[6] For example, if O_{k1}^{j1}, O_{k2}^{j2}, *and* O_{k3}^{j3} are three orders and a cluster has one vehicle then the masters test-assigns 7 separate and independent tasks (if the test-assignment is exhaustive) as power set of the three orders.

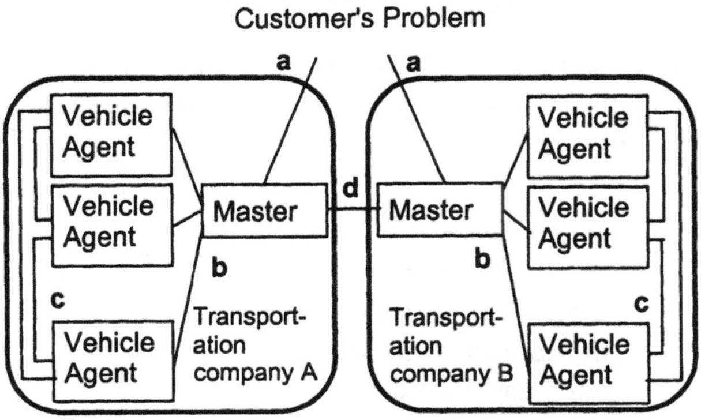

Figure 5-1. Four types of interactions between different agents in a multiagent model for logistics

- **Link c**: A vehicle may decide to split a task presented during the test-assignment and delegate the resulting sub-tasks to other vehicles. This link facilitates the collaboration among the vehicles.

- **Link d:** The masters negotiate with each other to distribute tasks among each other. The negotiation phase consists of a transportation company offering a bid for a set of tasks, and other companies accepting the bid or counter offering with another bid that minimizes the total bid. This negotiation phase is interleaved with the test-assignment phase link **b**), to reduce the negotiation space (number of deals/agreements/offers) for each master to decide on an offer. This link corresponds to the task distribution and negotiation.

The above four fundamental interaction links existing between different agents in the multi agent model of the transportation domain, are used to solve the stochastic transportation problem. We illustrate each interaction between the agents in the form of KQML messages. We also justify why the agents use a particular KQML performative based on the semantics described in (Satapathy and Kumara, 1999d) and thus avoid the ambiguity in deciding which performative to use.

3.1 KQML and Logistics Language (LogL)

A message in KQML that also specifies the illocutionary act associated with it (Labrou and Finin, 1997; Satapathy and Kumara, 1999d), can be represented as:

<Performative> :sender <word>
:receiver <word.>
:reply-with <word>
:in-reply-to <word>
:language <word>
:ontology <word>
:illocutionary_act <AnIllocutionaryAct>
:content <expression>

In this representation, <AnIllocutionaryAct> is a <word> representing 'assertive', 'directive', 'commissive', 'declarative', 'acknowledgement', 'interrogative', 'prohibitive' or 'permissive'. The semantics of each <performative> associated one of the illocutionary acts is described in detail in (Satapathy and Kumara, 1999d). The ':language' and ':ontology' keywords in KQML denote the language and the dictionary used in specifying ':content'. We use a language called LogL[7] to specify the ':content' of a KQML message. The BNF grammar of LogL may be expressed as:

< predicate> ({< type > < r|w|rw > < arg > < val > })

where, <predicate>, and <arg> are of <word> type whose context is explained in a dictionary. <val> represents the value of the <arg>, <type> represents the type of <arg> such as 'integer' or 'list' etc., <r|w|rw> represents if the <arg> is of input or output type. The content of a message in KQML can also be expressed in first order predicate logic. In this chapter, for simplicity, we express the contents in first order predicate logic.

3.2 KOMI, interactions in link a

The intention of a customer announcing the stochastic transportation problem is that the masters would distribute tasks among themselves and commit to the completion of those tasks. This complex intention could be mediated through a multi-step discourse between the masters and the

[7] LogL stands for Logistics Language. LogL is similar to Knowledge Interchange Format (KIF) used to specify ':content' of ~ KOML message. However. we feel BNF grammar of LogL is simpler and can be implemented easily.

customer. The suitable candidates for the customer initiating the discourse are those KQML performatives which are associated with *directive* and *interrogative* illocutionary intent. Those performatives are: achieve, insert, ask- if, ask-all, ask-one, and subscribe.

The content of a KQML message containing achieve performative is a ground formula f or an action a (Satapathy and Kumara, 1999d). The intention of the speaker of a KQML message with achieve performative is that the hearer takes an action which makes f realized (in a real-world). In the specific example, f or a is one of the tasks of T. However, the customer (i.e. the speaker) does not know which task to assign to which master apriori. In order to issue an achieve message with a task as its content, the customer must know that the master (i.e. the hearer) will commit to that task. Therefore, the customer may opt for some interrogative illocutionary act before issuing a directive act using achieve performative. The following options exist for a customer:

- *Option 1:* The customer can request each master to offer the bids for each task. In that case,

$$\text{Bid}(j,b),\ \forall_{j \in \mathbf{J}_T}, \forall\, T \in \mathbf{T} \tag{4}$$

where, Bid(j, b) implies that the bid b that is to be offered by the master for transporting a product j in $\mathbf{J}_T$. The bid b needs to be defined in logistics ontology so that all masters understand the concept of bid for the transportation of a product.

The customer may collect the bids for each task from all the masters and then decide the tasks to be assigned to the masters that minimizes the total bid and satisfies the constraint specified in equations (2) and (3). The customer may then assign the tasks to selected masters using achieve performative. However, this violates the intention of the customer. The customer intends that the masters would collaborate to decide the allocation of tasks. Therefore, we follow the second option for the customer to announce the problem.

- *Option* 2: The customer requests each master, to offer a bid for only one task, provided that she collaborates with all the other masters to decide on that task. The customer issues a message with ask-one performative to each master. The content of the message is as follows:

$$\exists T \in \mathbf{T} | \text{Bid}(j, b),\ \forall_{j \in \mathbf{J}_T}),\ \mathbf{r} \tag{5}$$

where, r denotes the constraint on the variable *b*. The constraint dictates that a group of masters G $\subseteq$ P negotiate using a negotiation protocol to distribute a set of tasks **T** such that final agreement x_f belongs to the set of agreements **X** specified by the customer. An agreement x is a quadruple < *j, p, T, b* >, where *j* is a product whose transportation is offered by the company *p* at the cost of *b*, by adopting the task *T* for its transportation. A finite set of agreements, X imposes abounded negotiation space. Now, we represent the constraint r in first order predicate logic as:

$$\mathbf{r} = \{b \in x_f \wedge \text{Negotiate Tasks } (\mathbf{G}, \mathbf{T}, \mathbf{X}, \mathbf{CMNP}, x_f) \quad (6)$$

The predicate 'NegotiateTasks' translates to: "find an agreement x_i which is the final agreement that is negotiated among a group of master G for a set of tasks T using a negotiation protocol called Cooperative Masters Negotiation Protocol (CMNP)". CMNP prescribes the negotiation rules, the constraints in equation (2) and (3). The terms such as 'NegotiateTasks', 'G' and CMNP are part of logistics ontology. Hence, a message containing the above terms in predicate logic must be interpretable by the masters.

The masters reply back to the customer with the bid (Bid(*j, b*) , $\forall_j \in \mathbf{J}_T$) for only one task *T* such that *T* and *b* for each j E $\mathbf{J}_T$ is specified the final agreement x_f, The masters use `tell` performative to reply back the bids and the selected task *T* specified in x_f as discussed in the semantics of `ask-one` performative (Satapathy and Kumara, 1999d). The master may also use `tell` with the negation of bid(j , *b)* as the content of the reply message, if she cannot satisfy the customer's specified constraint. The customer uses achieve performative to assign the selected task to each master who has replied with a `tell` performative. The customer may also add further constraints in **r** to obtain different replies from the masters, in subsequent iterations. This situation is beyond the scope of the current discussion in this chapter.

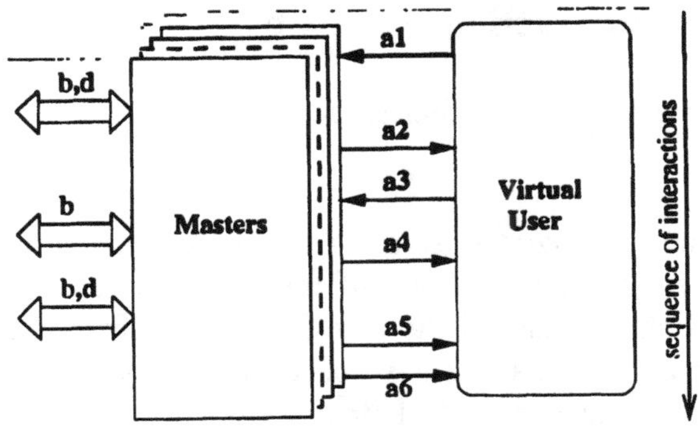

Figure 5-2. KQML Interaction between a customer and masters, link a.

A directive by the customer to achieve a task most likely leads to a commitment by the master. However, the master may not be able to keep the commitment in future. We follow a practical approach, where masters can retract their commitment to a task. The retraction can be strategic, but usually believed by other companies and the customer as the retraction due to the non-strategic causes. A master uses `untell` performative with *assertive* act to inform her retraction.

It must be noted that the retraction of a commitment simply means that a company's own resources will not be used to complete the committed task. Therefore, the master of the company who retracts the commitment to a task assigns that task to the masters of other companies (thus incurring a loss). The masters of other companies, to whom the task is assigned, inform the customer of their new commitment by sending a message containing tell performative associated with a *commissive* act. The interaction that leads to one master volunteering for another master's task is a part of link **d** interactions.

The link **a** is illustrated in figure 5.2. The content and the performative associated with each KQML message is listed as follows:

- **a1:** ask-one, interrogative, ($\mathbf{T}|\text{Bid}(j, b)$, $\forall_{j \in \mathbf{J}_T}$), **r**
- **a2:** tell, assertive, $\text{Bid}(j,b)$, $\forall_{j \in \mathbf{J}_{Ti}}$, T_i
- **a3:** achieve, directive T_i
- **a4:** tell, commissive, T_i
- **a5:** untell, assertive, T_i
- **a6:** tell, commissive, T_i

3.3 KQML interactions in link b

The expected action of the master upon receiving `ask-one` message (a1) is two fold:

1. Select an agreement x from the set of agreements X and offer the agreement to other masters of the group G. The selection of an agreement depends on the payoff expected from the agreements, and the agreements that have been offered by other masters. Each agreement specifies a task that must be solved by a company. Therefore, if the master of a company offers an agreement x, it implies that the master selects the task specified by x. We denote such a task[8] as T_x^p .

2. In order to select an agreement x, a master needs to know the cost of completing the task T_x^p so that he can calculate the expected payoff from the agreement x. The master creates the sub-tasks of the task T_x^p to be presented to the vehicles in the form of test-assignments. Unlike link a, where the customer presents the complete set of tasks to the group of masters to decide the distribution of tasks, the master does not present the complete set of sub-tasks to all the vehicles of his cluster to decide the final assignment. Instead, the master iteratively and selectively, presents the sub-tasks as test-assignment before the final assignment[9]. The final assignment of sub-tasks is accomplished based on (1) the result of all the test-assignments, and (2) the final agreement of the negotiation procedure.

In this section, we shall describe how a master selects the appropriate KQML performative and its content to achieve the second intention i.e. presenting sub-tasks and performing the final assignment. Presentation of a sub-task to a vehicle is a form of interrogative - the master creates a scenario identifier and requests a vehicle to feedback the cost and the probability (also called as *robustness of the plan)* of completing a sub-task The scenario identifier imposes the constraint that collaboration between the vehicles must be associated with that identifier. This implies that, if there exists a final agreement x_f then there exists at least a scenario identifier used by the

[8] It must be noted that we used notation T_n^p to denote any task $T_n \in \mathbf{T}$ assigned to the company *p*. If we preserve the notation, then we must use $T_{n\,x}^p$. However, to reduce the notational complexity we use T_x^p to denote any task $T \in \mathbf{T}$ that is specified by x for the company *p*.

[9] The rationale behind the master deciding the assignment as opposed to vehicles deciding the assignment themselves is that the centralized decision making is preferred than the distributed decision making, if there are large number of entities i.e. vehicles. The number of iterations or test-assignments given a selected task depends on the knowledge of reliability of the vehicles by the masters.

master p to present the subtasks of T^{p}_{xf} . Using this scenario identifier, the vehicles decide to collaborate. Therefore, the master issues a KQML message with ask-one as the performative that contains:

$$\text{Cost}(T_i, c) \wedge \text{Prob}(T_i, \pi), \text{r} \quad (7)$$

as its content. Cost(T_i, c) is a predicate which translates to "c is the cost of completing sub-task T_i". Prob(Ti, π) is a predicate which translates to "π is the probability of completing sub-task Ti". r denotes the set of constraints on the variable c and π:

$$\text{r} = \text{SolveTask}(\text{V}, T_i, \text{SCENE\_ID}(sp\_expr), c, \pi) \quad (8)$$

We explain the predicate symbol SolveTask in logistics ontology. SolveTask is a problem solving procedure to solve a task *Ti* by collaborating with a set of vehicles V such that all the vehicles in V are aware that scenario SCENE_ID(*sp_expr*) is being solved. In our specific example, *sp_expr* is defined as 'integer{ +integer}'and its interpretation is part of our logistics ontology.

The vehicle, to whom the ask-one message is issued, either asserts $Cost(T_i, c) \wedge Prob(T_i, \pi)$ using `tell` or asserts its negation using `tell`. The latter implies that the vehicle cannot find the probability and cost of completing the sub-task based on the information, that satisfies the constraint r. It must be noted that the vehicles do not (must not) perform any physical action to change the state of the world, and

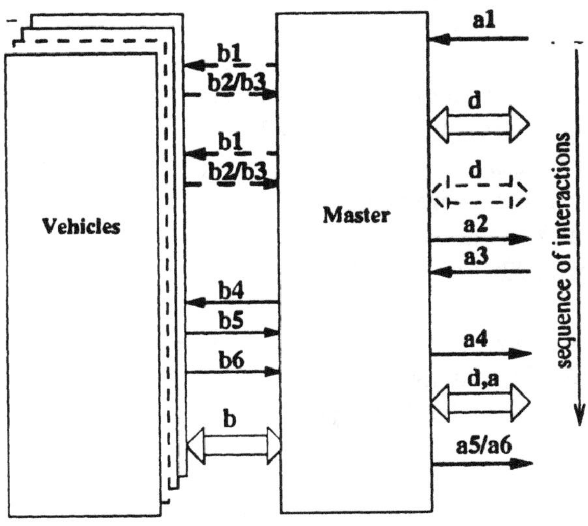

Figure 5-3. KQML Interaction between a master and his vehicles, link b

thereby change the solution of SolveTask. In other words, the expected actions due to an interrogative is preserved as explained in (Satapathy and Kumara, 1999d). The vehicle can also assert the same using untell performative, implying that it finds contradictory results about the *Cost*(T_i, c) based on its current information[10].

After several test-assignments and negotiation steps (link **d**), the masters finalize on an agreement x_f. If the customer approves x_f task T^p_{xf} is assigned to the master of the company *p*. We assume that the master has presented *at least* one set of sub-tasks generated from the task T^p_{xf}, to the vehicles as test-assignment. Upon the approval of the agreement x_f the master *p* assigns the sub-tasks to the vehicles that correspond to one of the previous test-assignments of the sub-tasks of T^p_{xf}.

The master issues achieve with *directive* illocutionary intent, containing sub-task T_i to the selected vehicles. The vehicle, to whom achieve message is issued, most likely commits to the sub-task. The vehicle uses the tell performative to commit to the assigned sub-task. We have developed a procedure using the fact that the subtask cannot be completed (due to nature). In that case, the slave declares its failure to its master by issuing a message that contains `untell` performative associated with *assertive* act. Unlike link **a**, another vehicle does not commit to the same sub-task. This unfulfilled sub-task needs to be re-assigned by the master (through test-assignment) or taken up by another master. We show various interactions between a master and his vehicles that are classified as link bin figure 5-3. The content and the performative associated with each KQML message is listed as follows:

- **bl:** `ask-one`, interrogative Cost (T_i, c) ^ Prob(T_i, π), **r**
- **b2:** `tell`, assertive, Prob(T_i, c_j) ^ Prob (T_i, π_j)
- **b3:** `untell`, assertive, Cost(T_i,c) ^ Prob(T_i, c) ^ Prob(T_i, π)
- **b4:** `achieve`, directive, T_i
- **b5:** `tell`, commissive, T_i
- **b6:** `untell`, assertive. T_i

The dotted arrows in figure 5-3 indicate that there is a possibility of several iterations of interactions (i.e. **bl, b2, b3** and link **d**) depending on the rate of convergence to arrive at a final agreement.

[10] This happens when the vehicle finds an optimal cost under the assumption that some goals (i.e. orders) may be dropped, which however, has not been guaranteed. Hence, the result is contradictory due to the assumption.

3.4 KQML interactions in link c

The interactions in link c are the interactions between two vehicles collaborating to solve a sub-task. The representation of a sub-task is same as the representation of a task, that is, a set of orders as defined in the constraint in equation (1). However, a sub-task presented or assigned to a vehicle is a subset of the task selected by its master, that meets the constraints in equations (2) and (3). The sub-task consisting of set of orders, is to be solved by the vehicle such that previously committed orders (i.e. goals) are not omitted. Therefore, the vehicles have following options:

1. Determine the cost and probability of fulfilling all the orders.

2. Delegate few orders to other vehicles.

3. Split an order into two orders and delegate one of them to another vehicle. An order consists of pick-up of an item from a source and its delivery at a destination. Therefore, splitting of an order implies determining an intermediate location which will serve as the destination of one order and source for the other order.

We have developed a procedure that lets a slave to determine a set of alternate routes including one master route for completing a sub-task presented. The procedure outputs the cost and probability of a robust plan[11]. While determining the routes, the procedure sometimes requires that the slave finds the orders or split the orders to be delegated. In the last two options i.e. delegating or splitting an order, a vehicle seeks collaboration with another vehicle. The vehicle's role is similar to that of the master, in delegating an order. However, we model the vehicles such that one vehicle does not have the knowledge of reliability of other vehicles. The vehicles may have the knowledge of plans of other vehicles to solve their committed sub-tasks[12]. Therefore, the vehicle announces an order to a set of vehicles. The vehicle issues a KQML message with ask-one performative to those set of vehicles. The content of this message is similar to the content of the test-assignment message by a master:

$$\mathrm{Cost}(T_i^{'}, c) \wedge \mathrm{Prob}(T_i^{'}, \pi), \mathbf{r}$$

[11] The plan is robust because it consists of several alternate plans that can be executed in case of *expected events.*

[12] Our modeling scheme allows a vehicle v to have the knowledge of plans of a set of vehicles whose plans intersect with the plan of vehicle v on a regional man. This facilitates collaboration, if the number of vehicles is large.

$$r = \text{SolveTask}(\mathbf{V}, T_i^{'}, \text{SCENE\_ID}(sp\_expr), c, \pi) \qquad (9)$$

where, **V** is a null set and $T_i^{'}$, is a task that contains only one order.

It is possible that two or more vehicles of the set of vehicles may reply to ask-one message with an assertive `tell` performative. The content of a tell message is: $Cost(T_i^{'}, C_j)$ ^$\text{Prob}(T_i^{'}, \pi_j)$.

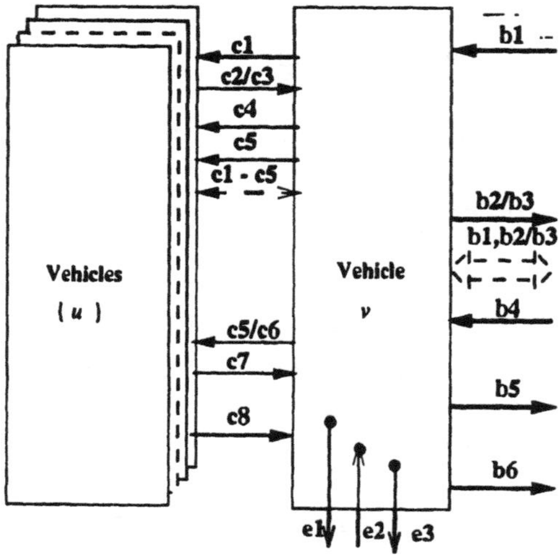

Figure 5-4. KQML Interaction between a vehicle and other vehicles, link c

The vehicle v who has issued ask-one may decide to consider only *one* of the `tell` messages towards the calculation of cost and probability of the sub-task T_i. The vehicles, who have replied with `tell` message, have not committed to the order or task $T_i^{'}$. Therefore, given another order or task, say $T_k^{'}$ of the same scenario (SCENE_ID(*sp_expr*) in Eqn. (9)), they may not consider $T_i^{'}$. On the other hand, the vehicle v cannot assign $T_i^{'}$ to a vehicle because it has not been assigned with task *Ti* by its master. Therefore, the vehicle v issues an `insert` message to only one vehicle u (that he selects) to make u believe that $T_i^{'}$ will be assigned to u pertaining to a scenario SCENE_ID(*sp_expr*), if a situation requiring the delegation of the task $T_i^{'}$ arises. The vehicle v (may) also issues `delete-one` message to the rest of the vehicles as a directive to remove the belief that $T_i^{'}$ will be assigned to those vehicles belonging to the scenario SCENE-ID(*sp_expr*). The content of these messages will be:

$$\text{ToBeAssigned}(T_i^{'}, \text{SCENE\_ID}(sp\_expr)) \qquad (10)$$

The predicate ToBeAssigned needs to be explained in logistics ontology.

When the vehicle v is assigned with task T_i by its master, the vehicle v also assigns $T_i^{'}$ to the selected vehicle (e.g. vehicle u). The vehicle v issues a KQML message with `achieve` performative to the vehicle u. The content of the message is $T_i^{'}$, In most cases, vehicle u commits to the order or task $T_i^{'}$. It is possible that vehicle u may be unable to keep the commitment to $T_i^{'}$ *in* future. In that case, the vehicle v uses `untell` to declare that it will no longer be able to complete the sub-task T_i (link **b6**), based on the information gathered from the sensors and the observation of the execution its own plans i.e. link **e2**. Links **el**, **e2** and **e3** are explained in section 3.6.

Figure 5-4 illustrates link c. The content and the performative associated with each KQML message is listed as follows:

- **cl:** `ask-one`, interrogative, Cost $(T_i^{'}, c)$ ^ Prob$(T_i^{'}, \pi)$, **r**
- **c2:** `tell`, assertive, $(T_i^{'}, c_j)$ ^ Prob$(T_i^{'}, \pi_j)$
- **c3:** `untell`, assertive $(T_i^{'}, c)$ ^ Prob$(T_i^{'}, \pi)$
- **c4:** `insert`, directive, ToBeAssigned
  $(T_i^{'}$,SCENE_ID(*sp_expr*))
- **c5:** `delete-one`, directive,ToBeAssigned
  $(T_i^{'}$,SCENE_ID(*sp_expr*))
- **c6:** `achieve`, directive, $T_i^{'}$
- **c7:** `tell`, commissive, $T_i^{'}$
- **c8:** `untell`, assertive, $T_i^{'}$

The dotted arrows in figure 5-4 indicate the possibility of several iterations of interactions.

3.5 KQML interactions in link d

As described in section 3.3, the expected actions of a master of a company p upon receiving `ask-one` message from the customer are: (1) the selection of an agreement x and (2) the test-assignment of the sub-tasks of T_x^p. The procedure to select an agreement and final assignment of the task is as follows:

1. A master p calculates the maximum expected payoff for task T_x^p for each x $\in$ X. The cost included in the calculation of the payoff is a probabilistic cost, which is obtained from the test-assignments.

2. The master *P* offers the maximum pay-off yielding agreement to other masters. The master *P* also receives offers from other masters. We assume a cooperative environment where masters also exchange the probability of failure of completing tasks. This probability may not be same as the probability of failure obtained from the test - assignment of these tasks by the masters to their vehicles. The probability of failure of tasks by one master p_1 may be used by other masters *{p2}* to prepare for an event where p_1 fails to complete the task. We have not investigated the issue of how this information is used for the purpose preparing for an event.

3. A master *p* may either accept the agreements offered by other masters or reject them. If the master *p* decides to reject one of the agreements received from another master, then it must offer another agreement whose total bid value $(\sum_p \in \mathrm{P} \sum_j \in \mathrm{J_T}{}^p_x (b \in \mathrm{x}))$ is less than the rejected agreement (Satapathy *and Kumara, 1999a).*

4. It must be noted that while a master is offering agreements, the expected pay-off from each agreement varies (Satapathy and Kumara, 1999a). This is due to the variation of the cost of executing the sub- tasks reported by the vehicles. The variation of the cost takes place due to the variation of the travel time and the service time that are used to determine the cost of solving the sub-task. Therefore, the steps 1, 2, and 3 are repeated until an agreement is accepted by all the masters.

In this section, we explain the KQML performatives that facilitate the interaction among the masters. Our task distribution procedure is a distributed approach where the masters offer the bids to each other, instead of declaring the bids to a central arbitrator. An agreement x captures the notion of a bid (Satapathy and Kumara, 1999a). Hence, in terms of the bargaining model, a bid (i.e. agreement) offered by a company is the final bid, if no other company can offer another bid so that the total value of the bid is reduced. In order to obtain bids from other masters, a master subscribes bids from other masters by sending a message that contains subscribe performative associated with *directive* act. The content of this message is another KQML based message containing ask-one performative, which is:

$$(\exists T \in \mathrm{T}|\mathrm{Bid}(j, b)\ \forall_{j \in \mathrm{J}_T} \wedge \mathrm{Prob}(\mathrm{T}, \pi)), \text{ r and}$$
$$\mathrm{r} = \{b \in \mathrm{x} \wedge \mathrm{x} \in \mathrm{X} \wedge (\mathrm{TotalBix(x)} < \mathrm{TotalBid}(\hat{x}))\} \quad (11)$$

where, TotalBid(x) is the total bid value of the agreement x. $\hat{x}$ is the agreement received from other masters that has the lowest total bid value.

A master p_1 offers a bid under two situations: (1) offer an initial bid to all other masters $\{p_2\}$ who have subscribed to bids from p_1, or (2) counter offer with a bid to the masters $\{p_2\}$ who have offered a bid to p_1, due to the subscription of the master p_1. The master p_1 issues a message containing tell performative with an *assertive* act to inform its offer or counter offer.

A master p_1 realizes that he has converged to a final agreement only when any of the following conditions occur:

- When all masters in a group of masters G have withdrawn subscription from him.
- When the bids sent by all other masters in G, is same as the bid that p_1 had sent.

The utterance of the `subscribe` performative implies that whenever a master finds a new bid that satisfies the constraint r, he asserts the bid to the subscribing master (refer to the semantics of `subscribe` performative in (Satapathy and Kumara, 1999d). The master to whom a `subscribe` message is issued, could have also issued a `subscribe` message to the other masters. Therefore, each master searches for a new bid upon receiving bids from other masters. When the master realizes that she has converged to a final agreement, she `unsubscribes` his subscription.

A master p commits to a task, if T^{p}_{xf} is not null. However, as mentioned in section 3.2, a master may still fail to complete the committed task. In that case, the master announces the unfulfilled task as a problem to be solved by other masters in his group G. The master acts like a customer to the group of masters in G. The master uses ask-one performative containing the following formula as the problem:

$$(\exists T_i \in T \,|\, \text{Bid}(j, b),\ \forall_j \in J_{Ti}),\ \text{r and}$$
$$\text{r} = \text{NegotiateTasks(G}, T, \text{X CMNP, x)} \tag{12}$$

where, T is the unfulfilled task, and X is formulated by the master who acts as the customer. This `ask-one` message is considered as the one originated from the customer. Therefore, the semantics of link d4, which represents this message, is same as the link at in section 3.2. The link d is illustrated in figure 5-5. The content and the performative associated with each KQML message is listed below:

- d1: `subscribe`, directive, *k*(*f*) where *k* is `ask-one` and *f* is $(\exists T_i \in \mathrm{T}\ |\mathrm{Bid}(j,b) \forall_{j \in \mathrm{J}_T} \wedge \mathrm{Prob}(T, \pi))$, r
- d2: `tell`, assertive, Bid(*j*, *b*) $\forall_{j \in \mathrm{J}_T,} T$
- d3: `unsubscribe`, directive, *k*(*f*) where *k* is `ask-one` and *f* is $(\exists T_i \in \mathrm{T}\ |\mathrm{Bid}(j,b) \forall_{j \in \mathrm{J}_T} \wedge \mathrm{Prob}(T, \pi))$, r

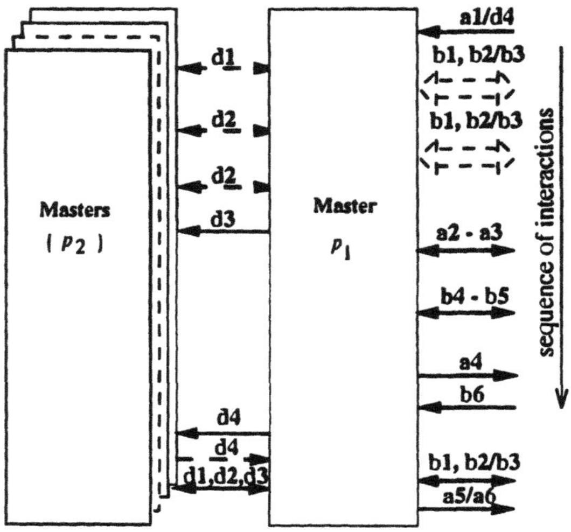

Figure 5-5. KQML Interaction among masters, link d

- d4: `ask-one`, interrogative, $(\exists T_i \in \mathbf{T}\ |\mathrm{Bid}(j,b) \forall_{j \in \mathbf{J}_T} \wedge \mathrm{Prob}(T, \pi))$.

3.6 All Other Interaction Links

In order to make the multiagent system with a master and slaves type of model operational in real-time, the agents need to interact with sensors and a plan executor. For experimental purposes, we let the masters and slaves use the sensor data (e.g. travel time, commitment reliability) as part of the agent program. However, we still demonstrate the interaction of the slaves with the plan executor, which in our case is a simulation program, using KQML based messages (Figure 5-4).

In figure 5-4, we present three links to (el, e2, e3) illustrate three primary KQML based interactions with the simulation interface program. It must be noted that the semantics of KQML messages as explained in (Satapathy and Kumara, 1999d), is not implemented in the simulation interface program because of its non-BDI structure.

- Link **el**: This link consists of sending KQML based messages with achieve performative to the simulation interface program. The content of this message is a plan. A plan is a data structure that describes the sequence of locations (including pick-up and delivery nodes), road links, and their arrival and departure times. This link also consists of sending a KQML message with achieve performative, but with a negation of a plan directing the interface program to stop simulating (executing) the plan.

- Link **e2**: This link consists of the simulation interface program asserting the arrival and departure times at various locations due to the simulation. The interface program uses a `tell` performative to assert the above information[13] and thus, act as a sensor. The interface program does not necessarily need a `subscribe` message to assert arrival and departure time at regular time interval using `tell` performative.

- Link **e3**: Based on the information from the simulation, if a slave determines that it can no longer satisfy the arrival and departure time requirements for the remaining pickup and delivery nodes, then it requests the simulation to stop plan execution. The message is similar to link el, but the content of the message relate to the negation of the plan. The slave declares (to the master) the remaining orders as unfulfilled sub-tasks.

4. MULTIAGENT MODEL

In this section, we describe the master-slaves model developed using the agent structure developed by Satapathy and Kumara (Satapathy and Kumara, 1999b). We describe the synthesis of a multiagent system, by combining the master and slave agents. The agent model is composed of a `reasoner` package and a *communication* package. As long as the same agent model is maintained, the communication package is generic to all agents. Different agents with different roles can be developed by enabling the `reasoner` class to use different belief, desire, and intention manipulation and search (M&S) methods. A set of `protected` BDI M&S methods such as `CheckBelief()`, `PlanGenerate ()` serve as virtual functions in the implementation of the agents. The details of these BDI M&S

[13] Most sensors also follow this type of KQML messages i.e. e2, asserting the status information about the world in which the agents are acting.

methods and their mapping to KQML performatives are discussed in (Satapathy and Kumara, 1999d).

4.1 The Model of a Master

The *belief* of a master of a transportation company is the set of vehicle route plans, vehicle commitment reliability[14], and the current status of all the vehicles of that company. The belief set of the master also contains the agreements offered by other companies. In this chapter we do not address the representational issues related to the belief of an agent. The *goal* of the master is the completion of the task of a chosen agreement. Therefore, at a given instance, a master has already committed to a set of goals i.e. tasks as he has already assigned the sub-tasks of those tasks to selected vehicles. Hence, the assignment of sub-tasks to a set of vehicles is considered as the *intention* of a master. Given a task, there are many possible ways of assigning sub-tasks of a task to its vehicles. Therefore, the set of actions is a set of possible ways of assigning its sub-tasks to its vehicles. A plan therefore consists of only one action. By executing this action (i.e. assigning sub-tasks to a set of vehicles), a master assures the accomplishment of goal (i.e. completion of a task) with certain degree of certainty. A plan generation scheme (`PlanGenerate()`) consists of iterative query (links **bl**, **b2** and **b3**) of the vehicles about the cost and probability of completing sub-tasks. It must be noted that a master not only tries to find a plan for a given task, but also tries to find a task that it must commit. In other words, it also searches for a goal (through negotiation).

Table 5-1 shows a list of various implementation methods that are invoked upon receiving a KQML message corresponding to an interaction link. The output of these implementation methods are the KQML messages that are sent to other masters, vehicles, and the customer. Each of these methods is either a direct implementation of a virtual BDI M&S methods or invoke a series of implementation methods of several virtual BDI M&S methods.

[14] The vehicle commitment reliability is a probability distribution over the deviation of cost and robustness of the plan reported by the slaves. This is used by the master to selectively present test-assignments.

Table 5-1. Different implementation methods of master and their input and output

The interactive links serving as input to a master	The Implementation methods of the `reasoner` object of a master	The Interaction links serving as output from a master	The related and associated virtual BDI M&S methods
a1, d4	`SolveTransport()`	**b1, d1**	`CheckBelief(), PlanGenerate()`
d1	`SolvesCMNPTransport()`	**d2**	`CheckBelief(), ReviseBelief()`
d2	`SelectNPresentTask()`	**b1, d2**	`CheckBelief(), ReviseBelief(), PlanGenerate()`
d3	`CheckForFinalTask()`	**a2**(to master or customer)	`RetractBelief(),PlanGenerate()`
b2/b3	`ReSelectNPresentTask()`	**d2, a2, b1**	`CheckBelief(), ReviseBelief(),PlanGenerate()`
a3 (from master or customer)	`Assign Tasks()`	**b4**	`ReviseIntention()`
b5	`CommitToATask()`	**a4** (to master or customer), **a6**	`ReviseGoal()`
b6	`UncommitNReconsiderTask()`	**a5, d4**	`CheckGoal(), GoalSkip(), Retract Goal, RetractIntention, PlanGenerate()`
a2 (from master)	`RedistributeTask()`	**a3** (to master)	`ReviseBelief(), PlanGenerate(), ReviseIntention()`

These methods contain algorithms such as negotiation, task selection and sub-task presentation, the details of which are not discussed in this chapter. Few important points to be noted in this table are: (1) None of these methods invoke any implementations of virtual M&S methods that manipulate intention and goal, until the customer has approved a task for the master (i.e. link **a3**). (2) During a sub-task failure reported by a vehicle of a transportation company, the master of the company assumes the roles of a customer in redistributing the sub-task among the group of masters. Hence, it can also receive KQML messages corresponding to links **a3**, **a2**, and **34**.

4.2 The Model of a Slave

The belief of a slave is a set of formulae that describe the following:

1. Road network
2. Travel time distribution of each road link over time.
3. Sensor reliability over the sensors reporting travel times.
4. The tasks presented/assigned by the masters and vehicles
5. Shortest route between two locations representing pick-up and delivery nodes of a task
6. The capacity of the vehicle

The desire (goal) of a slave is to arrive at the pick-up and delivery nodes within the arrival time interval as specified in a task specification. The intention is the set of plans consisting of sequence of *move* operations between the nodes of a road network and *load* and *unload* operations at the pick-up and delivery nodes, such that if the plan is executed successfully then it assures accomplishment of the goals.

Table 5-2 shows the list of different implementation methods of a slave that are invoked upon receiving a KQML message corresponding to interaction links, **b**, **c** and **e**. Each of these methods is either a direct implementation of a virtual M&S method or invokes a series of implementation methods of several M&S methods.

These methods implement algorithms such as stochastic PDP problem solving procedure, task splitting and delegation procedure, vehicle route plan evaluation procedure based on the sensor information (link **e2**), which are not discussed in this chapter. In table 5-2, it must be noted that, if a slave determines a possible failure of a plan based on the simulation information (link *e2)* or from the assertion from a collaborating slave (link **c8**), then the salve may skip few tasks depending on her relationship with the task assigning slave or master[15]. The slave may also choose to assert a task failure, and wait for a directive to un-achieve the failed task. A directive to un-achieve a task can also be expressed via links **b4** and **c6**. Hence, we have associated `GoalSkip ( )` and `RetractGoal ( )` methods with the `Achieve ( )` implementation method.

[15] These relationships are explained in (Satapathy and Kumara, 1999d) as relationships of trust and authority.

4.3 Putting It All Together

We presented a brief overview of belief, desire and intention of slaves and masters, and discussed different implementation schemes of important virtual methods. We have also discussed when these

Table 5-2. Different implementation methods of a slave and their input and output

The Interaction links serving as input to a slave	The Implementation methods of the reasoner objective a slave	The Interaction links serving as output from a slave	The Related and associated virtual BDI M&S methods
a1, d4	`Solve Task ()`	**b/b3 or c1, c2/c3**	`CheckBelief(), PlanGenerate()`
b4, c6	`Achieve ()`	**c6 or (e1 and b5), e1 and c7**	`ReviseIntention(), ReviseGoal(), GoalSkip(), RetractIntention(), RetractGoal()`
c2/c3	`ContinueToSolve Task()`	**c4/c5, c1 or b2/b3**	`ReviseBelief(), PlanGenerate()`
c4/c5	No Implementation	**NO Message Output**	`ReviseBelief() RetractBelief()`
c7	`CommitToTask()`	**e1 and b5**	`ReviseBelief(), ReviseIntention(), ReviseGoal()`
c8	`ReconsiderTask`	**b6, c6**	`ReviseBelief() CheckGoal(), GoalSkip(), RetractIntention(), RetractGoal()`
e2	`Evaluate Task`	**c8, b6, c6**	`ReviseBelief(), CheckGoal(), GoalSkip(), RetractIntention(), RetractGoal()`

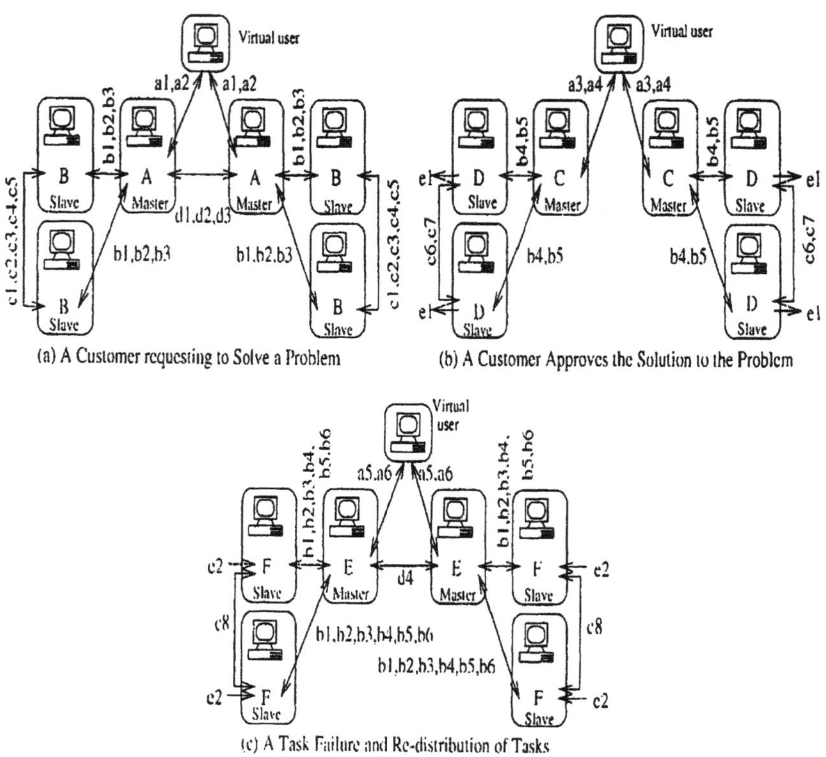

Figure 5-6. The Multiagent Model illustrating the Agent Interactions

methods are invoked. In this section, we combine the interactions and generate various scenarios in which the multiagent system operates. The important scenarios are:

1. A customer requesting to solve a stochastic transportation problem
2. A customer approving the final agreement, provided by a group of masters
3. A task failure reported by a simulation interface program or by a vehicle leading to the redistribution and re-assignment of tasks

In figure 5-6, which is self explanatory, we illustrate various interactions between masters and slaves corresponding to these three scenarios. This figure shows two transportation companies; each containing only two slaves, and a customer. Each of these scenarios is described by only a subset of interaction links discussed in section 3.

5. COMPARISON WITH OTHER RESEARCH

The master and slave model presented in this chapter has many similarities with the approach presented by Fischer in (Fischer et al., 1996). However, there are some vital differences, which we believe, are of importance in terms of optimality, timeliness, and robustness of the solution. The differences are as follows:

- In the approach presented by Fischer, a manager (i.e. a master) requests all truck agents (i.e. slaves) to submit a bid for a task[16], and then the manager assigns the task based on the bids. The approach considers extended Contract Net Protocol, where they allow the truck agents to submit bids for part of the orders in terms of the weight or volume of the products to be transported. We significantly differ from this approach. In our approach, the master performs selective test- assignments of tasks as opposed to requesting for bids for tasks[17]. This is because a customer's problem is composed of several orders, where each order is indivisible in terms of weight or volume. A near-optimal solution in terms of the cost is possible, if the master requests all the slaves to submit bids for all the tasks of power set of orders separately. However, the number of such requests and their response becomes exponential with increasing number of orders constituting a customer's problem. It must be noted that an optimal solution is also not possible through the sequential assignment of *temporal grant* and *definitive grant* of orders as discussed in Fischer's Approach. Furthermore, Fischer does not consider collaboration among the truck agents *I* slaves in terms of splitting of orders by finding intermediate goals. Our approach considers the collaboration among trucks as it happens in real-world.

- Fischer considers a horizontal cooperation among the managers of transportation companies, which is more similar to a third-party task assignment in real-world. A manager of a transportation company 'A' declares the capacity available by his vehicles to another company 'B', and the manager of company B considers company A as another vehicle. Thus, the manager of the company 'B' ultimately decides if an (part of) order will be assigned to company A. We take a different stance which is more close to the real-world approach.

[16] A task is a set of orders. A task can be considered as consisting of one order in Fischer's approach. In other words, a customer's problem is composed of only one order.

[17] We presume that the task assignment though selective test-assignment will improve the timeliness of assignment.

The masters of all the companies have equal autonomy in deciding which task to commit to. The masters negotiate on peer-to-peer level based on CMNP, which is based on game theoretic approach. One may consider the negotiation based on CMNP as using Contract Net Protocol, where each master is also an arbitrator with respect to all other masters, for one customer problem. The master-slaves model presented in this chapter has the provision for a master submitting false information (bid) to other masters, and thus winning a task that he may be incapable of solving it. At the same time, a master also considers the possibility of failure in completing the task bid by other masters, in bidding competitively (but constrained by CMNP) in the whole negotiation process. Fischer's approach does not consider these vital aspects of real-world problem solving.

- Finally, the master-slave type model is based on a BDI model of an agent and uses KQML based messages. This facilitates a robust information infrastructure for multiagent models in transportation problems in specific, and in general for supply chain problems.

6. CONCLUSIONS

In this chapter, we define a transportation problem of practical importance, containing uncertainties in travel times, and the availability of products for transportation. We present a master-slaves type of model to solve this transportation problem where each master and slave is modeled using a BDI agent model and interact using KQML based messages. This model addresses most of the requirements mentioned in section 1, such as (1) dynamics in real-time problem solving, (2) autonomy of the vehicles in developing their own plans, (3) collaboration among the vehicles by taking possible task failure into consideration, and (4) negotiation among the transportation companies without much of intervention by the customer. Furthermore, The model has the provision to incorporate the manufacturers and supplier's decisions[18] in the task selection and assignment process, and thus integrating all enterprises of supply chain domain from a practical perspective. Lastly, the approach presented in this chapter addresses most of the important factors of real-world transportation problems compared to the existing literature.

[18] These decisions relate to declaring the probability of availability of goods, and the probability of producing finished items, which can be changed by the producers and suppliers during the planning and execution process.

ACKNOWLEDGEMENTS

The authors wish to thank Army Research Office (Grant DAAHO4-96-1-0464) and National Science Foundation (DMI-0075584) for their generous support in pursuing this research.

REFERENCES

Benerecetti, M., Giunchiglia, F., and Serafini, L. (1997). Model checking multiagent systems. Technical Report IRST-9708-07, Istituto per la Ricerca , Povo (Trento), Italy, http://mambo.cs.unitn.it/bene/pub.html.

Burmeister, B., Haddadi, A., and G., M. (1997). Application of multi-agent systems in traffic and transportation. *JEE Transactions on Software Engineering,* 144(1): 51-60.

Burmeister, B. and Sundenneyer, K. (1992). Cooperative problem solving guided by intentions and perception. In Werner, E. and Demazeau, Y., (editors), *Decentralized AIJ - Proceedings of the Third European Workshop on Modeling Autonomous Agents in a Multi-Agent World (MAAMAW-91),* pp. 77-92. Elsevier Science Publishers B. V.: Amsterdam, The Netherlands

Dean, T. and Greenwald, L. (1992). A formal description of the distributed transportation problems. Technical Report CS-92-14, Department of Computer Science, Brown University, Providence, RI, USA.

Dror, M., Laporte, G., and Trudeau, P. (1989). Vehicle routing with stochastic demands: properties, and solution frameworks. *Transportation Science,* 23(3): 166-176.

Fischer, K., Muller, J. P., and Pischel, M. (1996). Cooperative transportation scheduling: An application domain for DAI. *Applied Artificial Intelligence,* 10(1): 1-34.

Golden, B. L. and Assad, A., (editors) (1988). *Vehicle routing: Methods and studies.* Elsevier Science Publishers B. V.: Amsterdam, The Netherlands.

Labrou, Y. and Finin, T. (1997). A proposal for a new KQML specification. Technical Report TR CS-97-03, Computer Science and Electrical Engineering Department, University of Maryland, Baltimore County, Baltimore, MD, USA.

Laporte, G., Louveaux, F., and Mercure, H. (1992). *Transportation Science,* 26(3): 161-170. The vehicle routing problem with stochastic travel

Ljunberg, M. and Lucas, A. (1992). The OASIS traffic management system. Technical Report 28, Australian Artificial Intelligence Institute, Melbourne, Australia. Also appears in Second Pacific Rim International Conferences on Artificial Intelligence.

Sastry, S., Meyer, G., Tomlin, C., Lygeros, J., Godbole, D., and Pappas, G. (1995). Hybrid control in air traffic management systems. In *Proceedings of the 1995 IEEE Conference in Decision and Control,* pp. 1478-1483, New Orleans, LA.

Satapathy, G., Kumara, S., and Moore, L. M. (1998a). Distributed intelligent architecture for logistics (DIAL*). International Journal of Expert Systems with Applications,* 14:409-424.

Satapathy, G. and Kumara, S. R. T. (1999a). Negotiation for transportation tasks with stochastic payoffs. In *Workshop on Agent based Decision-Support for Managing the Internet Enabled Supply-Chain, Third International Conference on Autonomous Agents, Seattle, WA, USA*, pp. 101-107.

Satapathy, G. and Kumara, S. R. T. (1999b). Object oriented design based agent modeling. In *Fourth International Conference on Practical Applications of Agents and Multiagent (PAAM99),* London, UK.

Satapathy, G. and Kumara, S. R. T. (1999c). Robust vehicle route planning for pick-up and delivery orders using real-time vehicle-to-vehicle collaboration. Working Paper Series, Department of Industrial and Manufacturing Engineering, Pennsylvania State University, University Park, PA, USA.

Satapathy, G. and Kumara, S. R. T. (1999d). Semantics of KQML messages based on illocutionary acts and BDI model of an agent. In *The Sixth International Workshop on Agent Theories, Architectures, and Languages (ATAL-99).*

Satapathy, G., Kumara, S. R. T., and Moore, L. M. (1998b). Software agents in logistics. In *Encyclopedia of Library and Information Science.* Merkel Dekker, Inc, NY.

Wellman, M. P. (1993). A market-oriented programming environment and its applications to multicommodity flow problems. *Journal of AI Research,* 1: 1-22.

Chapter 6

IMPROVING SCALABILITY OF E-COMMERCE SYSTEMS WITH KNOWLEDGE DISCOVERY[1]

Sigurdur Olafsson
Iowa State University

Abstract: The efficiency of many data driven e-commerce system may be compromised by an abundance of data. In this chapter we discuss how knowledge discovery and data mining techniques can be useful in improving the scalability of data driven e-commerce systems. In particular we focus on improving scalability via dimensionality reduction and improving the information view experienced by each user. To address these issues, we cover several common data mining problems, including feature selection, clustering, classification, and association rule discovery and present several scalable methods and algorithms that address each of those problems. Numerous examples are included to illustrate the key ideas.

Keywords: Data mining, feature selection, clustering, classification, association rules, support vector machines, scalability, e-commerce

1. BACKGROUND

This chapter surveys tools and techniques for knowledge discovery as they relate to improving scalability of complex data driven systems and e-commerce systems in particular. The main motivation behind this work is that the scalability of such systems can be limited by an abundance of data, much of which may be redundant or irrelevant, which in turn can cause the system performance to degenerate. Furthermore, the users of such systems may experience information overload or be overwhelmed if too much data is presented in an unstructured manner. In this paper we argue that a more scalable approach would automatically reduce the amount of information to a manageable size regardless of the total amount of system traffic and data, thus making the decision making time independent of the load. In other words, in the ideal design a given user will always have a similar amount of information, regardless of how many other user are generating data. We discuss how by drawing on tools from knowledge discovery and data

[1] This work was supported by the National Science Foundation under grant DMI-0075575.

mining, automatic decision support can be developed to improve the overall efficiency of the system and reduce the information overload for its users. Thus, by incorporating the knowledge discovery techniques surveyed in this paper, the scalability of data driven e-commerce systems can be improved.

The interdisciplinary area of data mining and knowledge discovery draws on well-established fields such as statistics, artificial intelligence, and optimization. However, in recent years knowledge discovery has experienced an explosion in interest from both industry and academia. The primary reason for this surge in attention is that databases in many modern enterprises have become massive, containing a wealth of important data, but traditional business practices often fall short in transforming such data into relevant knowledge that can be used to gain a competitive advantage. Thus, to obtain an advantage, data mining draws on a variety of tools to extract previously unknown but useful knowledge from massive databases.

This process of extracting useful information from large databases consists of several steps and a complete description is beyond the scope of this chapter. For good introductory texts to data mining process we refer the reader to Han and Kamber (2001) and Witten and Frank (2000). Briefly, the process usually includes the following steps. It typically starts by identifying the sources of data within the enterprise and integrating this data from numerous legacy databases. This preparation may also include manipulation of the data to account for missing and incorrect data, and often involves construction of data warehouses and data marts (Immon, 1994). After the data has been prepared a model is induced to describe the target concept of interest. This may for example be looking for a correlation in the data, finding natural clusters, or determining how the data can be classified according to some specific variable. Depending on the target concept an appropriate data mining algorithm is selected. After the model has been induced it is usually necessary to go through a process of data abstraction to identify relevant patterns, understand what knowledge has been gained, and determine what actions to take within the enterprise as a result of this knowledge.

As indicated by the above description, all data mining starts with a set of data that is commonly termed the training set. This training set consists of instances describing the observed values of certain variables that we refer to as features. These instances are then used to learn the target concept, and depending upon the nature of this concept different learning algorithms are applied. One of the most common is classification, where a learning algorithm is used to induce a model that classifies any new instances into one of two or more known categories. The primary objective may be for the classification to be as accurate as possible, but accurate models are not necessarily useful or interesting and other measures such as simplicity and novelty are also important. In addition to classification, other common concepts to be learned include association rules discovery, numerical prediction, and natural clustering of the instances.

In recent years, knowledge discovery tools have been extensively applied to various data driven systems, and in particular for many e-commerce applications (Ansari et al., 2000; Kosala and Blockeel, 2000). For example, given the increased business emphasis on mass customization and personalization (Pine and Gilmore, 1999), intelligent recommender systems have received considerable attention. Such systems typically suggest products to their customers, as well as providing information that helps customers decide which products to buy. Early recommender systems were generally based on simple nearest-neighbor collaborative filtering algorithms and this is still the most common approach in practice (Resnick et al., 1994; Shardanand and Maes, 1995). However, other learning methods have also been used, including Bayesian networks (Breese et al., 1998), classification methods (Basu et al., 1998), association rules (Lin et al., 2000), regression (Vucetic and Obradovic, 2000), and intelligent agents (Good et al., 1999). Schafer et al. (2001) survey commercial recommender systems and identify the technologies employed.

The remainder of this paper is organized as follows. In the next section we introduce a case study involving online auctions of recyclable products that will be used throughout to illustrate relevant techniques. Sections 3 and 4 contain the actual survey of knowledge discovery tools and constitute the main part of this paper. In Section 5 we revisit the case study in light of how the data mining tools introduced can be used for scalability improvements. Finally, we conclude in Section 6 with a general discussion of the importance of these tools to scalability.

2. CASE STUDY: ONLINE AUCTIONS FOR RECYCLABLE PRODUCTS

To illustrate the knowledge discovery tools surveyed in this chapter, we will use a case study for a design of a e-commerce auction system that is currently being developed to support reverse logistics operations. A detailed description of this system can be found in Ryan, Min, and Olafsson (2001). The basic idea of this system is to create an e-hub that brings together a fragmented market of recyclers, demanufacturers, and manufacturers, and allows these enterprises to trade via online auctions. The dynamics of the system are that manufacturers have computers on stock that they want to sell and a demanufacturer can buy those computers and disassemble them into CPUs, memory boards, and plastic shells. The memory boards are sold to a party outside of the system but the CPUs and the shells can be sold via the same auction system as the computers. A recycler can buy shells and melt them into plastic. Finally, the manufacturers are in the market to buy both CPUs and (melted) plastic to make computerized coffeemakers. Apart from these basic functions any participant may act as a broker and buy and sell any item.

This system is driven by large amounts of product and transactional data, and there are several ways in which knowledge discovery can be used to improve its scalability. For example, a recommender system can be used to determine which auctions a user will be interested in, explain why the auction is interesting, and notify potentially interested users. This improves overall efficiency of the system and enables it to effectively deal with higher traffic loads. Secondly, products, and hence auctions, that are complimentary can be identified and grouped together (bundling). Users that display interest in one auction can then be notified of others in the same bundle. Finally, many of the products are essentially similar in the sense that a user that is interested in one product is likely to be interested in several other equivalent products. Thus, clustering together similar products into product groups can reduce the dimensionality of the database.

To gain insights into the performance of this system, experiments were conducted where the behavior of manufacturers, demanufacturers, and recyclers were simulated and data regarding their bidding was gathered. The features for the data gathered for data mining purposes is shown in Table 1. The data has 11 nominal features and 20 numeric or continuous features, in addition to the class feature that indicates if the user participated in the auction. A total of 2592 instances were obtained.

Table 1. Features for the Recycling Data

Nominal Features	Numeric Features
Item	Price for Last *X* Auction*
Item Contains CPU	Quantity for Last *X* Auction*
Item Contains Shell	Time Since Participants Last *X* Auction*
Item Contains Plastic	Price for Participants Last *X* Auction*
Round	Quantity for Participants Last *X* Auction*
Category of Participants Last *X* Auction*	
Participant Name	
Participant Type	

*$X \in$ {Computer, CPU, Shell, Plastic}

Numerous knowledge discovery tools can be applied to support this reverse logistics auction system, and in the next two sections we survey the tools that can be used to address the scalability of such data driven systems. In Section 3 we start by reviewing supporting knowledge discovery technologies that can be used to reduce the dimensionality of the data, such as (a) reducing the number of data points via clustering and (b) dimensionality reduction by using feature selection to eliminating redundant or irrelevant features. In Section 4 we survey tools that can be used to directly improve the operations of the system.

3. THE CURSE OF DIMENSIONALITY

One of the primary concerns when it comes to scalability of data driven system is the dimensionality of the data. By dimensionality we mean the number of variables, or features, that are maintained in the database. To motivate the knowledge discovery techniques, we consider two distinct dimensionality problems that are often encountered in date driven systems.

Example: A Data Warehouse.

A data warehouse is a subject oriented database designed for decision support (Immon, 1996). The purpose of the operational (transactional) databases and the data warehouse is vastly different, so many of the features contained in the transactional databases are irrelevant to the purpose of the warehouse and using all of those will significantly degrade the performance and scalability of the warehouse (Gupta, 1997). The usual practice is to assume that the warehouse designers know which features are relevant and manually choose those to be included in the warehouse. This approach assumes significant knowledge on the part of the designers and does not allow for automatic adapting to changes. □

Example: Online Information Retrieval.

The World Wide Web contains a wealth of information, but its size and lack of structure make utilizing this system challenging. In particular, suppose we are interested in classifying Web documents with the words and possibly functions of words taken as the features. Clearly the number of possible words is enormous and the effectiveness and scalability of an automatic Web document classification system is restricted by this large dimensionality. □

In this section we discuss two knowledge discovery techniques that can be used to address the two dimensionality problems presented above as well as many other high dimensionality challenges. We first discuss the use of feature selection methods for identifying relevant features, and second discuss clustering and how it can be used to reduce the number of features in an unsupervised learning setting.

3.1 Feature Selection

An important problem in knowledge discovery is analyzing the relevance of the features, usually called feature selection. This involves a process for determining which features are relevant in that they predict or explain the data, and conversely which features are redundant or provide little information (Liu and Motoda, 1998). Such feature selection is commonly used as a preliminary step preceding a learning algorithm. This has numerous benefits. By eliminating many of the features it becomes easier to train other learning methods, that is, computational time is reduced. Also, the resulting model may be simpler, which often makes it easier to interpret and thus more useful in practice. It is also often the case that simple models are found to generalize better when applied for prediction. Thus, a model employing fewer features is likely to score higher on many interestingness measures and may even score higher in accuracy. Finally, discovering which features should be kept, that is identifying features that are relevant to the decision making, often provides valuable structural information and is therefore important in its own right.

The literature on feature selection is extensive within the machine learning and knowledge discovery communities. Some of the methods applied for this problem in the past include genetic algorithms (Yang and Honavar, 1998), various sequential search algorithms (see e.g., Aha and Bankert, 1996; Caruna and Freitag, 1994), correlation-based algorithms (Hall, 2000), evolutionary search (Kim, Street, and Menczer, 2000), rough sets theory (Modrzejewski, 1993), randomized search (Skalak, 1994), branch-and-bound (Naranda and Fukunaga, 1977), mathematical programming (Bradley, Mangasarian, and Street, 1998), and the nested partitions method (Olafsson and Yang, 2001). These and other feature selection methods are typically classified as either *filtering methods*, that produce a ranking of all features before the learning algorithm is applied, and *wrapper methods*, that use the learning algorithm to evaluate subsets of features. As a general rule, filtering methods are faster whereas wrapper methods usually produce subsets that result in more accurate models. Some feature selection methods evaluate one feature at a time and either include or eliminate this feature, whereas for other methods an entire subset of features is evaluated together. Wrapper methods always fall into the latter category.

3.1.1 An Information Gain Filter

A simple method for evaluating the value of a feature is the information gain used by the ID3 decision tree induction algorithm (Quinlan, 1986). The basic idea of this approach starts by evaluating the entropy of a data set D relative to classification into some c classes:

$$E(D) = -\sum_{i=1}^{c} p_i \log_2(p_i),$$

where p_i is the proportion of the data set D that belongs to class i. The idea of entropy stems from information theory and it can be interpreted as the number of bits of information it takes to encode the classification of a member of the data set. Now suppose a feature A has been fixed to one of its values $v \in \text{Values}(A)$ resulting in a subset D_v of the data set corresponding to all the instances where A has this value. The entropy $E(D_v)$ of this subset D_v can be calculated according to the same formula. The *information gain* of the feature is the difference between the original entropy and the weighted sum of the entropy when the feature is fixed in turn to each of its possible values:

$$Gain(D, A) = E(D) - \sum_{v \in \text{Values}(A)} \frac{|D_v|}{|D|} \cdot E(D_v).$$

The interpretation of the information gain is that this is how many fewer bits of information are required to encode a classification if the value of the feature A is known. Using the concept of information gain we can assign importance values to each feature. That is, a feature with high information gain is considered more important that a feature with low information gain. Thus, a simple feature selection filter evaluates the information gain of each feature and select those features that have the highest gain. □

3.1.2 Correlation-Based Feature Selection

The entropy-based filter evaluates each feature individually, but there are also many methods that evaluate subset of features instead of a single filter. For example, Hall (2000) proposes using the following measure to evaluate the quality of a subset S of features, where there are a total of n features:

$$Merit(S) = \frac{n\bar{r}_{cf}}{\sqrt{n + n(n-1)\bar{r}_{ff}}}$$

$\bar{r}_{cf}$ = Average feature to class correlation

$\bar{r}_{ff}$ = Average feature to feature intercorrelation

The intuitive idea of this measure is that good features are highly correlated with the class feature, whereas there should be little correlation with other features. We note that this method is a filter method, as the learning algorithm itself is not used to evaluate the quality of the subset. □

The feature selection problem is generally difficult to solve. The number of possible feature subsets is 2^n, where n is the number of features, and evaluating every possible subsets is therefore prohibitively expensive unless n is very small. Furthermore, in general there is no structure present that allows for efficient search through this large space, and a heuristic approach, that sacrifices optimality for efficiency, is typically applied in practice. Thus, most existing methods do not guarantee that the set of selected features is the optimal in any sense.

In the recycling auctions system introduced in Section 2, feature selection can be used to determine which features are important for a user to consider when deciding if to participate in a given auction. In particular, for the recycling auction data introduced in Section 2 above, feature selection can be used to reduce the number of features. For example, the correlation based feature selection with a genetic algorithm search yields the following five features:

a) Time_Of_Participants_Last_Computer_Auction
b) Price_For_Participants_Last_Shell_Auction
c) Time_Of_Participants_Last_Plastic_Auction
d) Category_Of_Participants_Last_Plastic_Auction
e) Participant_Type

These features can then be used by an induction algorithm such as a decision tree or a support vector machine to obtain a model that classifies when an auction is of interest to a specific participant (see Section 4 below).

Example: Data Warehouses.

One of the key issues in designing a data warehouse is to determine which features should be included in the warehouse (Gupta, 1997). In practice this is typically assumed to be the responsibility of the designer, that is, the designer is assumed to know which features are relevant. However, feature selection may be useful to automate this process and result in dimensionality reduction and a more scalable warehouse (Last and Maimon, 2000). □

3.2 Clustering

Another knowledge discovery task that can sometimes be used for dimensionality reduction is clustering of similar data object, where unlike in typical feature selection the primary feature of interest is unknown. In this context a cluster is simply a group of data objects that are similar to each other but dissimilar to other data objects. As an example of dimensionality reduction via clustering, consider the idealized situation illustrated in Figure 1 below. In this figure there are a total of fifteen instances, representing fifteen different computers that have been recycled, but clustering analysis reveals that they can be naturally divided into three groups or clusters depending on their CPU. Thus, instead of working with each product individually they can be processed based on product group, presumably improving the overall scalability of the system.

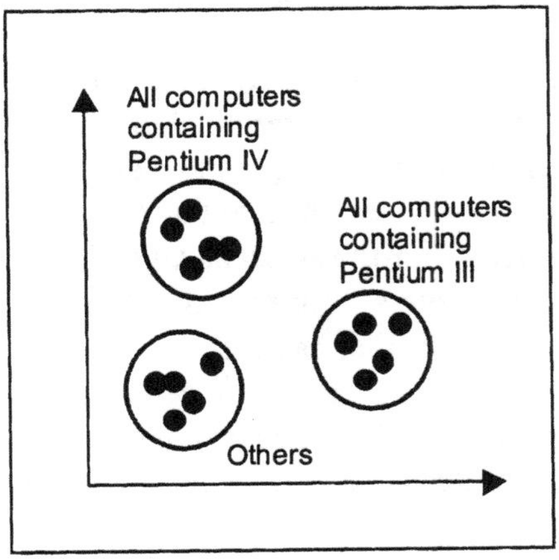

Figure 1. Dimensionality Reduction using Clustering

Clustering algorithms can be roughly divided into two categories: *hierarchical clustering* and *partitional clustering* (Jain, Murty, and Flynn, 1999). In hierarchical clustering all of the instances are organized into a hierarchy that describes the degree of similarity between those instances (e.g., a dendrogram). Such representation may provide a great deal of information, but the scalability of this approach is questionable as the number of instances grows. Partitional clustering, on the other hand, simply creates one partition of the data where each instance falls into one cluster. Thus, less information is obtained but the ability to deal with a large number of instances is improved. There are, however, many other characteristics of clustering algorithms that must be considered to ensure scalability of the approach. For example, most clustering algorithms are *polythetic*, meaning that all features are considered simultaneously in tasks such as to determine the similarity of two instances. However, as the number of features becomes large this may pose scalability problems and it may be necessary to restrict attention to *monothetic* clustering algorithms that consider features one at a time. Most clustering algorithm are also non-incremental in the sense that all of the instances are considered simultaneously. However, there are a few algorithms that are *incremental,* which implies that they consider each instance separately. Such algorithms are particularly useful when the number of instances is large and keeping the entire set of instances in memory poses scalability problems.

To illustrate these concepts let's consider two specific clustering algorithms that have been widely used: the *k*-means algorithm and the CobWeb algorithm.

3.2.1 Clustering Using the *k*-Means Algorithm

The classic *k*-means algorithm (MacQueen, 1967) remains the basis for some of the most commonly used clustering algorithms. This is a partitional algorithm; that is, each instance is assigned to a single cluster. Furthermore, it is both polythetic and non-incremental so that all features and instances are considered simultaneously. For the *k*-means algorithms we must specify the number of cluster, for example a 2-means algorithm assumes that there will be two clusters. The algorithm then proceeds as follows:

- Randomly select k instances as a starting centroid.
- Assign each instance to the closest centroid, resulting in k clusters.
- Based on these assignments recalculate the centroid of each cluster.
- Repeat.

Thus, the algorithm moves the centroids in an iterative manner until some stopping criterion is met, which is usually that no or little further change in the centroids occurs. The *k*-means algorithm is simple but fairly effective and scalable and has been applied to large practical applications.

3.2.2 Clustering Using the CobWeb Algorithm

The CobWeb algorithm is an incremental hierarchical clustering algorithm proposed for nominal data (Fisher, 1987). The basic idea is to determine the best location of each instance incrementally based on the category utility measure:

$$CU(C_1, C_2, \ldots, C_k) = \frac{\sum_l \Pr[C_l] \sum_i \sum_j \left(\Pr[a_i = v_{ij} \mid C_l]^2 - \Pr[a_i = v_{ij}]^2 \right)}{k}$$

Here k is the number of clusters that are denoted C_1, C_2,…, C_k. There is some apriori probability $\Pr[a_i = v_{ij}]$ of a feature a_i taken a give value v_{ij} and a conditional probability $\Pr[a_i = v_{ij} \mid C_l]$ given that the cluster of this instance in known. For the clustering C_1, C_2,…, C_k to be useful it should reveal something about the values of the features, so the conditional probability should be higher for valuable clustering. Thus,

$$\sum_i \sum_j \Pr[a_i = v_{ij} \mid C_l]^2 - \Pr[a_i = v_{ij}]^2$$

measures the value of each cluster and its weighted average measures the overall value of the clustering. To avoid overfitting, that is, favoring clusters that have very few instances, the category utility measure divides by the total number of clusters. The CLASSIT algorithm extents this approach for numerical data (Gennari et al., 1990).

The *k*-means and CobWeb are two clustering algorithms that have been found to be quite effective in practice. Many other algorithms exist but there are still many challenges that call for further development of clustering algorithms. A key challenge is the scalability of the clustering algorithm with respect to number of instances or data objects. In this respect, incremental algorithms such as CobWeb are the most scalable, but their performance may suffer as a result and in many cases such incremental algorithms are sensitive to the order in which instances are considered. Another way of dealing with large number of instances is to use random sampling to select a subset of instances and apply the clustering algorithm to this subset. This is the basic idea of the CLARA algorithm that combines random sampling and what is called *k*-median, which is similar to *k*-means but uses the median instead of the mean (Kaufman and Rousseeuw, 1990). CLARANS is a further extension that uses dynamic random sampling and has been found empirically to be quite scalable (Ng and Han, 1994). High dimensionality also often poses difficulties for clustering and here monothetic algorithms are the most scalable. Such algorithms rarely produce high quality cluster, however. Other issues that are subject to active research include better ability to handle mixed numerical and nominal data, incorporating domain knowledge into clustering, and the interpretability and validation of cluster.

Example: Online Information Retrieval

Information Retrieval (IR) from static databases is a well established field, but new challenges arise when applying IR to a dynamic environment such as the Internet. In particular, the number of simultaneous users and number of documents far exceed that of most traditional databases. Thus, examination of every document may not be feasible and a reasonable approach is to cluster similar types of documents that can be treated as a single unit (Rasmussen, 1992; Kobayashi and Takeda, 2000). For this type of clustering similarity measures can for example be (a) terms contained in the documents, (b) similar citations, and (c) the context in which the documents occur. □

4. EXPEDITING THE SYSTEM OPERATIONS

Dimensionality reduction is a critical issue for improving scalability but there are other ways in this knowledge discovery can be applied to more directly improve the efficiency of a data driven system. In particular, effective and scalable decision support can reduce the amount of time required by each user and limit the number of queries and requests. In this section we focus on inductive learning methods and their application to making recommendations to users, which presumably will speed up their decision process and improve scalability.

4.1 Decision Trees

Classification is the process of building a model that describes a priori specified classes or concepts based on data objects with known classifications, and then using this model to classify new data objects where the class is unknown. The primary performance measure of interest is the accuracy of the classification model, but induction speed, robustness with respect to noisy or incomplete data, scalability, and interpretability are also critical issues. In our context we are primarily interested in classifying information as either relevant or irrelevant so that only relevant information is presented to a particular user, and thus improving the scalability of the system.

Various induction algorithms have been proposed for classification, but the most common approach is top-down induction of decision trees. Here a node in the tree represents a test for a particular feature, each branch represents the outcome of this test, and the leaf nodes are the classes for the class features. A typical decision tree algorithm works in a greedy top-down manner by determining the value of, or a range for, one feature at a time. Thus, one feature is selected and the entire training data set is partitioned or split according to the value of this feature.

As an example, considering the auction data where there are 31 features that can be used to induce a decision tree and one class feature that take two values: participate, don't participate. At the top of the tree one feature, say "Item", is selected and the data split according to this feature. If the feature is numeric, say "Time Since Participants Last Computer Auction", then the node splits according to range of values (see Figure 2). Following this scheme, the features are selected one at a time until a stopping criterion is satisfied.

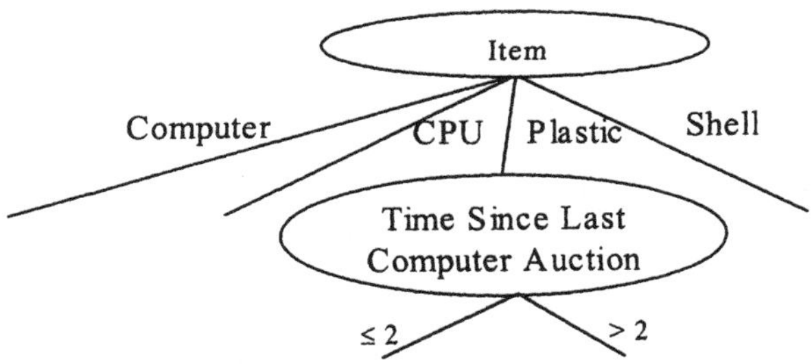

Figure 2. Part of a Simple Decision Tree

One of the main issues is how to determine the order in which features are selected and a measure of the quality of features is thus required. A commonly used measure is the information gain (see Section 3.1.1), which is used by the ID3 algorithm (Quinlan, 1987), and the related gain ratio used by the C4.5 algorithm (Quinlan, 1993).

As mentioned above, one way in which decision trees can be useful in improving performance of data driven system is in classifying information as either interesting or not interesting to a potential user. If a user accesses the system he or she may need to work through large amounts of data to evaluate its relevance and may in the process require multiple complex and time consuming queries. If the information can be organized such that each user is only presented with a limited view of the data that is determined to be relevant, the decision making process can be greatly facilitated and this may imply a more scalable system that can handle larger volumes of traffic.

An example of a decision support system that improves the view and type of information experienced by a user are the recommender systems introduced in Section 1 above. One way of constructing such a recommender is to use decision trees (Basu et al., 1998). Say for example that we want to construct a decision tree to recommend if an auction should be recommended to a manufacturer in the recycling auction. Instead of using all 32 features, which would result in a fairly complicated tree, we use NP-wrapper feature selection method (Olafsson and Yang, 2001) to reduce the number of features and then apply the C4.5 decision tree induction algorithm (Quinlan, 1993) to induce the tree. The resulting tree is shown in Figure 3.

From the tree we can read the following decision rules for a specific manufacturer that we can label as, say, M:

- **IF** item = computer **THEN** recommend = yes.
- **IF** item = CPU **THEN** recommend = yes.

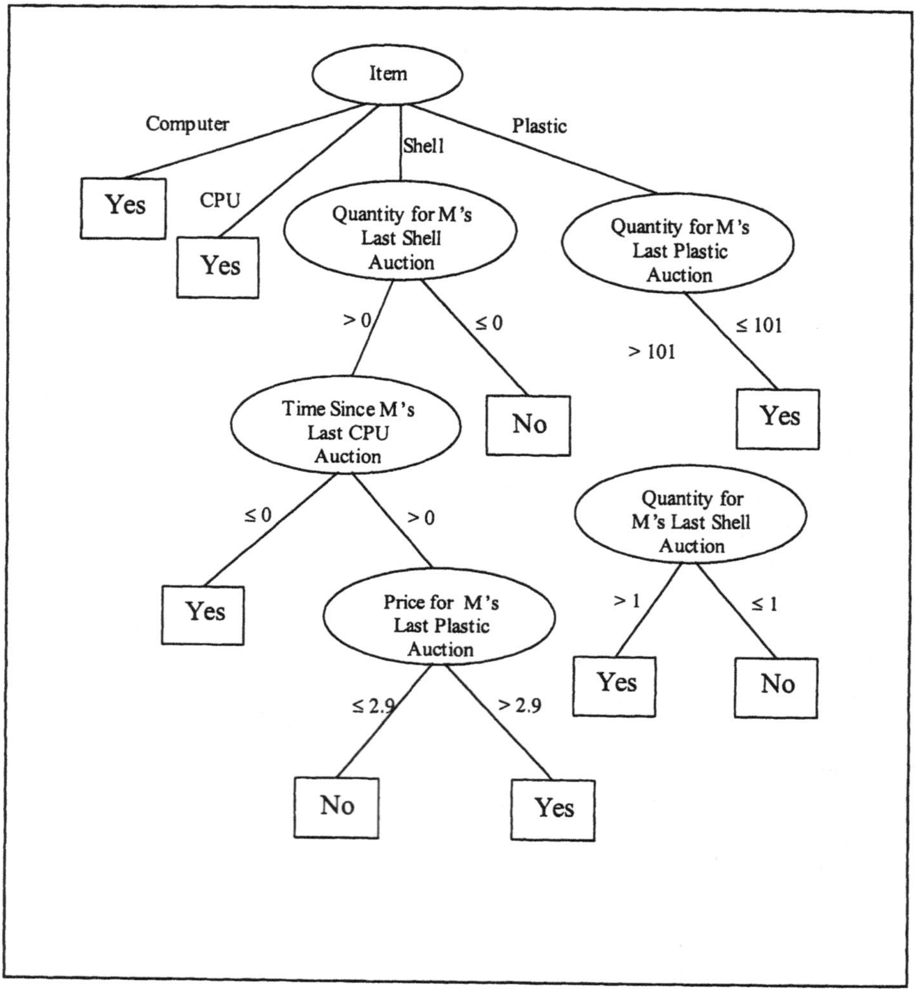

Figure 3. Decision Tree for Auction Recommendation

- **IF** item = shell **AND** quantity for M's last shell transaction > 1 **AND** time since M's last CPU auction ≤ 0 **THEN** recommend = yes.
- **IF** item = shell **AND** quantity for M's last shell transaction > 1 **AND** time since M's last CPU auction > 0 **AND** price for M's last plastic auction > 2.9 **THEN** recommend = yes.

- **IF** item = plastic **AND** quantity for M's last plastic auction ≤ 101 **THEN** recommend = yes.
- **IF** item = plastic **AND** quantity for M's last plastic auction > 101 **AND** quantity for M's last shell auction >1 **THEN** recommend = yes.
- **OTHERWISE** recommend = no.

Let us consider how these rules can be interpreted. As a primary function, a manufacturer sells computers and buys CPUs and plastic to make coffeemakers. Thus, at first glance one might expect a manufacturer to be always interested in computer, CPU, and plastic auction, and never be interested in shell auctions. The model reveals a somewhat more complicated situation that is still quite intuitive. As expected, auctions involving computers and CPUs should always be recommended but plastic auctions are only recommended some of the time and even shell auctions are sometimes recommended. So why are plastic auction not always recommended? Intuitively, a reason might be that the manufacturer has sufficient supply of plastic and does not need to purchase any further supplies. Thus, in the decision tree, if the manufacturer did not participate in the last plastic auction, the next plastic auction is not recommended. On the other hand, manufacturer does not have any primary use for shells so why should shell auctions be recommended? Intuitively, such auctions should only be recommended if the manufacturer wants to acts as a broker that is to buy and sell shells. The primary indicator of this in the decision tree is if this particular manufacturer participated before in shell auctions, that is, M's quantity for the last shell auction is bigger than zero.

Beyond decision trees and decision rules there is a large number of other methods that can be used for classification. For example, statistical methods such as naïve Bayes (Duda and Hart, 1973) and Bayesian networks (Heckerman, 1996) are commonly used in practice. Another frequently used approach is neural networks (Ripley, 1996). Reviewing all classification methods is beyond the scope of a short survey paper and we will limit ourselves to describing one additional method, which is called support vector machines.

4.2 Support Vector Machines

Support vector machines (SVM) trace their origins to the seminal work of Vapnik and Lerner (1963) and Mangasarian (1964), but have only recently received the attention of much of the knowledge discovery and machine learning communities. Detailed expositions of SVM can be found in the book by Vapnik (1995) and in the survey papers of Burges (1999) and Bennett and Campell (2000); and in this section we only briefly introduce some of the key ideas.

Consider Figure 4 that illustrates the concept of support vector machines for the problem of determining if an auction should be recommended. Each instance in the training set is labeled as either interesting and should hence be recommended (filled circles), or not interesting (triangles).

The problem here is to determine a best model for separating interesting and non-interesting auctions. If the data can be separated by a hyperplane H as in Figure 4, the problem can be solved fairly easily. To formulate it mathematically, let us assume that the class feature y_i takes two values -1 (not interesting) or $+1$ (interesting). We assume that all features other than the class feature are real valued and denote the training data, consisting of l data points, as $\{(\mathbf{x}_j, y_j)\}$, where $j = 1,2,\ldots,l$, $y_j \in \{-1,+1\}$ and $\mathbf{x}_j \in \mathbf{R}^d$. The hyperplane H can be defined in terms of its unit normal $\mathbf{w}$ (see Figure 4) and its distance b from the origin. In other words, $H = \{\, \mathbf{x} \in \mathbf{R}^d : \mathbf{x} \cdot \mathbf{w} + b = 0\}$, where $\mathbf{x} \cdot \mathbf{w}$ is the dot product between those two vectors.

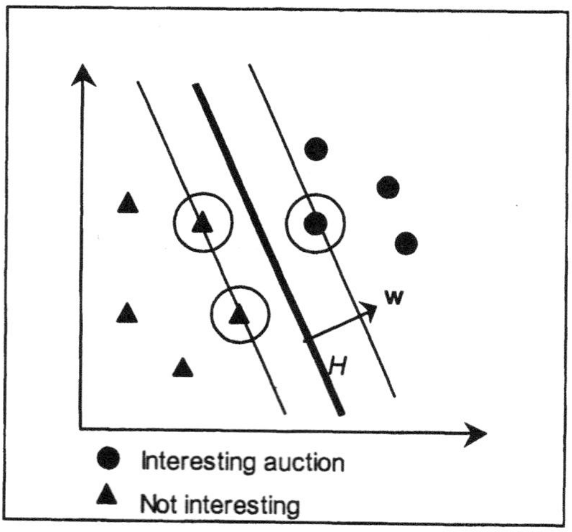

Figure 4. Support Vector Machine for Classifying Auctions

If a separating hyperplane exists then there are in general many such planes and we define the optimal separating hyperplane as the one that maximizes the sum of the distances from the plane to the closest positive example (recommend) and the closest negative example (not recommend). Intuitively we can imagine that two hyperplanes, parallel to the original plane and thus having the same normal, are pushed in either direction until the convex hull of the sets of all instances with each classification is encountered. This will occur at certain instances, or vectors, that are hence called the support vectors. Those support vectors are identified with circles in Figure 4 for the simple auction classification example. This intuitive procedure is captured mathematically by requiring the following constraints to be satisfied:

$$\mathbf{x}_i \cdot \mathbf{w} + b \geq +1, \forall i : y_i = +1$$
$$\mathbf{x}_i \cdot \mathbf{w} + b \leq -1, \forall i : y_i = -1.$$

With this formulation the distance between the two planes, called the margin, is readily seen to be 2/τwτ and the optimal plane can thus be found by solving the following mathematical optimization problem:

$$\begin{array}{ll} \max\limits_{\mathbf{w},\mathbf{b}} & \|\mathbf{w}\|^2 \\ \text{subject to} & \mathbf{x}_i \cdot \mathbf{w} + b \geq +1, \forall i : y_i = +1 \\ & \mathbf{x}_i \cdot \mathbf{w} + b \leq -1, \forall i : y_i = -1. \end{array}$$

When the data is non-separable this problem will have no feasible solution and the constraints for the existence of the two hyperplanes must be relaxed. One way of accomplishing this is by introducing error variables ε_j, for each data point $\mathbf{x}_j$, $j = 1,2,\ldots,l$. Essentially these variables measure the violation of each data point and using these variables the following modified constraints are obtained:

$$\mathbf{x}_i \cdot \mathbf{w} + b \geq +1 - \varepsilon_i, \forall i : y_i = +1$$
$$\mathbf{x}_i \cdot \mathbf{w} + b \leq -1 + \varepsilon_i, \forall i : y_i = -1$$
$$\varepsilon_i \geq 0, \forall i.$$

As the variables ε_j represent the training error, the objective could be taken to minimize τwτ^2/2 + $C\cdot$ ($\sum_j \varepsilon_j$), where C is constant that measures how much penalty is given. However, instead of formulating this problem directly it turns out to be convenient to formulate what is called the Wolfe dual of the problem (Fletcher, 1987).

$$\begin{array}{ll} \max\limits_{\alpha} & \sum\limits_i \alpha_i - \frac{1}{2}\sum\limits_{i,j} \alpha_i \alpha_j y_i y_j \mathbf{x}_i \cdot \mathbf{x}_j \\ \text{subject to} & 0 \leq \alpha_i \leq C \\ & \sum\limits_i \alpha_i y_i = 0. \end{array}$$

The solution to this problem are the dual variables α, and to obtain the primal solution, that is, the model classifying instances defined by the normal of the hyperplane, we calculate

$$\mathbf{w} = \sum_{i:\mathbf{x}_i \text{ support vector}} \alpha_i y_i \mathbf{x}_i .$$

The benefit of using the Wolfe dual is that the constraints are much simpler and easier to handle, and that the training data only enters through the dot product $\mathbf{x}_i \cdot \mathbf{x}_j$. This latter point is important when we consider extending this approach to non-linear model.

Thus far we have only considered linearly models. However, the above SVM approach can be extended to non-linear models in a very straightforward manner using kernel functions $K(\mathbf{x},\mathbf{y})=\Phi(\mathbf{x})\cdot\Phi(\mathbf{y})$, where $\Phi:\mathbf{R}^d \rightarrow \mathrm{H}$ is a mapping from the d-dimensional Euclidean space to some Hilbert space H. This approach was introduced to the SVM literature by Cortes and Vapnik (1995) and it works because the data $\mathbf{x}_j$ only enters the Wolfe dual via the dot product $\mathbf{x}_i \cdot \mathbf{x}_j$, which can thus be replaced with $K(\mathbf{x}_i,\mathbf{x}_j)$. The choice of kernel determines the model. For example, to fit a p degree polynomial the kernel can be chosen as $K(\mathbf{x},\mathbf{y}) = (\mathbf{x}\cdot\mathbf{y} + 1)^p$. Many other choices have been considered in the literature but we will not explore this approach further.

4.3 Association Rule Discovery

A classification model using either a decision tree or a support vector machine model can be applied for recommendation or other information selection if sufficient data exists regarding each user's prior history so that a useful model can be induced based on this data. For a new or an infrequent user such data is not likely to exist, however, so a recommendation cannot be made solely on this basis. In such situations association rule mining may be useful to induce rules such as the following: If a user is interested in Auction A then the user is also interested in Auction B with 90% confidence (Lin et al., 2000). Alternatively, associations may be induced between users, for example that Manufacturer M5 is interested in the same auctions as Manufacturer M1 with 90% confidence.

This type of induction is the domain of association rule mining that aims to discover interesting correlation or other relationships in large databases where it is unknown which variables or features will be included in the relation. Association rule mining algorithms typically seek relationship that have high support (that is, are based on large number of data instances) and high confidence (that is, are highly consistent with the historical data). For example, for a rule $A \Rightarrow B$ the support is the probability that both A and B occur, while the confidence is the conditional probability that B occurs given that A occurs:

$$\text{support}(A \Rightarrow B) = \Pr(A \cup B),$$
$$\text{confidence}(A \Rightarrow B) = \Pr(B \mid A).$$

Typically the user specifies a minimum support and/or a confidence level. In addition to requiring a minimum support and confidence, it is also necessary to assess if a discovered rule is interesting, that is, if anything new has been learned. Thus, a major issue is determining the interest value of the rules.

Another key issue is the selection of algorithms. The *Apriori algorithm* (Agrawal and Srikant, 1994; Mannila, Toivonen and Verkamo, 1994) is the most used association rule discovery algorithm. This algorithm starts by generating what is called frequent item sets. An item is an attribute value pair and a *k*-item set is a set of *k* such pairs. A frequent item set is a set that occurs at least as often as the required minimum support. The Apriori algorithm starts by generating all frequent item sets and then uses these item sets to generate association rules that also have the required confidence. Most other association rule discovery algorithms follow a similar process, but in mining large-scale data, the scalability of the algorithm will be critical. Many authors have therefore addressed improving the efficiency of the Apriori algorithm (Agrawal, Aggerwal, and Prasad, 2000) and other more efficient algorithms such as CLOSET have been suggested (Pai, Han, and Mao, 2000). As generating item sets dominates the computational time of most association rule discovery algorithms this task has received considerable attention (Han, Pei, and Yin, 2000).

As an illustration, consider the application of the Apriori algorithm to the nominal features of the auction data set. The association rules discovered for this application are all straightforward or trivial and do not seem to add much to our understanding of the data. However, they do give the flavor for what type of knowledge might be discovered from a more complex set of data. Some examples of the rules discovered are:

1. **IF** Category_Of_Participants_Last_Computer_Auction = sell **AND** Participant_Type = Manufacturer **THEN** Category_Of_Participants_Last_Plastic_Auction = buy (1.00)
2. **IF** Round = 1 **THEN** Category_Of_Participants_Last_Cpu_Auction = none **AND** Category_Of_Participants_Last_Shell_Auction = none **AND** Category_Of_Participants_Last_Plastic_Auction = none (1.00)

3. **IF** Category_Of_Participants_Last_Computer_ Auction = sell **AND** Category_Of _Participants_Last_ Plastic_Auction = buy **THEN** Participant_Type = Manufacturer (0.93)

The confidence of each rule is given in brackets after the rule. The first two rules have confidence of 100% whereas the last has confidence of 93%.

5. IMPROVING SCALABILITY OF AN ONLINE AUCTION SYSTEM

The knowledge discovery techniques reviewed in this chapter can be used to improve the scalability of the online auction system described in Section 2 above. A direct way in which scalability can be improved is for example inducing a decision tree model based on past transactions to determine if an auction is relevant to a particular user. To evaluate the potential benefits of doing this we induced a C4.5 decision tree (see Section 4.1 for discussion). The data set contains 2592 instances, each labelled based on if the particular user participated in the auction or not. Of those, the user participated in 1493 auctions or 58% of all auctions. Without a recommender, every user must therefore consider every auction but will only participate in an average of 58% of those auctions. With the decision tree based recommender, 1612 auctions or 62% of the auctions are recommended. Thus, if a user only considers those auctions that are recommended, the number of auction considered reduces from 2592 to 1612 or by 38%. This should significantly reduce the time needed by each user.

Another way in which the decision making process of each user can be facilitated is by displaying all relevant information and only the relevant information. For example, the auction data contains 32 feature all of which are potentially relevant, but displaying all these features may be to complex and it is unlikely that all are relevant to all decisions. Again, knowledge discovery tools can help by selection the most important features. Recall from Section 3.1 that the following features are found using one feature selection tool to be the most important:

a) Time_Of_Participants_Last_Computer_Auction
b) Price_For_Participants_Last_Shell_Auction
c) Time_Of_Participants_Last_Plastic_Auction
d) Category_Of_Participants_Last_Plastic_Auction
e) Participant_Type

Of course each participant will be aware of their own type, but this indicates that it is important to prominently display the time since the last computer and plastic auctions, as well as the clearance price of the last shell auction in which the user participated. With this information a user may be able to quickly determine his or her interest in a given auction.

Other techniques that could be used to improve performance of this system include bundling of auctions using association rule discovery (see Section 4.3) and clustering of similar products, e.g., computers that are essentially the same, to reduce dimensionality (see Section 3.2). Such techniques may greatly improve scalability. For example, if all essentially different types of computers have already exist in the system then adding further products will not effect performance as they will simply be added to one of the existing groups.

6. CONCLUSIONS

In this chapter we have surveyed numerous knowledge discovery problems and techniques. These tools can be used to improve the scalability of data driven systems such as most e-commerce systems. For example, many such systems suffer from high dimensionality that degrades the overall system performance and limits it capacity for expanding. To allow such systems to scale up, automatic feature selection can be applied to reduce the number of dimensions by eliminating redundant, irrelevant, or otherwise unnecessary features. Another way to reduce the dimensionality of the data is to use automatic clustering to group together similar features that can then be treated as one. Many clustering algorithms exist and we considered a few with a special attention to scalable algorithms.

Another way in which data mining can be utilized is as decision support that simplifies the views of the users, allowing them to obtain the necessary information quickly, in a structured manner, and with less time consuming queries and browsing. To this end models can be induced that allow the system to predict what information is relevant to a specific user. For example, a model can be used to identify which auctions are relevant in an online auction system. In some cases, in particular when sufficient data exists regarding a user's past history, classification models may be appropriate and there are many available algorithms to induce such models. In this chapter we considered two such induction algorithms: decision trees and support vector machines. If less is known about users then association rule discovery may be more appropriate.

In summary, the potential applications of data mining to improving scalability of e-commerce system are plentiful. To date, only a few such applications have been explored in detail and there are many open issues that remain in how these tools are best utilized.

REFERENCES

Agrawal, R. and Srikant, R. (1994). Fast algorithms for mining association rules. In *Proceedings of the 1994 International Conference on Very Large Data Bases* (VLDB'94), 487-499.

Agrawal, R., Aggerwal, C. and Prasad, V.V.V. (2000). A tree projection algorithm for generation of frequent itemsets. *Journal of Parallel and Distributed Computing.*

Aha, D.W., and Bankert, R. L. (1996). A comparative evaluation of sequential feature selection algorithms. In D. Fisher & J.-H. Lenz (Eds.), *Artificial Intelligence and Statistics* V. New York: Springer-Verlag.

Ansari, S., R. Kohavi, L. Mason, and Z. Zheng (2000). Integrating e-commerce and data mining: architecture and challenges. In *Proceedings of ACM WEBKDD 2000.*

Basu, C., Hirch, H., and Cohen, W. (1998). Recommendation as classification: using social and content based information for recommendation. In *Proceedings of the National Conference on Artificial Intelligence.*

Bennett, K.P. and C. Campbell (2000). Support vector machines: hype or hallelujah? *SIGKDD Explorations* **2**(2), 1-13.

Bradley, P.S., Mangasarian, O.L., and Street, W.N. (1998). Feature selection via mathematical programming, *INFORMS Journal on Computing*, **10**(2), 209-217.

Breese, J., Heckerman, D., and Kadie, C. (1998). Empirical analysis of predictive algorithms for collaborative filtering. In *Proceedings of the 14th Conference on Uncertainty in Artificial Intelligence.*

Breiman, L., Friedman, J., Olshen, R. and Stone, C. (1984). *Classification and Regression Trees.* Wadsworth International Group, Monterey, CA.

Burges, C.J.C. (1998). A Tutorial on Support Vector Machines for Pattern Recognition. *Knowledge Discovery and Data Mining*, **2**(2).

Caruana, R., and Freitag, D. (1994). Greedy feature selection. In *Proceedings of the Eleventh International Conference on Machine Learning*, pp. 28-36. New Brunswick, NJ: Morgan Kaufmann.

Cortes, C. and V. Vapnik (1995). Support vector networks. *Machine Learning* **20**, 273-297.

Duda, R. and P. Hart (1973). *Pattern Classification and Scene Analysis.* John Wiley & Sons, New York.

Fisher, D. (1987). Improving inference through conceptual clustering. In *Proceedings of the 1987 AAAI Conference*, 461-465, Seattle, WA.

Fletcher, R. (1987). *Practical Methods of Optimization.* John Wiley and Sons, New York.

Gennari, J.H., P. Langley, and D. Fisher (1990). Models of incremental concept formation. *Artificial Intelligence* **40**, 11-61.

Good, N., Schafer, J.B., Konstan, J.A., Borchers, A., and Sarwar, B. (1999). Combining collaborative filtering with personal agents for better recommendations. In *Proceedings of the National Conference on Artificial Intelligence.*

Gupta, V.R. (1997). *An Introduction to Data Warehousing*, System Services Corporation, Chicago, IL.

Hall, M.A. (2000). Correlation-based feature selection for discrete and numeric class machine learning. *Proceedings of the Seventeenth International Conference on Machine Learning*, Stanford University, CA. Morgan Kaufmann Publishers.

Han, J. and Kamber, M. (2001). *Data Mining: concepts and techniques.* Morgan Kaufmann, San Francisco, CA.

Han, J., Pei, J. and Yin, Y. (2000). Mining frequent patterns without candidate generation. In *Proceedings of 2000 ACM International Conference on Management of Data (SIGMOD'00)*, 1-12.

Heckerman, D. (1996). Bayesian networks for knowledge discovery. In *Advances in Knowledge Discovery and Data Mining*, 273-305, MIT Press, Cambridge, MA.

Hipp, J., Güntzer, U. and Nakhaeizadeh, G. (2000). Algorithms for association rule mining - a general survey and comparison. *SIGKDD Explorations* **2**, 58-64.

Immon, W.H. (1996). *Building the Data Warehouse*. John Wiley & Sons, New York.

Kaufman, L. and P.J. Rousseeuw (1990). *Finding Groups in Data: and introduction to cluster analysis*. John Wiley & Sons, New York.

Kim, Y.S., Street, W.N, and Menczer, F. (2000). Feature selection in unsupervised learning via evolutionary search. In *Proceedings of the 6th ACM SIGKDD International Conference on Knowledge Discovery and Data Mining*.

Jain, A.K., Murty, M.N. and Flynn, P.J. (1999). Data clustering: a review. *ACM Computing Surveys* **31**, 264 - 323.

Kobayashi, M. and Takeda, K. (2000). *Information retrieval on the web: selected topics*.

Kosala, R. and H. Blockeel (2000). Web mining research: a survey. *SIGKDD Explorations* **2**, 1-15

Last, M. and Maimon, O. (2000). Automated dimensionality reduction of data warehouses. In Jeusfeld, Shu, Staudt, and Vossen (eds), *Proceedings of the International Workshops on Design and Management of Data Warehouses*.

Lauritzen, S.L. (1995). The EM algorithm for graphical association models with missing data. *Computational Statistics and Data Analysis* **19**, 191-201.

Lin, W., S.A. Alvarez, and C. Ruiz. (2000). Collaborative recommendation via adaptive association rule mining. In *Proceedings of ACM WEBKDD 2000*.

Liu, H. and Motoda, H. (1998). *Feature Selection for Knowledge Discovery and Data Mining*. Kluwer, Boston.

MacQueen, J. (1967). Some methods for classification and analysis of multivariate observations. In *Proceedings of the 5th Berkeley Symposium on Mathematical Statistics and Probability*, 281-297.

Mannila, H., Toivonen, H. and Verkamo, A.I. (1994). Efficient algorithms for discovering association rules. *In Proceedings of AAAI'94 Workshop on Knowledge Discovery in Databases (KDD'94)*, 181-192.

Modrzejewski, M. (1993). Feature selection using rough sets theory. In P.B. Brazdil, editor, *Proceedings of the European Conference on Machine Learning*, pp. 213-226.

Mangasarian, O.L. (1965). Linear and nonlinear separation of patterns by linear programming. *Operations Research*, **13**, 444-452.

Narendra, P.M., and Fukunaga, K. (1977). A branch and bound algorithm for feature subset selection. *IEEE Transactions on Computers*, **26**(9), 917-922.

Ng, R. and J. Han (1994). Efficient and effective clustering method for spatial data mining. In *Proceedings for the 1994 International Conference on Very Large Data Bases*, 144-155.

Olafsson, S. and J. Yang (2001). Intelligent partitioning for feature relevance analysis. *Working Paper*, Industrial Engineering Department, Iowa State University, Ames, IA.

Pai, J., Han, J. and Mao, R. (2000). CLOSET: An efficient algorithm for mining frequent closed itemsets. In *Proceedings of 2000 ACM-SIGMOD International Workshop on Data Mining and Knowledge Discovery (DMKD00)*, 11-20.

Pine, B.J. and Gilmore, J.H. (1999). *The Experience Economy*. Harvard Business School Press, Boston, MA.

Quinlan, J.R. (1987). Induction of decision trees. *Machine Learning* **1**(1), 81-106.

Quinlan, J.R. (1993). *C4.5: Programs for Machine Learning*. Morgan-Kaufmann, San Mateo, CA.

Rasmussen, E. (1992). Clustering algorithms. In Frakes, W., Baeza-Yates, R (eds.), *Information Retrieval*, 419-442. Prentice-Hall, Englewood Cliffs, NJ.

Resnick, P. Iacovou, N. Suckak, M., Bergstrom, P, and Riedl, J. (1994). Grouplens: an open architecture for collaborative filtering of netnews. In *Proceedings of ACM CSCW'94 Conference on Computer-Supported Cooperative Work*, 175-186.

Ripley, B.D. (1996). *Pattern Recognition and Neural Networks*. Cambridge University Press, Cambridge, UK.

Ryan, S.M., K.J. Min, and S. Olafsson (2001). Experimental study of scalability enhancements for reverse logistics e-commerce. In Prabhu, Kumara and Kamath (eds.) *Scalable Enterprise Systems - An Introduction to Recent Advances* (submitted).

Schafer, J.B., Konstan, J., and Riedl, J. (2001). Electronic commerce recommender applications. *Journal of Data Mining and Knowledge Discovery*, **5** (1/2), 115-152.

Setiono, R., and Liu, H. (1997). Neural network feature selector. *IEEE Transactions on Neural Networks*, **8**(3), 654-662.

Shardanan, U. and Maes, P. (1995). Social information filtering: algorithms for automating 'word of mouth'. In *Proceedings of ACM CHI'95 Conference on Human Factors in Computing Systems*, 210-17.

Skalak, D. (1994). Prototype and feature selection by sampling and random mutation hill climbing algorithms. *Proceedings of the Eleventh International Machine Learning Conference*, pp. 293-301, New Brunswick, NJ: Morgan Kauffmann.

Vapnik, V. (1995). *The Nature of Statistical Learning Theory*. Springer, N.Y.

Vapnik, V and A. Lerner (1963). Pattern recognition using generalized portrait method. *Automation and Remote Control*, **24**.

Vucetic, S. and Z. Obradovic. (2000). A regression-based approach for scaling-up personalized recommender systems in e-commerce. In *Proceedings of ACM WEBKDD 2000*.

Witten, I.H. and Frank, E. (2000). *Data Mining: practical machine learning tools and techniques with Java implementations*. Morgan Kaufmann, San Francisco, CA.

Yang, J., and Honavar, V. (1998). Feature subset selection using a genetic algorithm. In H. Motada and H. Liu, editors, *Feature Selection, Construction, and Subset Selection: A Data Mining Perspective*, Kluwer, New York.

Chapter 7

A SCALABLE SUPPLY CHAIN INFRASTRUCTURE RESEARCH TEST-BED

Z.-Y. Zhao, M. Ball and C.-Y Chen
Robert H Smith School of Business
University of Maryland
College Park, MD 20742

Abstract: Two classes of systems, enterprise resource planning (ERP) systems and supply chain management (SCM) systems, are now widely used to support the coordination of activities across major organizational components and supply chain partners. We collectively refer to such systems, as well as closely related systems and tools, as supply chain infrastructures (SCIs). In this chapter, we provide a review of various SCI architectures, describe in detail a scalable SCI research test-bed that we developed and define an integration architecture and model. In our test-bed, we use middleware and a collaboration-based architecture to achieve "application scalability". In addition, we illustrate, by embedding an advanced available to promise (ATP) model and a simulation model within our test-bed, how to achieve "business scalability". This chapter also describes these two classes of decision models in detail and summarizes on-going research projects based on our test-bed.

Keywords: Supply chain infrastructure test-bed, application scalability, business scalability, available to promise (ATP), SCOR-based simulation

1. BACKGROUND

1.1 Introduction

Supply chain management is the planning, organizing, and controlling of activities inside and outside an organization that enable the value chain to make products and provide services to the customer (APICS Dictionary, 1998). Recently, two classes of software systems, enterprise resource planning (ERP) systems and supply chain management (SCM) systems, have been broadly adopted for managing supply chains. The "traditional" overall objective of ERP systems is to provide an integrated transaction processing fabric inside an organization. This enhances organizational performance by providing broader information availability, reducing information inconsistency and improving transaction-processing efficiency. The implementation of an ERP requires that many business processes be organized, documented and, to a degree, standardized. In many cases, benefits also accrue directly from these activities. SCM systems, on the other hand, are aimed at providing a higher level of decision support related to activities that involve the coordination and control of inter-organizational production and distribution processes. With the shifting focus of supply chain managers from a plant, a warehouse or a company to multiple companies and their interactions, standalone ERP and SCM systems are incapable of supporting today's business dynamics. ERP and SCM systems are generally maturing their capabilities and their features have begun to overlap. A class of reliable, cost-effective and synchronized ERP and SCM systems becomes vital for today's business in order to better support complicated internal players and collaborate with trading partners along the total supply chain. We collectively refer to such systems, as well as closely related systems and tools, as supply chain infrastructures (SCIs). In fact, the SCIs have been one of the fastest growing trends in business today. Some studies have indicated that as much as 50% to 60% of the potential benefits from effective supply chain management may not be achieved without this integrated infrastructure (Miller and Berger, 2001).

With the emergence of the global economy, the rapid growth of information systems technology and the increasing use of the IT as an essential part of supply chain management, global corporations are increasingly changing the way they do business. As businesses place more focus on providing the highest levels of customer service, product flexibility (configuration-to-order) and volume flexibility, the SCI must be able to effectively handle high-mix-low-volume products characterized by numerous features, options, sizes and colors (Markland 1997). Meanwhile, the SCI should also be capable of easily adapting business processes in response to varying customer demands whether there is a single order with a simple configuration or thousands of orders with complicated configurations. We refer to this kind of capability as *business scalability*, which evaluates how quickly the SCI business processes such as order fulfillment, customer service, delivery mechanism, etc. can be scaled to accommodate large and variable inflows of customer orders with potentially complicated product configurations. Another fundamental change companies face today is the "deconstruction" of existing supply chains. Traditional vertically integrated organizations are giving way to more flexible decentralized networked forms of coordination. Many organizations are changing their strategies to a more proactive approach of building a standardized architecture that enables fast integration of business processes and corresponding business application systems. However, this goal is very difficult to achieve in the heterogeneous technical environments with multiple partner-supported standards and protocols (Anant and Pandya, 2000). To support this trend, the SCI must constitute a scalable framework capable of integrating a wide variety of business systems and state-of-art tools. We refer to this capability as *application scalability* – it is concerned with how easily and quickly an SCI can be scaled up to integrate many enterprise applications and systems. *Scalable SCIs* possess both application scalability and business scalability.

The broad application of SCIs leads to a series of research issues. For example, the management of application scalability; guaranteeing and improving the business scalability; carrying out and assessing architecture transformation of SCIs; embedding and evaluating various decision models within scalable SCIs. In order to support such research we have developed a scalable supply chain test-bed at the Robert H. Smith School of Business at the University of Maryland. In developing this SCI we were faced with challenges related to both application scalability and business scalability.

1.2 Literature Review

SCIs are developing and evolving at a fast pace based on both the advance of information technology and the demands of business requirements. Based on current technology, we can identify three categories of SCIs: 1) connectivity-based infrastructure, 2) database-based infrastructure and 3) middleware-based infrastructure. The connectivity-based infrastructure employs various EAI (Enterprise Application Integration) technologies, e.g. agents, the web, transaction servers or other connectors to bind various ERP and SCM systems together within and between organizations to perform electronic data exchange. The data flow in the infrastructure may come in many flavors such as flat file, EDI (Electronic Data Interchange), HTML (Hyper-Text Markup Language), and XML (eXtensible Markup Language) (Sokol, 1995). This is the most typical and straightforward infrastructure and there exist many examples from practice. For example, BizTalk server, which was launched by Microsoft and is supported by a wide range of organizations such as SAP and CommerceOne, can be categorized as this kind of infrastructure with XML framework for electronic commerce and application integration (see www.microsoft.com/biztalk/default.html). The webMethods integration server provides various data-connectivity capabilities for business applications that enable companies to connect many ERP and SCM systems like SAP, i2, or J.D. Edwards and other top applications (see www.webMethods.com). AMR Research Inc. has proposed the enterprise commerce management (ECM) model to tie enterprise resource planning (ERP) systems with private exchanges and Web technologies to create an open, multi-software, multi-vendor-based supply chain management infrastructure (Songini, 2001). Under this SCI category, EAI provides a very effective layer to link different applications. Linthicum identifies different levels of EAI including the data level, application level, method level and user interface level (Linthicum, 2000). Puschmann and Alt define and analyze the underlying topological structures: EAI-like Federated architecture, Point-to-Point architecture, Hub-and-Spoke architecture, etc (Puschhmann and Alt, 2001). Lee and Whang address the data model of connectivity-based SCIs from an information sharing perspective and define an "Information Transfer Model" (Lee and Whang, 2001). Justified by today's dynamic, unpredictable business environment and the considerable associated demands for process flexibility, a relatively novel stream of research argues for the use of intelligent agents, which offer the potential and capability to work as connectors in SCIs. At Penn State, a connectivity-based test-bed (IMEInc.com) was constructed using distributed-agent technology to serve as an enterprise integration infrastructure of various manufacturing facilities (see www.ie.psu.edu/cmei/home.htm). However, agent technology remains relatively immature for these problems even though there are many successful applications (Mehra and Nissen, 1998; Nissen, 1999; Nissen, 2000).

For database-based infrastructures, a central database or data warehouse is designated as the information platform, and various systems within and between organizations are connected to this database to share predefined information. Lee and Whang (1999, 2001) present an "information hub" architecture to tie different systems together. The "information hub" is a data warehouse supported by one or more databases where multiple enterprise application systems can interact in pursuit of supply chain integration. SAP developed the SAP Business Information Warehouse (BW) to support data-level integration of supply chain systems (Vidovic et al., 1999). BW is a comprehensive data warehouse solution that supports the representation and processing of large volumes of data. Lee and Whang described two data models for this category of SCIs: a "Third-Party Model" and an "Information Hub Model" (Lee and Whang, 2001).

In order to accomplish total supply chain management, many researchers and practitioners argue that SCIs should be able to support business process level integration (Linthicum, 2001; Chorafas, 2001). Hence, the concept of using middleware as a communication and control platform within SCIs is emerging and is attracting more and more research interest. In a middleware-based infrastructure, the message broker (or information bus) is used for data transactions and the integration manager for business processes execution (Britton, 2001; Bernstein, 1996). The middleware provides inter-operational services such as transaction, directory, persistence, event-handling, messaging and process execution based on an architectural standard such as CORBA (Common Object Request Broker Architecture -- see www.omg.com) and DCOM (Distributed Common Object Management -- see www.microsoft.com). Presently, many sophisticated middleware solutions are available such as Integration Manager and activeDB from TIBCO (see www.tibco.com), OpenWorld from iBann (see www.iBann.com), WebConnect from Manugistics (see www.manu.com), etc. Mercator Software provides an intelligent B2B integration solution for product intensive enterprises using the Mercator integration broker (see www.mercator.com). IBM Supply Chain Integration Enabler (SCIE) is a very comprehensive architecture. The foundation of the architecture is IBM's MQ Series messaging broker that facilitates e-Business and advanced applications integration(see www.ibm.com). MySAP.com e-business platform is another example of middleware that empowers interaction among multiple heterogeneous applications based on a communication standard (HTTP and XML), and a connectivity standard (DCOM and CORBA) (see www.mySAP.com). S. Linthicum, the CIO at Mercator Software, Inc., has summarized the B2B application integration methodology and middleware technology in his book (Linthicum, 2001).

The connectivity-based infrastructure represents a straightforward and scalable approach to linking application systems within and between organizations. However, the difficulties of using the approach arise relative to standardization among the various connectors and the time required to develop individual connectors. In addition, this approach may be inefficient if the connectivity architecture is not properly designed. The database-based approach, on the other hand, can provide for efficient data transaction and a scalable to link multiple applications. But for this class of infrastructures there can be difficulties in achieving business processes integration. A middleware-based infrastructure is the most sophisticated both at the data-level and the business-process-integration level. However, it is not realistic to expect full integration with outside organizations with middleware solutions. In recent years, more and more research has proposed the creation of supply chain community consisting of a number of critical trading partners. The trading partners are integrated by collaboration-based software solutions to facilitate the exchange of data and decisions. Several companies (e. g. American Software and Syncra) have developed an information system that facilitates knowledge sharing and collaborative decision-making in the spirit of CPFR (Collaborative Planning, Forecasting and Replenishment – see Syncra, 1998). It is always challenging to seamlessly blend a middleware-based infrastructure with collaboration solutions to achieve business efficiency and collaboration along an entire supply chain. In this paper, we propose and introduce an SCI using a middleware-based infrastructure integrating a collaboration solution to achieve "application scalability" within and between organizations. Obviously, this infrastructure also reduces the complexity of software integration, since it introduces the middleware as the integration node for software packages inside an organization, and the collaboration solution for software packages outside the organization.

Clearly, an SCI is not the only supporting architecture in supply chain management. The business scalability of an SCI is still highly dependent on effective decision models, which in turn are based on effective supply chain information flow. To illustrate and investigate decision model integration, we have embedded two advanced decision models within our SCI test-bed: 1) advanced ATP models, and 2) simulation models based on the Supply Chain Operations Reference (SCOR) model. Our ATP decision model can handle a wide range of customer orders with complicated product configurations, and coordinating production activities through dynamic linkages among multiple suppliers and multiple customers. Our SCOR-based simulation models can deal with complex global supply chain decisions via "what-if" analyses of different scenarios. The SCOR model was developed and has been endorsed by the Supply-Chain Council (SCC). According to Hakanson (1999), the SCOR model "enables companies to communicate supply chain issues, measure their performance objectively, identify performance improvement objectively, and influence future supply chain management software development".

There are many articles that discuss the needs and features for ATP decision models (for example, Lee and Billington, 1995; Zweben, 1996; Fordyce and Sullivan, 1999; Robinson and Dilts, 1999 and SAP AG, 1999). However, only a very limited number of papers present quantitative models to support order promising and fulfillment decisions. Taylor and Plenert (1999) introduce a heuristic technique called Finite Capacity Promising (FCP) which keeps track of traditional ATP quantities to generate feasible due dates to promise customer orders. Kilger and Schneeweiss (2000) describe the concept of APS-based "Allocated ATP" motivated by seat-class allocation in airline yield management. They also illustrate how to use a search procedure along time, customer, and product dimensions of "Allocated ATP" quantities to promise customer orders. Balakrishnan and Geunes (2000) define and study the impact of "demand substitution" and "reactive substitution", which address the complexity of product configurations on the order fulfillment process. Ball et al. (2001) models component incompatibility constraints commonly seen in ATP execution. Chen et al. (2000, 2001) present two Mix-Integer-Programming based models to support effective and efficient order promising and fulfillment for a wide range of customer orders and product configurations.

Computer simulation may provide the only approach for attacking the more complex and comprehensive supply chain issues that need to consider strategic, tactical and operational level decisions (Ganeshan, 1998; Swaminathan, 1998). Research on using computer simulation for supply chain decision support has been ongoing over the past 40 years (Mason-Jones and Towill, 2000; Zhang 2000). However, little research has been carried out on SCOR-based simulation models. In order to standardize and facilitate the process of building the supply chain models, the SCOR model is used as the standard to establish the basic supply chain simulation building blocks. Since the SCI should contain all important supply chain data, we propose effective decision-support models based on the SCOR model. Our simulation models can directly retrieve data from the SCI and then feed the results back into the SCI as appropriate transactions or database updates.

In the next section, we describe our middleware and collaboration based scalable SCI research test-bed and its components, as well as the integration architecture and integration models. The ATP decision models and the SCOR-based simulation models are discussed in Section 3. We summarize several research projects that are being carried out based on our research test-bed in Section 4.

2. A SCALABLE SCI TEST-BED ARCHITECTURE

The scalable supply chain test-bed includes several ERP components, an SCM component, a middle-ware component, a simulation component and a decision-model component. The purpose of incorporating a simulation capability is to create one or more virtual and operating organizations. Thus, one may observe and experiment with these organizations in carrying out research projects. The goal of our scalable supply chain test-bed is to investigate the following issues:

- Analyze and improve supply chain scalability;
- Evaluate the impact on organizational performance of new decision support models;
- Determine the limitations to scalability introduced with new architectures and implementation techniques;
- Compare the performance of alternative supply chain architectures, organizational structures and managerial policies;
- Evaluate the manner in which (human) decision makers interact with various SCI components.

2.1 Test-bed Components

Figure 1 illustrates the test-bed components. These components are closely integrated to provide the ability to conduct a variety of SCI experiments. In the test-bed, the ERP component contains multiple ERP instantiations for individual supply chain entities (like supplier, manufacturer, retailer, etc.). Under the current implementation, these ERP instantiations are implemented using multiple configurations of the Oracle ERP. In the future we plan to also employ multiple ERP systems such as SAP, J. D. Edwards, etc. The SCM component, which employs software from Manugistics, is designed to integrate all the ERP instantiations. The integration of the SCM component and the ERP components forms the SCI architecture. The simulation component includes a discrete-event simulation environment, and a library of supply chain simulation models. The visualization component is implemented by software for activity animation and information illustration. Finally, the middleware component, which makes use of TIBCO software, consists of an information bus, a message broker, data adaptors and a variety of APIs.

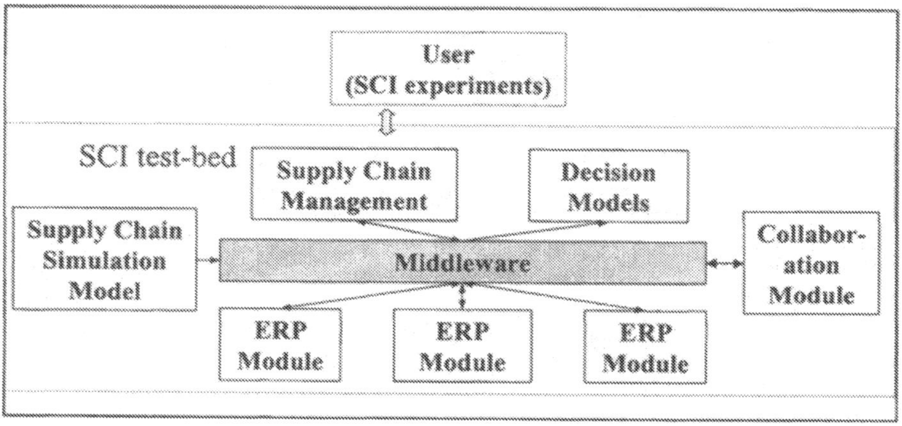

Figure 1. Components of the Scalable SCI test-bed

Components and functionality: The major functionality embodied in each of the components is described below.

- ERP Component: The ERP component includes a variety of planning and execution functions to handle business processes inside an enterprise. Each ERP instantiation represents an autonomous management unit for each individual supply chain entity. Each ERP instantiation is able to perform a group of typical ERP functions such as order management, inventory, business planning, production planning, shop-floor control, logistics management, cost evaluation, quality control, purchasing, human resource management, and so on. All these functions have the ability to process detailed business transactions that provide critical information to support high-level decision-making. In our current SCI test-bed, a web-enabled Oracle ERP package from *Oracle Corporation* is installed. We have configured instantiations with this package for different supply chain entities that include suppliers, manufacturers, distributors and retailers.

Even though the Oracle ERP package has a suite of SCM functions, we activated only the ERP functions (Bartholemy et al., 1999).

- SCM Component: While the ERP component processes detailed transactions within one enterprise, the SCM component focuses primarily on inter-enterprise planning and decision-making. In fact, the SCM scope covers the entire supply chain, from acquiring raw materials, to manufacturing products, and finally to distributing finished goods to the end consumers. Typical examples of SCM functions are supply chain demand forecasting, supply chain network analysis, global order fulfillment, distribution planning, deployment, supply chain execution, etc. In our current SCI test-bed, a commercial SCM package from *Manugistics Inc.* is employed (Manugistics, 1999a). This SCM package consists of three related modules: 1) "*SCPO*" for supply chain planning and optimization, 2) "*Deployment*" for transportation management and product deployment, and 3) "*Demand Planning /EE*" for demand forecasting across the supply chain.

- Simulation Component: The integrated ERP and SCM components provide the functionality to manage a physical supply chain system. However, these components are required to work interactively with a real-life physical supply chain to retrieve live data and deploy various schedules. Meanwhile, the data and deployment in simulation should contain specific configurations of particular business scenarios in order to support different SCI experiments. For example, the comparison of decision policies under different business scenarios, or the evaluation of organizational transformation across different industries. Therefore, the discrete-event simulation component was created to provide a "virtual supply chain" which generates a continuous transaction data stream and executes supply chain decisions. Another reason to use computer simulation in an SCI test-bed is to capture complicated dynamics and processes of real-life supply chains, which are often hard to represent with "static" ERP and SCM components. The simulation component performs two roles: data generation and operations execution. In its data generation role, the simulation generates transaction data such as customer orders, production discrepancy and transportation delay. These transactions are transmitted to the ERP and SCM components. In its operations execution role, the simulation carries out and executes various ERP and SCM schedules by simulating real-life physical processes such as goods delivery, fabrication and material sourcing. The outcomes of these operational decisions are also fed into the ERP and SCM components, which update the schedule deployment status. The simulation component is currently implemented with *Arena©* simulation software from *System Modeling Corporation* (Kelton et al., 1998). The SCOR model is utilized as the standard to establish our basic supply chain simulation building blocks. In this implementation, these blocks include Make, Source, Delivery, Plan (Plan Supply Chain, Plan Source, Plan Make and Plan Delivery) and the performance matrix blocks (SCOR, 1999).

- Decision Model: The sophisticated decision models are able to effectively match fluctuating customer demands with available supply chain resources to guide complicated procurement, production, distribution, and even marketing activities. Due to the ever-increasing complexity of product configuration, process design, and supply chain strategy, more advanced decision models are definitely required to take full advantage of the real-time information flow along the supply chain to support responsive scalable decision makings. In the SCI test-bed, we have implemented two classes of advanced decision models: ATP models and simulation-based decision models.

- Middleware: In the SCI test-bed, each component discussed above requires information or results from other ones. Middleware serves as an information backbone to tie together disparate components, namely the ERP component, the SCM component, the simulation component and the visualization component. The middleware is able to support functional calls or data transactions across multiple platforms and software packages. Additionally, corresponding APIs for each package are required to accomplish the integration together with the middleware. In our current test-bed, a middleware package from *TIBCO Software Inc.* is implemented and configured. This middleware package consists of several modules: TIB/Rendezvous, TIB/Adapter for Active Database, TIB/Integration Manager and TIB/Message Broker (TIBCO, 1999). In addition, the APIs from Oracle ERP, Manugistics SCM and *Arena* simulation software were used when direct calls to the software packages was appropriate. Our experience shows that the difficulty of using middleware to accomplish the integration is the synchronization and coordination of the functional calls and the data transactions since the former is driven by business process while the latter is by actual events. Meanwhile, the complexity of integrating one more application package in our test-bed mainly depends on the users' understanding and flexibility of the APIs provided by the software vendor.

- SCM Collaboration: While the Middleware component manages intra-organization integration, the collaboration solution provides collaborative decision-making process and information sharing among the focal company and its suppliers and customers. Recently, the most popular collaboration model available has been the CFPR model that defines a set of business processes that entities in a supply chain can use for collaboration on a number of buyer/seller functions, towards overall efficiency in the supply chain (CPFR, 2000). In our current SCI test-bed, a commercial Web-based collaboration package, "*Network/Collaborate (v6.0)*", from *Manugistics Inc.* is employed (Manugistics, 1999b).

Test-bed Generalization: The SCI test-bed is designed to be a general-purpose tool, in the sense that it should be able to support SCI experiments in different business scenarios and industrial configurations. Since different industries have different characteristics and business processes, the SCI test-bed should have the flexibility to incorporate knowledge specific to a particular industry or business. However, it is time-consuming and cost-ineffective to reconfigure the SCI components from one industrial setting to another. Therefore, the SCI test-bed should provide a library of configurations for each component. In this way, the user can select a desired configuration of each component in the test-bed with minimal reconfiguration effort to perform a customized SCI experiment. Presently, our test-bed is only configured for the discrete parts, electronic manufacturing industry. Additional industrial cases will be studied to enrich the test-bed configuration library.

Facilitating User Decision Making in the Test-Bed: The SCI test-bed is designed to include the user as an integral part of the supply chain control system. This is accomplished in the following manner: The simulation component dynamically generates live data, such as customer orders, to the ERP and SCM components. Then, the user takes actions to manage the supply chain by executing functions from the integrated ERP and SCM. The simulation component periodically monitors and retrieves data such as production plans and shipment schedules from the ERP and SCM components. Given these plans and schedules, the simulation component simulates the execution processes and feeds back results to the ERP and SCM components. During an experiment, the user can monitor progress and critical information such as material movements, orders, and inventory levels using the visualization component. The visualization component is responsible for consolidating and presenting the required information from other components. By computing and evaluating metrics, the performance of the supply chain software and decision models and human decision-making can be judged. In this way we hope to investigate issues at the interface between organizational structure, management policies and technology infrastructure.

A critical issue to be addressed is synchronization of all software (and human) components. Usually the simulation clock is much faster than its real-life counterpart so that the simulation can cover a statistically large set of events. With this in mind, the global time clock should be based on the clock of the simulation component. At the same time the global clock must be slowed down, or suspended, when human decisions are required. In between it would operate at a fast rate. In this way, the user could, for example, make decisions covering a year's operation of the supply chain within a half-day experimental session.

2.2 SCI Test-bed Architecture

In our scalable SCI test-bed, we deployed a general supply chain architecture that includes a focal company with multiple suppliers and customers. The focal company has one headquarters for global supply chain management, multiple factories to produce products and multiple retailers to serve customers. The suppliers and customers are independent organizations that have their own ERP or enterprise systems. However, they are supply chain business partners linked by common end products. As we discussed earlier, the approach to integration within the focal organization (intra-organization) is different from the approach outside the organization (inter-organization). In our test-bed, the intra-organization integration employs middleware technology, while the inter-organization uses collaboration techniques as shown in Figure 2.

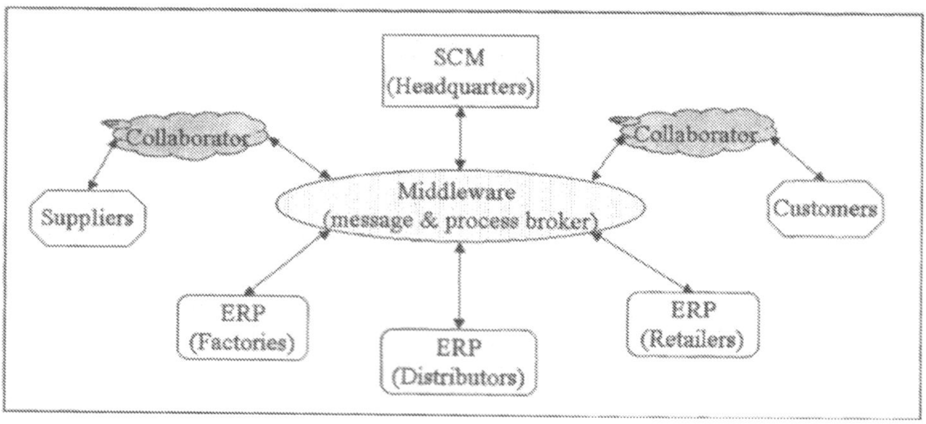

Figure 2. Integration Architecture of the Scalable SCI

For intra-organization integration, the middleware ties the SCM system with factory ERP systems and retailer ERP systems. The middleware serves as an information backbone to transact data and control business processes among disparate ERP systems and SCM systems. To support this integration, corresponding application programming interfaces (APIs) for each software package are required. The benefits of this integration are effective business processes execution, efficient information sharing and automatic error-proof data transactions. Unlike the intra-organization integration, the inter-organization integration employs collaboration solutions to create an extended supply chain community involving suppliers and customers in addition to the focal organization. The collaboration process can bring together suppliers, customers and the focal company toward achieving decision consensus with shared goals. Clearly, this kind of integration can improve the efficiency of the entire supply chain and reduce uncertainty along the supply chain. Figure 2 illustrates the integration architecture. One point that should be stressed is that this integration architecture achieves application scalability through the use of both middleware and collaboration systems. On the one hand, heterogeneous ERP and SCM systems within an organization can be added by simply building connectors with the middleware. On the other hand, suppliers and customers belonging to other organizations can also be integrated by simply building connectors with the collaboration systems (via Middleware) to transfer only the necessary data between the focal company SCM/ERP systems and suppliers' or customers' application systems.

2.3 SCI Test-bed Integration Data Models

The various modules in the scalable SCI test-bed are required to be tightly integrated to work as a whole. However, the integration of these complex packages has raised several interesting issues including the integration data model and integration business model. Implementing our test-bed required significant investigation into the capabilities of each of the packages, and a careful design of both the integration data model and business model. These two models are the guidance (references) for configuring the Middleware to integrate different test-bed modules. As we addressed in the literature review, the middleware-based intra-organization integration can take place at two levels: one is at the data level and the other is at the business-process level. For data-level integration, all distributed data stores along the supply chain should be properly connected. Specifically, data is retrieved from one application database, for example an ERP database, to update another database like the SCM database. This kind of integration is relatively easy to implement without even using advanced connectors in the middleware. However, this level of integration cannot enforce logical business rules and processes to accomplish synchronized business integration. To support this integration, we employ an *integration data model*, which defines general integrated data transactions across different applications. An example data model diagram can be illustrated in Figure 3.

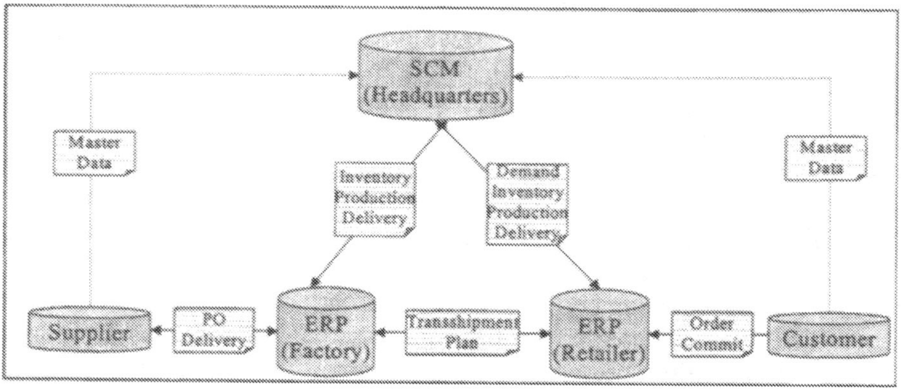

Figure 3. Example Integration Data Model

In Figure 3, the supply chain management package sits in the headquarters to manage the whole supply chain, while the ERP systems, as well as other enterprise systems, are used for individual factories, distribution centers or retailers. The detailed data transformation in the Integration data model is summarized in Table 1.

Table 1. Example Data Transformation in Integration Data Model

Data Linkage	Data Transferred
Retailer ERP to and from Headquarters SCM	▪ Sales master info ▪ Forecasted demands (long-range) ▪ Firmed demands (short-range) ▪ Urgent sales orders ▪ Urgent sales orders commitments
Headquarters SCM to factory ERP	▪ Supply chain master info ▪ Distribution requirement plan (DRP) ▪ Master production schedule (MPS) ▪ Key parts transshipment schedule ▪ Targeted production performance for factory ▪ Targeted business performance for factory ▪ Supplier info of key parts ▪ Firm demand commitment to sales

Factory ERP from Headquarters SCM	▪ ERP master info ▪ DRP deployment, acknowledge & confirmation ▪ MPS deployment, acknowledge and confirmation ▪ Inventory ▪ Production execution plan (long-range) ▪ Firm production schedules (short-range) ▪ Scheduled receipt of key parts ▪ Up-to-date actual receiving of key parts ▪ Up-to-date factory production performance ▪ Up-to-date factory business performance
Factory ERP to and from retailer ERP	▪ Key parts transshipment schedule ▪ Resource sharing exchange
Supplier to and from factory ERP	▪ Supplier master info ▪ Purchase order (PO) of key parts ▪ PO commitment
Customer to and from retailer ERP	▪ Customer master info ▪ Sales orders ▪ Sales order commitment

We have developed and endorsed a complete integration data model that can be found on our website "http://www.rhsmith.umd.edu:8001/ckim/nsf_sci.htm".

2.4 SCI Integration Business Models

One approach to business-process-level integration is through the development of an integrated application suite based on application-to-application interfaces. However, such an approach may not be practical since different applications may be distributed globally and have autonomy to a certain extent. Alternatively, the business-event-driven approach triggers applications based on business events, which avoids the need for developing new code for each transaction class. To support this capability, an *integration business model,* which describes the general business process relationship and carefully designs the business process sequence, is required. The integration process model defines the execution sequence of different applications or processes in the integrated scalable supply chain infrastructure that can be implemented in the middleware. It corresponds to the business process execution logic across the integrated supply chain. A typical example is the supply chain planning process. Following supply-chain-level demand forecasting, distribution requirement planning (DRP) for all suppliers, wholesalers, retailers and customers, and master production scheduling (MPS) for all production factories should be executed. Then, the DRP results will drive the transportation execution and delivery plan; and the MPS results should be passed on to factory-level ERP systems to start the material requirement planning (MRP) and capacity requirement planning (CRP) for individual locations. The MRP and CRP will push the material flow from supplier to factories and from factories to retailers. All the subsequent activities are triggered by the completion of events of previous activities in the predefined sequence described in the integration process model. Figure 4 illustrates an example integration business model.

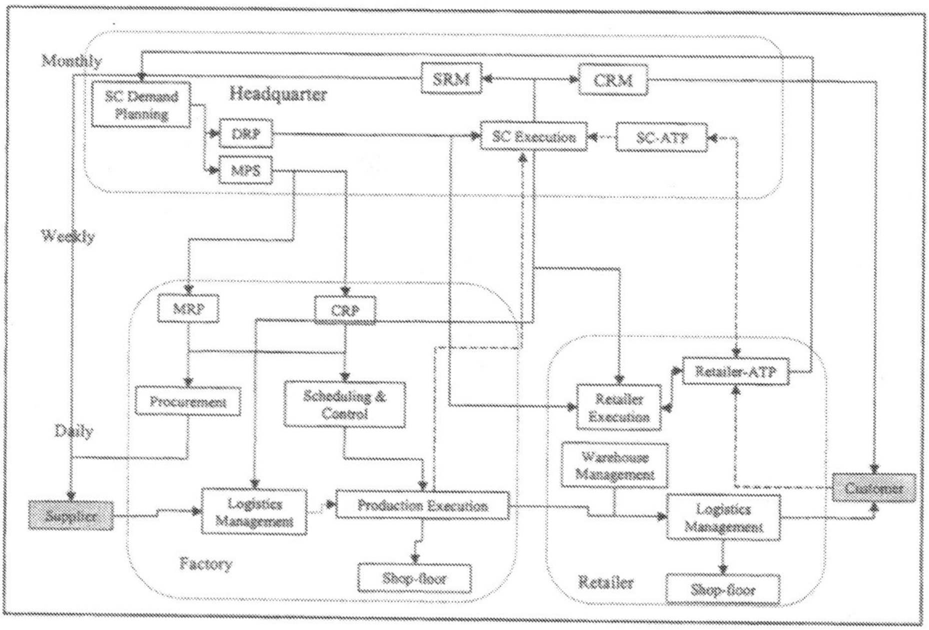

Figure 4. Example of Integration Functional Model

From Figure 4, one can observe the two major information flows in the system: push-based and pull-based. The push information flow (denoted by the solid lines) starts from the supply chain demand forecasting and pushes the production and logistics processes until demand is satisfied. This process consists of supply chain level DRP for suppliers, factories and retailers, and MPS for factories, and enterprise-level MRP, CRP and shop floor scheduling. All these activities are driven by the forecasted demands. The pull information flow (denoted by the dashed lines) starts from the customer orders. This process consists of order management, supply chain execution, production control and logistics execution. All these activities are driven by the customer orders. A variety of ATP mechanisms such as supply chain ATP, retailer ATP and production ATP play an important role in effectively matching demands with supplies on the push-pull boundaries. We have constructed and documented an integration process model based on our test-bed which can be found on our website "http://www.rhsmith.umd.edu:8001/ckim/nsf_sci.htm".

3. ADVANCED DECISION MODELS

As discussed earlier, a comprehensive perspective of supply chain infrastructures should include decision models as well as the networked information systems. In this section, we will discuss two advanced decision models and show how these models fit in a scalable supply chain infrastructure. Before we introduce the characteristics of these advanced decision models, we would like to review the key requirements of decision models in a scalable supply chain infrastructure.

Without advanced decision models in place, a company cannot take full advantage of a scalable supply chain infrastructure and operate efficiently and effectively as intended. This implies that the embedded decision models have to be "scalable" as well. As the business environment is moving toward customer-centric, we see both the variety and the complexity of products increase significantly. Decision models are therefore required to be able to easily accommodate an ever-increasing number of products and configurations. At the same time, because of the Internet, companies have access to more and more suppliers and can be approached by many customers in real time. The ability to handle an enormous amount of suppliers and customers becomes another important requirement. Nonetheless, the scalable infrastructure itself makes the business environment more dynamic and unpredictable. These requirements imply the need for advanced decision models that can help an enterprise understand the possible consequences of its business actions and allow it to review and finely tune its decisions accordingly.

To investigate issues related to embedding decision models in SCIs, we have developed and included two kinds of advanced decision models in our SCI research test-bed: 1) Available-to-Promise (ATP) decision models, and 2) SCOR-based simulation models. Our ATP decision models address the needs of handling complicated product families and coordinating production activities through dynamic linkages between multiple suppliers and multiple customers. To study complex interactions among different decisions across a supply chain, we rely on simulation-based models through "what-if" analyses of different scenarios. A set of SCOR-based simulation templates (building blocks) has been developed. We can take advantage of these building blocks to mimic any kind of supply chain architecture quickly and perform simulation studies easily. Eventually, all of these advanced decision models will be linked with the other core components in the test-bed to form a "truly" scalable supply chain infrastructure. More details of each model will be discussed below.

3.1 Available-To-Promise Decision Models

From a systems integration point of view, an Available-to-Promise (ATP) system is an advanced execution-based decision model that dynamically allocates and reallocates the planning-based resource availability to satisfy actual customer orders. Especially in a make-to-order (MTO) or assemble-to-order (ATO) environment, the last step of putting together a final (customized) product is postponed until a customer requested configuration is actually realized. This pull strategy guarantees that the system produces the right products at the right time, overcoming deficiencies in make to stock (MTS) production planning where customer needs may not be adequately addressed. However, in order to speed up the entire order fulfillment process, most of the companies adopt a push strategy like MRP to prepare and stock required subassemblies or components, in advance, based on demand forecasts. Due to the relatively longer lead times to produce or procure these subassemblies or components, by the time actual customer orders arrive, there is usually almost no economical way to raise their availabilities. In this case an advanced ATP system is needed to effectively allocate and reallocate the limited component availabilities on the push-pull boundary to individual customer orders.

To help a company pursue "business scalability," we have developed two such pull-based models: 1) quantity quoting ATP, and 2) due-date and quantity quoting ATP. In both models, we explored the properties of these pull-based ATP models and developed constraints to align short-term objectives with long-term objectives in a rolling horizon simulation. It is these advanced decision models that will allow us to effectively and dynamically link multiple customer orders with a variety of limited production resources through complicated production and configuration constraints. In the first model (based on Maxtor Hard Disk Drive production), customers specify a predetermined due date. In addition, customers also provide acceptable upper and lower bounds on desired quantity and an acceptable group of suppliers for each required component. If the desired due date cannot be met with a quantity in the acceptable range, the order is denied. The ATP model has to determine which orders to accept at what quantity, taking into account current resource availability and previously promised orders. The objective of the ATP model is to maximize the net revenue over a predefined planning horizon. Net revenue is calculated as sales revenue minus inventory holding cost (for raw materials, work-in-process and finished goods), denial penalties, and capacity under-utilization penalties. For detailed formulation please refer to (Chen et al., 2000) and (Ball et al., 2001). This mixed-integer-programming model is run periodically on a rolling horizon basis, as shown in Figure 5. Under this approach, subsequent ATP executions have the opportunity to reallocate some of the resources previously allocated based on the arrival of new competing orders. When we ran experiments with different sizes of batching interval (the length of time between successive

ATP executions), we found that, when the material availabilities are relatively tight, a larger batching interval tends to yield higher tangible net revenue (excluding denial and capacity penalties) over the simulation periods. However, this growing trend becomes dominated by the significant revenue loss caused by missed orders whose due dates are too close to be fulfilled due to the delay associated with a lengthy batching interval. We conducted a sensitivity analysis to evaluate the impact of order quantity flexibility and configuration flexibility on the tangible net revenue figures. The results show that the impact of order quantity flexibility is higher than that of configuration flexibility when material availability is relatively tight, but lower with looser material availability.

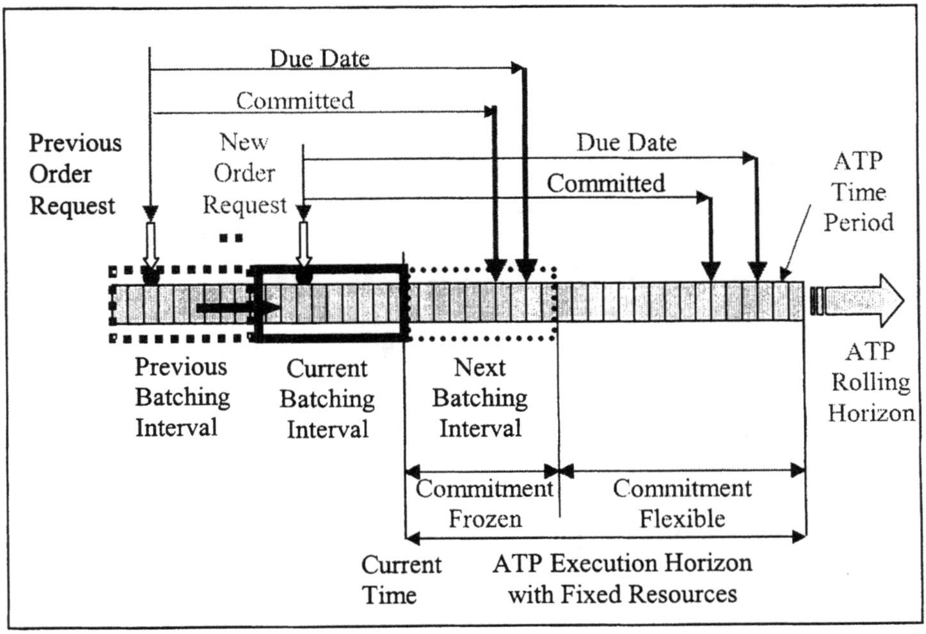

Figure 5. ATP Execution Rolling Horizon

In the second model, we explore another ATP decision – due date quoting – in addition to the quantity quoting as in the first model. In this case, customers specify not only a range of acceptable quantities but also a window of acceptable delivery dates. To promise an order, we have to assign a due date and a quantity within the acceptable due date window and quantity range. A similar denial penalty will be charged for each denied order. This model also maximizes the net revenue (see Chen et al., 2001 for details). After applying this ATP model to the Toshiba notebook computer supply chain, we observe that the effect of batching interval size is more significant in this model than in the previous (quantity quoting only) model. We believe that the increased product complexity and resource tightness could partially contribute to this result. In addition, the way that the ATP model assigns due dates may lead to a more significant increase in tangible net profit when we expand the batching interval. In order to maximize net revenue, the model assigns the earliest possible due date for each order to save on inventory costs. This results in a lack of flexibility for reallocating resources committed to earlier orders. In the first model, since a specific delivery date was not promised, the order commitment due-date and production flexibility was maintained (up to a specified limit). The decrease in profitability of using the model with shorter batching intervals reveals the weakness of greedy order promising models without look-ahead mechanisms. Therefore, we see a significant benefit of using optimization-based order promising with longer batching intervals.

Although there are benefits to adopting longer batching intervals from the (short-term) net revenue maximization perspective, one must consider the implication of longer batching intervals from a customer service perspective. Assuming customer orders arrive randomly over time, a customer is expected to wait for one half of the batching interval (plus some constant communication and processing delay) to obtain a confirmed response of a promised quantity and/or delivery date. Longer response times correspond to a decrease in the level of customer service (especially in the e-commerce era). Customers might be on-line or on-the-line waiting for an answer; therefore, long batching intervals might not be acceptable at all. This is the reason that many order promising systems adopt a "standard lead time" strategy, which promises a standard due date for all customers without really knowing whether they can keep the promise. Although these two models could be efficiently executed in our simulation experiments, we believe that there could be scalability problems for implementation within large-scale enterprises. The efficiency of decision support systems based on these models will depend not only on the ability to quickly solve the optimization models, but more importantly on the ability to access and update the relevant SCI systems that supply data to the models and that execute the decisions specified by the models. Our approach to addressing these issues involves the development of push-pull ATP models.

A push-pull model consists of an integration of push-based planning ATP with pull-based execution ATP to dynamically guide the allocation and reallocation of resources in nearly real time. In general, the push-based ATP model allocates resource availability among a number of demand classes, in advance, based on demand forecasts. Later, after orders are received, a pull-based ATP model allocates the available resources in each demand class against actual orders within that class (see Kilger and Schneeweiss, 2000 and Taylor and Plenert, 1999 for the general concept). We are now investigating integration issues for these push-pull models to improve not only their effectiveness and reliability but also the "scalability" (in terms of computation speed) of coordinated order promising and order fulfillment activities.

To successfully execute an ATP model for a complex supply chain, lots of data are required from SCM/ERP systems. Also, ATP execution results are required to feedback into the SCM/ERP systems to support further order fulfillment decisions. In general, data required from the test-bed to ATP models include customer orders, customer preferences, cost structure, material compatibility, material availability, capacity availability, bill of materials (BoM), material substitution, etc. ATP execution results like customer order commitment, material utilization plan, production capacity utilization plan, etc. are required to feedback into the SCM/ERP.

3.2. Simulation-Based Decision Models

One of the objectives of our research is to make use of computer-simulation technology to perform various "what-if" decision evaluations within the context of SCIs to pursue "business scalability" across dynamic global supply chains. Since global supply chains can be very complex, involving a variety of explicit and implicit business rules, the use of detailed mathematical programming models may not be feasible. Computer simulations can enhance decision quality by allowing the evaluation of options under consideration. However, to effectively support decision-making (particularly execution-level) it must be possible to quickly generate a simulation model consistent with the problem faced. Hence, in our research, supply chain simulation templates were developed and directly linked with ERP and SCM systems for effective data input. These templates provide "off-the-shelf" model building capabilities. They can be seen as libraries containing predefined building blocks, which enable users to create different simulation models easily and quickly. Since the SCOR model is widely accepted as a standard for modeling supply chain processes, it was chosen as a reference model in building our supply chain templates. Our ultimate objective is that the simulation templates should be directly linked to the supply chain test-bed. Figure 6 illustrates the concepts underlying simulation-based decision-support in our scalable supply chain test-bed.

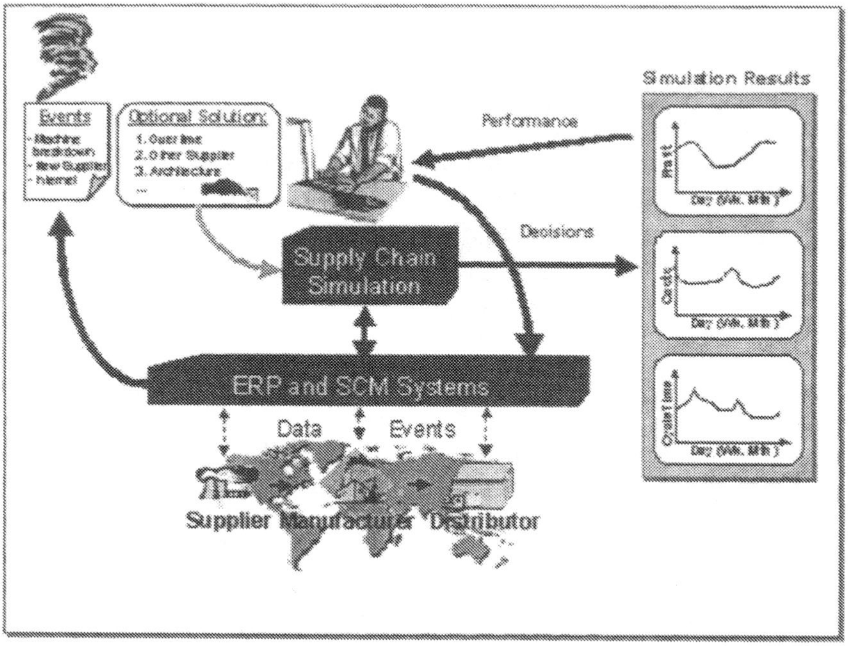

Figure 6. Concept of Simulation-Based Decision Models

In Figure 6, the ERP and SCM systems jointly contain data representing the supply chain status and its history. Should an unexpected event occur, for example a delay in a supplier shipment, the manager needs to react and develop decision options that address the problem. In order to evaluate the results of different options, the manager could invoke an embedded simulation model, and configure it based on the current business problem.

The SCOR model takes a closer look at the procedures and operations in a supply chain as the cross-industry standard for supply-chain management. The SCOR model makes use of standardized processes in supply chains and common sets of definitions to enable common models to be used for different industries and to create the opportunity for comparisons among them. This model integrates the well-known concepts of business process reengineering, benchmarking, and process measurement into a cross-functional framework.

The concept behind the SCOR model is to break down the different supply chain processes into standardized procedures. The SCOR version 3.1 contains three levels of detail (SCOR, 1999). The top level defines the scope and content for the SCOR model and consists of the management processes: Plan, Source, Make and Deliver. In SCOR, the management processes are subdivided into process categories. The Plan process is split up into a P1 (Plan Supply Chain), P2 (Plan Source), P3 (Plan Make) and a P4 (Plan Deliver) process category. The remaining three management processes each consists of three types of process categories. The first is "Make-to-Stock", so S1 (Source Stocked Products), M1 (Make to Stock) and D1 (Deliver Stocked Products). This process category contains all the functions and operations for those companies or parts of a company that do not directly react to customer demand. Of course the customer always lies as the main focus, but the sourcing, production and delivery are executed according to either forecasts or customer orders and then pushed through the supply chain towards the end customer.

We developed sixteen SCOR modules (see Keilen, 2001 and Spille, 2001) using ARENA software according to the SCOR process elements. This, of course, did not represent the complete set of SCOR models. A "Warehouse" template was developed to provide a conduit between any two of the other templates. This module was added due to the fact that there is no specific warehouse mentioned in the SCOR model. The templates have been connected with the ERP and SCM systems database for importing information (e.g. when to produce, what quantity, which product, where, and according to which bill of material), and writing performance measures into the database. With these configurable building blocks, we can quickly and easily put together a supply chain simulation model of interest and conduct "what if" analyses. This advanced decision model can help a company face the dynamic business environment as well as develop innovative business strategies. Another level of "business scalability" can then be achieved.

4. SUPPORT FOR OTHER RESEARCH PROJECTS

There are several ongoing research projects based on our research test-bed, which are described below.

4.1 Performance Scalability in Supply Chain Infrastructures

Technology changes have impacted the business processes that govern the supply chain in many significant ways (Saaksjarvi, 1999 and Jutla and Ma, 1999). A network of information servers, which may be distributed across multiple organizations, is replacing the physical material supply chain. In addition, supply chain management functions are becoming Web-enabled. The supply chain business model is changing from supplier-centric towards customer-centric. Retailers, distribution centers, manufacturers and suppliers are establishing extensive supply chain networks to better support end-customer service.

Figure 7 illustrates a multi-echelon supply chain configuration. At each level of retailer, distribution center, manufacturer and supplier, there could be multiple players. A manufacturer or supplier may further extend the "internal" supply chain if they themselves participate in an "external" supply chain. Depending on the extensiveness of the internal and external supply chains, a request generated in one chain could spawn multiple requests in another chain. The *scalability problem for* the supply chain infrastructure is that it must meet quality of service requirements, while the number of clients and requests increase, and as the extensiveness of the supply chains increases, in a dynamic environment such as the Internet. Quality of service has many dimensions as follows:

- The *end-to-end latency (delay)* between the time that a request is generated and an answer is obtained.
- The *data quality* related to the value of the response.
- The ability to meet performance targets in dynamic and unpredictable environments.

We describe some typical scenarios of processing requirements within the extended supply chain (Handfield and Nichols, 1999). In a supply chain distribution network, a retailer may need to consult manufacturers to obtain data on inventory, shipment dates, etc. in order to make a commitment on a customer order. In a supply chain make network, a manufacturer may consult multiple information servers from the retailer network to consolidate historical sales orders to obtain a demand forecast. The manufacturer may also consult multiple servers from the sourcing network to query supplier inventory data and shipment data.

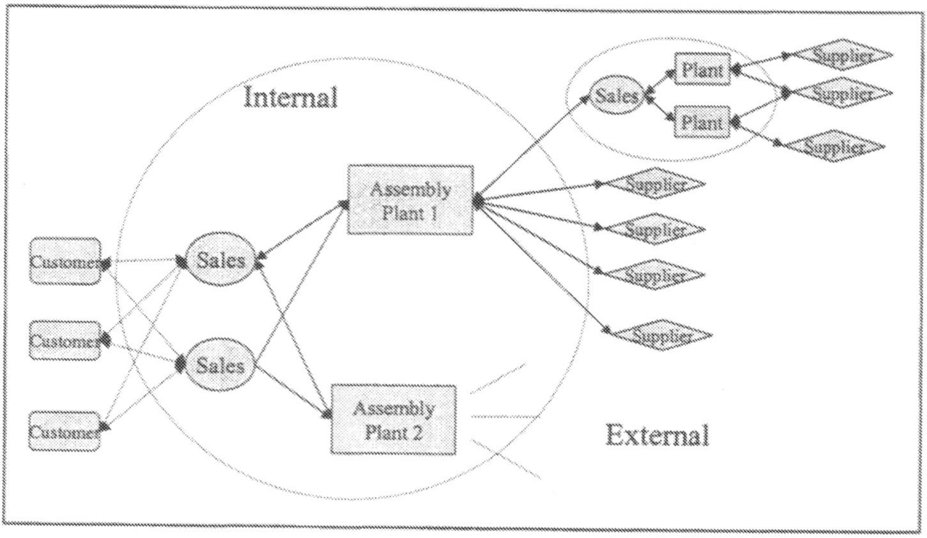

Figure 7. Multi-Echelon Extended Supply Chain

As the supply chain becomes more extensive, it will rely more heavily on wide area networks, and its services will increasingly become Web-enabled. There is related research in wide area applications. Adaptive query evaluation techniques have been developed to overcome initial delays and bursty data delivery in wide area networks (Hellerstein, 2000). Query optimization techniques have been developed for mediators that process queries on multiple remote sources (Zadorozhny and Raschid, 2001b). There is also research on optimization to meet performance targets in noisy environments (Chu et al., 1999; Zadorozhny et al., 2001a; Zadorozhny et al., 2001c).

Scalability in meeting quality of service requirements is being investigated in a simulation environment with multiple clients and servers, representing retailers, distributors, manufacturers and suppliers. Our simulation considers the impact of the following factors:

- The topology of the chain: This includes depth and fanout at each level, and the impact of noise (variance).
- Behavior to client requests: This includes real-time response or off-line response to aggregated requests. A server can respond immediately to a request, or forward the request to a varying number of servers at the next level, or both.
- Level of knowledge sharing: This includes knowledge about the quality associated with servers (low latency or low noise (variance) or high data quality) as well as priority associated with client requests.

Our simulation is used to monitor end-to-end latency of the first response to request as well as average latency. We also consider the variance in latency and the ability to meet target latency (Zadorozhny and Raschid, 2001a).

Based on initial simulation results, we observe the following:

- The latency of the first response to a request *reduces when we increase the fanout* from 1 to 2, i.e., we benefit when a server forwards a request to 2 servers at the next level rather than only one.
- However, when we increase fanout from 2 to 3, the latency of the first response shows some asymptotic behavior.
- If we consider 50% of the requests, the latency is similar to the latency with fanout = 2. However, for 100% of the requests, the latency is similar to fanout = 1.
- This indicates that there is only a limited benefit of increasing the fanout from 2 to 3 compared to increasing the fanout from 1 to 2. Thus, there is a limitation on the benefit represented by a reduced latency of the first response to a request, as fanout increases.

A simulation-based study on the impact of knowledge sharing is currently being performed. Here too we expect to observe some benefits as well as some asymptotic behavior that identifies the limitations of knowledge sharing. Our objective is to determine potential business strategies based on performance scalability results as follows:

- Reducing the latency of the response to a request with increasing fanout can lead to strategies to improve the performance of the ATP task.
- The benefits of knowledge sharing can be reflected in developing strategies such as when to use the *best supplier* or how much benefit can be expected from associating *priority* with client requests.

4.2 Toshiba Global Supply Chain Model

We implemented a global supply chain model to support production-planning decisions in the context of a complex geographically dispersed supply chain. This project was carried out in cooperation with Toshiba Japan. The MPL optimization package, using Cplex as the solver, was employed to implement the optimization model. All necessary data, e.g. lead-times, Bill of Materials and unit costs, were imported directly from the Toshiba ERP Database.

The project's test case was the Toshiba notebook computer global supply chain, which consists of 5 PCB factories, 4 PC final assembly factories, and 8 sales subsidiaries. These facilities are globally distributed, e.g. including locations in Japan, Philippines, United States and Germany. Toshiba manufactures PCBs in different countries, then ships them to other factories as input to PC final assembly, and finally ships the finished goods to sales subsidiaries. The mixed integer programming production planning model produced monthly global production plans that take into account variations in manufacturing costs, transportation costs, including tariff and currency exchange rates, penalties for missing order targets, as well as other factors.

The model implemented is a multi-period, multi-echelon and multi-product system. The time horizon is 6 periods (months). This model is a mixed-integer-programming model that contained 1.8 million decision variables, and 1.5 million constraints. The major constraint categories include: supply chain capability, material, material substitution and compatibility, production capacity, business priority, material transshipment, duty constraints and currency constraints. The objective function is to maximize revenue minus total cost including raw material cost, production cost, transportation cost, duty cost, lost sales penalty and lead time penalty. Alternate objective functions were also employed in order to allow comparison of business strategies. These alternate objective functions were: minimize lost sales and maximize fill rate. Thus, in addition to supporting production-planning decisions, this model enables the comparison of alternative strategies/policies.

4.3 Impact of Internet on Supply Chain Architectures

Articles based on anecdotal evidence in the practitioner and the academic literature suggest that the Internet, as a distribution channel tool, has positive effects in transaction and product distribution operations in the supply chain

(Andrews and Trites, 1997). Our ongoing experiments aim at providing theoretical and empirical validity to these claims by studying transaction and product distribution efficiencies associated with Internet-based purchases by end consumers.

It is hypothesized that Internet-based purchases by end consumers reduce the costs resulting from the exchange of product ownership across multiple echelons of the supply chain by streamlining the flow of transactions (Bailey, 1998). Thus, it is expected that for a given product, "Internet-enhanced" supply chains generate increasingly lower transaction costs as the distance between the demand and the supply entities decreases in the supply chain network.

It is also hypothesized that Internet-based purchases by end consumers enable greater product distribution efficiencies. In this line of argument, it is expected that the use of the Internet, as a technological interface for the fulfillment of customer orders, enables the realization of *statistical* economies in inventory-carrying costs (Zinn et al., 1989) and *scope* and *scale* economies in transportation and ordering costs (Blumenfield et al, 1985). These efficiency improvements are generated by consolidated order processing functions (Spector, 2000), increased inventory centralization, greater inventory sharing among stock keeping entities in the supply chain (Krishnan and Rao, 1965), and greater end-consumer-order processing frequencies. These Internet-enabled enhancements are expected to result in greater reliance on direct distribution through the postponement of Inventory location to upstream supply chain echelons until after end-consumer orders are placed (Bucklin, 1965).

We are building, within our test-bed, models and a simulation environment to study the effects that Internet-based purchasing by end consumers has on product distribution efficiency. The analysis is being performed under different levels of integration of physical and virtual supply chain operations. It is aimed at identifying substitution and harmonizing mechanisms and strategies between *physical* operations, such as inventory management and material sourcing, and *virtual* operations, embodied by Internet-driven transactions between retailers and end consumers and retailers and suppliers.

5. CONCLUSIONS

In this paper, we have summarized the results from our ongoing project on scalable supply chain infrastructures and the current status of the supply chain test-bed implemented in the R. H. Smith School of Business at the University of Maryland. This test-bed can support a variety of research projects. Projects, which involve the development of decision support models, are enriched within the test-bed environment by the challenges and opportunities of interfacing with the SCI data sources. The many dimensions of scalability pose a variety of research challenges, including developing and choosing a software architecture, efficiently implementing decision models and efficiently executing supply chain queries in wide area networks. The test-bed can also serve as the basis for a variety of behavioral experiments related to how SCIs impact the effectiveness of decision-makers and how organizations should adapt to SCIs. In this chapter, we have described research initiatives in these directions and, hopefully, have given an indication of the breadth of the potential research in this area.

6. ACKNOWLEDGEMENTS

This work is supported by the National Science Foundation under grant number DMI9908137. Matching support is being provided by the Robert H. Smith School of Business and the University of Maryland Institute for Advanced Computer Studies (UMIACS). Software, hardware and systems support for the associated laboratory is being provided by Sun Microsystems Inc., Oracle Corporation, Manugistics Inc., TIBCO Software Inc. and Electronic Data Systems (EDS).

7. REFERENCES

Anant, B. and Pandya, D. Why We Need Extensible B2B Infrastructures. EbizQ (www.b2b.ebizq.net), 2000.

Andrews, J., and Trites, G. Net Sales. CA Magazine, 1997;130:12-15.

Bailey, J.P. *Intermediation and Electronic Markets: Aggregation and Pricing in Internet Commerce*, PhD dissertation, Massachusetts Institute of Technology, Cambridge, MA, 1998.

Balakrishnan, A., and Geunes, J. Requirements Planning with Substitutions: Exploring Bill-of-Materials Flexibility in Production Planning. Manufacturing & Service Operations Management, 2000;2(2): 166-185.

Ball, M. O., Chen, C. Y. and Zhao Z. Y. Material Compatibility Constraints for Make-to-Order Production Planning. Under revision for publication in *OR Letters*, available at http://bmgt1-notes.umd.edu/faculty/km/papers.nsf, 2001.

Ball, M. O., Ma, M., Raschid, L. and Zhao Z. Y. Supply Chain Infrastructures: System Integration and Information Sharing. *ACM SIGMOD Record*, 2002;31:1.

Bartholemy, B., Bloch, R. and Myers, T. *Implementing the Primary Oracle Manufacturing and Supply Chain Management Applications Release 11*, Oracle Corporation. 1998;1-4,

Bernstein, P. Middleware: A model for distributed system services. Communications of the ACM 39,1996;1:86-98.

Blumenfield, D.E., Burns, L.D., Diltz, J.D., and Daganzo, C.F. Analyzing Tradeoffs between Transportation, Inventory, and Production Costs on Freight Networks. Transportation Research, 1985;19B:361-380.

Britton, C. IT Architectures and Middleware: Strategies for Building Large, Integrated Systems, Addison-Wesley, 2001

Bucklin, L.P. Postponement, Speculation, and the Structure of Distribution Channels. Journal of Marketing Research, 1965;2:26-32.

Chen, C.-Y., Zhao, Z.-Y. and Ball, M. O. A Model for Batch Advanced Available-to-Promise. To appear in Production and Operations Management, 2000.

Chen, C.-Y., Zhao, Z.-Y. and Ball, M. O. Quantity and Due Date Quoting Available-to-Promise. Information Systems Frontiers, 2001;3(4):477-488.

Chorafas, D. N. Integrating ERP, CRM, Supply Chain Management, and Smart Materials. Auerbach Publications, 2001.

Chu, F., Halpern, J. and Seshadri, P. Least Expected Cost Query Optimization: An Exercise in Utility. Proceedings of the Symposium on the Principles of Database Systems, pages 138-147, 1999.

CPFR. Collaborative Planning, Forecasting and Replenishment. Voluntary Inter-industry Commerce Standards (VICS) Association (www.cpfr.org), 2001.

Fordyce, K.J. and Sullivan, G.A. Supply Chain Management, Decision Technology, and e-business Information Technology at IBM Microelectronics. MicroNews, 1999;1.5:4 (http://chips.ibm.com/micronews/).

Franklin, M., Mihaila, G., Raschid, L., Urhan, T., Vidal, M.E. and Zadorozhny, V. Searching and Querying Wide-Area Distributed Collections. Proceedings of the First National (Russian) Conference on Digital Libraries, St. Petersburg, Russia,1999.

Hakanson, B. (Supply Chain Council). Supply Chain Management: Where Today's Businesses Compete. White Papers of Montgomery Research Inc. (on http://www.ascet.com/ documents.asp?d_ID=28), 1999.

Handfield, R.B. and Nichols, E.L. Jr. *Introduction to Supply Chain Management*, Prentice Hall: New Jersey, 1999.

Hellerstein, J., Adaptive Query Processing: Technology in Evolution, *IEEE Data Engineering Bulletin,* 2000;23:2.

Jutla, D. and Ma, S. WebTP: A Benchmark for Web-based Order Management Systems," Proceedings of the 32nd Hawaii International Conference on System Science, 1999.

Keilen, J., "A SCOR Based Simulation Model for Supply Chain Decision Support", Master's Thesis, Aachen University of Technology, Aachen, Germany, 2001.

Kelton, W. D., Sadowski, R. P. and Sadowski, D.A. *Simulation with Arena*, McGraw-Hill,1998.

Kilger, C. and Schneeweiss, L. "Demand Fulfillment and ATP," in *Supply Chain Management and Advanced Planning: Concepts, Models, Software and Case Studies*, H. Stadtler and C. Kilger (eds.), Springer, Berlin, Germany, 2000;135-148.

Krishnan, K.S. and Rao, V. K. Inventory Control in N Warehouses. Journal of Industrial Engineering, 1965;16:212-215.

Lee, H.L. and Billington, C. The Evolution of Supply-Chain-Management Models and Practice at Hewlett-Packard. Interfaces, 1995;25(5):42-63.

Lee, H. L. and Whang, S. Information sharing in a supply chain. International Journal of Technology Management, 2000;20:3/4.

Lee, H. L. and Whang, S. Supply chain integration over the Internet. CommerceNet, January 8, 2001.

Linthicum, D. *Enterprise Application Integration*. Addison-Wesley, 2000.

Linthicum, D. S. Binding companies before you take on B2B Integration: Understanding supply chain management. eAI Journal, June 2001.

Manugistics. *Supply Demand Planning and Distribution Planning*. Manugistics Inc., 1999(a).

Manugistics. *Networks/Collaborate*. Manugistics Inc., 1999(b).

Markland, R. E. Going International--What Are You Doing and How Are You Doing? Decision line, 1997;28(4):6.

Mehra, A. and Nissen, M. E. Case Study: Intelligent Software supply chain agents using ADE. Proceedings from the AAI Workshop on Software Tools for Development Agents, 1998.

Miller, E. T. and Berger, W. D. *Totally Integrated Enterprises: A framework and Methodology for Business and Technology Improvement*. St. Lucie Press, 2001.

Nissen, M. E. Agent-based supply chain integration. Journal of Information Technology Management Special Issue on E-Commerce in Procurement and the supply chain, 1999.

Nissen, M. E. Supply Chain Process and Agent Design for E-Commerce. Proceedings of the 33rd Hawaii International Conference on System Science, 2000.

Puschmann, T. and Alt, R. Enterprise Application Integration – The Case of the Robert Bosch Group. Proceedings of the 34th Hawaii International Conference on System Science, 2001.

Robinson, A.G. and Dilts, D.M. OR & ERP: A Match for the New Millennium? OR/MS Today, 1999;26(3):30-35.

Saaksjarvi, M. In Search of Business Success on the Web: The Dilemma of Defensive Functionality. Proceedings of the 32nd Hawaii International Conference on System Science, 1999.

SAP AG. Functions in Detail – SAP Advanced Planner & Optimizer: Global Available-to-Promise (http://www.sap.com/), 1999.

SCOR. *Supply-Chain Operations Reference model (SCOR), Plan / Source / Make / Deliver*, Version 3.1. Supply-Chain Council, Pittsburgh (PA) (http://www.supply-chain.org), 1999.

Sokol, P. From EDI to Electronic Commerce. McGraw-Hill Inc., 1995.

Songini, M. L. AMR touts E-business supply chain concept, Computerworld, Framingham; May 28, 2001; 35(22):10.

Sokol, P. *From EDI to Electronic Commerce: A Business Initiative*. McGraw-Hill: New York, NY, 1996.

Spector, R. *Amazon.com - Get Big Fast: Inside the Revolutionary Business Model that Changed the World*, New York: Harper Business, 2000.

Spille, J. "A SCOR Based Simulation Model to Analyze the Impact of Fluctuations in Demand on Supply Chains", Master's Thesis, Aachen University of Technology, Aachen, Germany, 2001.

Swaminathan, J. M., Smith, S. F. and Sadeh, N. M. Modeling Supply Chain Dynamics: A Multiagent Approach. Decision Sciences, 1998; 29(3):607-632.

Syncra Software Inc. Collaborative Planning, Forecasting, & Replenishment (CPFR). One Cambridge Center, Cambridge MA 02214, 1998.

Taylor, S. G. and Plenert, G. J. Finite Capacity Promising. Production and Inventory Management Journal, 1999;40(3):50-56.

TIBCO. *TIB/MessageBroker: A TIB/Active Enterprise Connectivity Component*. TIBCO Software Inc., 1999.

Vidovic, A., Engelbart, M. and Murphy, K. SAP Business Information Warehouse on the AS/400 System. International Technical Support Organization on www.redbooks.ibm.com, 1999.

Zadorozhny, V., Raschid, L., Bright, L., Urhan, T. and Vidal, M.E. Efficient Evaluation of Queries in a Mediator for WebSources. Under review, 2001a.

Zadorozhny, V. and Raschid, L. Query Optimization to Meet Performance Targets for Wide Area Applications, Under review, 2001b.

Zadorozhny, V., Raschid, L., Zhan, T. and Bright, L. Validating an Access Cost Model for Wide Area Applications, *Proceedings of the CoopIS Conference,* 2001c:371-385.

Zhang, J.G. The Development of the Internet based Extended Enterprise Supply Chain Management Simulator, Unpublished Document, CIMRU, NUI, Galway, Ireland, 1998.

Zinn, W., Levy, M., and Bowersox, D.J. Measuring The Effect of Inventory Centralization/ Decentralization. *Journal Of Business Logistics*, 1989; 10:1-14.

Zweben, M. Delivering on Every Promise: Real-Time Sales Order Promising of Supply Chain Resources. APICS – The Performance Advantage, 1996(6):3 (http://www.apics.org /magazine/).

Chapter 8

PUBLISH SUBSCRIBE MIDDLEWARE *

Lukasz Opyrchal
Department of EECS
The University of Michigan
Ann Arbor, MI 48109, USA
lukasz@eecs.umich.edu

Atul Prakash
Department of EECS
The University of Michigan
Ann Arbor, MI 48109, USA
aprakash@eecs.umich.edu

1. Introduction

Publish subscribe (pub-sub) middleware has become very popular in building large scale distributed systems. Application domains that benefit from publish subscribe architectures include supply-chain management, financial applications, e-commerce applications, workflow systems, and information distribution systems.

A number of research and commercial middleware systems support the publish subscribe paradigm. Publish subscribe provides a simple communications abstraction which enables development of large scale distributed applications. Different publish subscribe products provide different sets of features and present varying levels of flexibility to the application developer. Many of today's distributed ecommerce applications require varied security guarantees which have to be integrated with high performance and flexible message delivery systems.

*This work is partially supported by the National Science Foundation under Grant Nos. 0082851, 0075447, and 9873025. Any opinions, findings and conclusions or recommendations expressed in this material are those of the author(s) and do not necessarily reflect the views of the National Science Foundation (NSF).

This section defines publish subscribe in general and describes specific variants of the paradigm. The remainder of the chapter is organized as follows. Section 2 describes the evolution of publish subscribe systems from first subject-based implementations to current content-based prototypes. Section 3 presents design issues facing content-based systems designers. Section 4 describes a number of event multicast algorithms based on IP Multicast. Section 5 presents a more detailed coverage of security issues. Section 6 presents a short summary of issues discussed in this chapter.

1.1 What is a publish subscribe system

The publish subscribe paradigm provides an asynchronous communications abstraction between *publishers* and *subscribers*.

- *Publisher*: an entity that produces and publishes messages (events) to be received by subscribers.

- *Subscriber*: an entity that receives messages (events) published by publishers.

A few other terms related to publish subscribe systems are:

- *Broker*: a server which processes messages received from publishers and forwards them toward interested subscribers (also called *message broker*).

- *Event*: a message published by a publisher.

- *Subscription*: an expression indicating interest in certain types of events. Subscriptions are expressed in a language specific to a publish subscribe system and are used to determine which events should be delivered to particular subscribers.

Subscribers indicate their interest in messages by means of subscriptions. Different publish subscribe systems use different subscription languages that allow more or less specific subscriptions. Publishers publish new messages and the publish subscribe system performs matching to determine which subscribers are interested in a particular message.

A publish subscribe system receives messages from publishers and in turn forwards them to interested subscribers based on their subscriptions. Subscribers need not know who produces messages and publishers need not know who receives published messages. An example of a publish subscribe system is presented in figure 8.1.

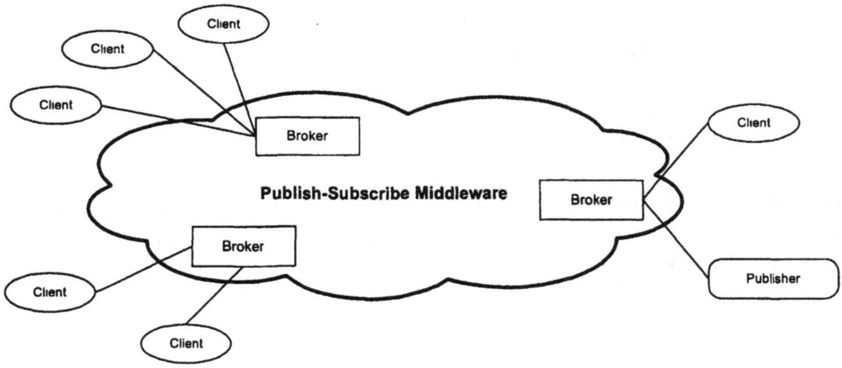

Figure 8.1. An example of a publish subscribe system

1.2 Classification of publish subscribe systems

There are two main dimensions along which publish subscribe systems can be categorized. One is the expressiveness of the subscription language. There are two main classes of subscription languages:

- subject-based subscriptions
- content-based subscriptions

There are many varieties within each of those classes and they are described in the following sections. The second dimension along which publish subscribe systems can differ is whether the event matching/delivery mechanism is centralized or distributed.

1.2.1 Subject-based.

Channels, Subjects, Groups. The earliest publish subscribe systems were subject-based. In those systems, each unit of information (event) is classified as belonging to one of a fixed set of *subjects* (also known as channels, groups, or topics). Publishers are required to label each event with a subject; consumers subscribe to all events within a particular subject. For example a subject-based pub-sub system for stock trading may define a group for each stock issue; publishers may post information to the appropriate group, and subscribers may subscribe to information regarding any issue. Several academic and industrial strength solutions have adopted this paradigm [5, 11, 18].

Subject Hierarchies. Newer systems have expanded the subject paradigm into hierarchies of subjects. Instead of simple group names, subjects have taken the following form:

subject1.subject2.subject3

Wildcards have been introduced in such systems to broaden the subscription language. To subscribe to all events in the following 2 subjects in the standard group system, two subscriptions are needed.

stocks.quotes.ibm

stocks.reports.ibm

With the introduction of wildcards, it is possible to use only one subscription to receive all events sent to both subjects:

stocks.*.ibm

A number of commercial publish subscribe middleware products have adopted the subject hierarchy paradigm [29, 19, 28, 20].

1.2.2 Content-based. The problem with subject-based pub-sub is that in large systems one has to pre-define large numbers of subjects (subject hierarchies) or the subscribers have to filter out large numbers of un-needed messages. The subject-based subscription language isn't expressive enough in many cases. The fact that many subscribers receive un-needed events also contributes to increased bandwidth usage and possible network overload.

Consider an example of a stock quote delivery application. A subject based solution could assign each ticker to a separate subject. Then every quote is published with the appropriate subject heading. Subscribers interested in particular issues can subscribe to the appropriate subjects. A problem arises if a subscriber is interested in large trades (i.e. 1000 or more shares) instead of trades in a particular issue. To accomplish this in a subject based system, the subscriber would have to subscribe to all issues and then filter out smaller trades. The large amount of un-needed messages arriving at the subscriber could easily overload it.

An emerging alternative to subject-based systems are *content-based messaging systems* [26, 25, 3, 10, 13, 15, 17]. These systems support an event schema defining the type of information contained in each event (message).

For example, applications interested in stock trades may use the event schema

[issue: string, price: dollar, volume: integer]

A content-based subscription is a predicate against the event schema, such as

(issue = "IBM" & price < 120 & volume > 1000)

Only events that satisfy (match) the subscription predicate are delivered to the subscriber.

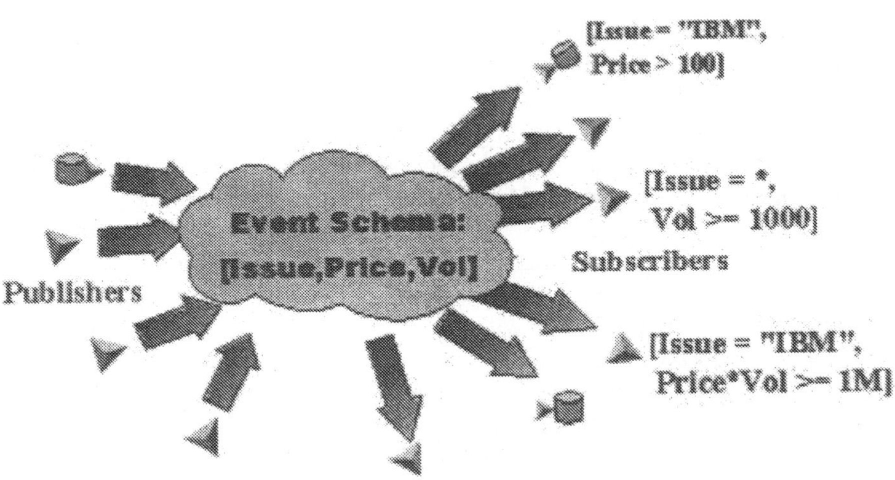

Figure 8.2. An example of a content-based system.

With content-based subscription, subscribers have the added flexibility of choosing filtering criteria along multiple dimensions, without requiring pre-definition of subjects . In the stock trading example, a subject-based subscriber could be forced to select trades by issue name because those are the only subjects available. In contrast, a content-based subscriber is free to use any orthogonal criterion, such as volume, or indeed a predicate on any collection of criteria, such as issue, price, and volume (figure 8.2).

In fact, it is possible to match events based not only on the event schema but also on the entire content of an event or even sequences of events [13].

There is a trade-off between the expressiveness of subscription language and efficiency of matching and delivery algorithms. Most existing content-based systems use an event schema (as described above) and the subscription language allows simple predicates involving logical and conditional operators.

2. Evolution of Publish Subscribe Systems

This section describes the evolution of publish subscribe systems from the earliest subject-based implementations to the current commercial middleware solutions and content-based prototypes.

2.1 Internet News

Internet News (USENET News) [11] is perhaps the most familiar example of the publish subscribe paradigm. In this system, publishers are people who send messages (similar to electronic mail) to *news groups*. Subscribers are people who read those messages. The system distributes messages to 180,000 machines and 7 million users [24]. The Internet News uses the simple subject-based paradigm where every subscriber receives all messages sent to the particular subscribed groups.

Even though the needs of an inter-process communication system differ substantially from those of end-user system such as Internet News, the Internet News system proves that the publish subscribe paradigm can be highly scalable. Decoupling of publishers and subscribers makes the system scalable and supports delivery of data from a source to a large number of subscribers.

The USENET system consists of a distributed network of servers. Servers are arranged in a tree infrastructure with *slave* and *master* servers. Messages are distributed *downwards* from master servers to slave servers. New messages posted to a local server are first sent up to a master server and then distributed down to other slave servers. Internet News system experiences long latencies and low throughput which makes it unsuitable for inter-process communication.

2.2 Group Communication Systems

Group communication systems are in a way similar to subject-based publish subscribe systems. Each group is a subject. A major difference is that in most group communication systems publishers and subscribers (*group members*) are tightly coupled as opposed to loose coupling in publish subscribe systems.

A very well known group communication system is Isis. Isis is a group communication system for building reliable distributed applications [5]. A fundamental abstraction of the Isis toolkit is the *process group*. A process group is a set of processes each of which maintains the same membership information for the group. The system provides communication primitives for joining, leaving, and sending messages to process groups.

Isis uses a model of execution known as *virtual synchrony.* Informally, the virtual synchrony model provides atomic delivery[1] of messages to the members of a group and ensures that these members maintain consistent knowledge of the group's membership. Isis has been successfully used in commercial applications like *trading floor* data delivery systems but it doesn't scale to wide-area distributed applications.

A number of other group communication systems exist. Often, they provide different reliability or ordering guarantees. Some examples include Totem [1], Transis [16] and follow on projects to Isis: Horus [30] and Ensemble [14].

2.3 Linda

Linda system, developed at Yale university, supports a model based on *generative communication* [9]. This model can be used to implement a publish subscribe system through its *tuple space.* Messages can be published by inserting a tuple into the tuple space. To subscribe to messages, a process simply needs to read from the tuple space by providing a template tuple that matches against the desired messages. This model can be used to implement subject-based as well as content-based publish subscribe system.

The generality of the model can make some of the operations expensive to support in a wide-area setting with large tuple spaces without constraining the system's behavior. In particular, the *in*[2] operator may have to perform an exhaustive pattern match against the entire tuple space.

2.4 The Information Bus

Like group communication systems, the Teknekron Information Bus (TIB) [20] uses the simple subject model of publish subscribe. The system focuses on providing a highly reliable service for distributing high volumes of data. TIB integrates self describing objects into its publish subscribe model. Data objects are annotated with subjects and sent by publishers to repositories and subscribers.

The basic design of the Information Bus is shown in figure 8.3. Messages are typically transmitted by a publisher using hardware broadcast with an overlaid retransmission mechanism that ensures reliable and ordered delivery. If the number of subscribers subscribing to a subject is small, point-to-point communication is used.

[1] Atomic delivery requires that all group members deliver all messages in the same order.

[2] The *in* operator searches the tuple space for matching tuples. It returns the first matching tuple.

2.5 Tibco Randezvous

Many of today's commercial e-commerce solutions use publish subscribe middleware as their communications backbone. One of the first commercial systems to use the publish subscribe paradigm is Tibco Randezvous messaging infrastructure [29]. This system uses the subject hierarchy model. It provides scalable and reliable message delivery with basic security features based on subjects. It features distributed architecture for better scalability and performance. Tibco Randezvous supports both publish subscribe and request/reply modes of communication. Unfortunately, due to the commercial nature of this system, it is hard to obtain more details about design specifications and performance characteristics of this system.

2.6 Content-based systems

Many emerging distributed applications could benefit from more expressive subscription languages than the subject hierarchy paradigm. In large applications, many subscribers receive un-needed events due to the fact that subject-based subscriptions are too coarse to precisely define which events a subscriber is interested in. Publish subscribe systems based on the content-based paradigm began to appear, filling the need for more expressive subscriptions.

First content-based systems were centralized. All subscribers and publishers were connected to one server (*broker*). In such systems, a broker receives messages from publishers, performs a match against all existing subscriptions and forwards messages to the appropriate subscribers. A centralized server based design doesn't work well in large distributed applications and it was quickly shown that those systems didn't scale well. Some examples of those early prototypes are: Yeast [15] and Gem [17].

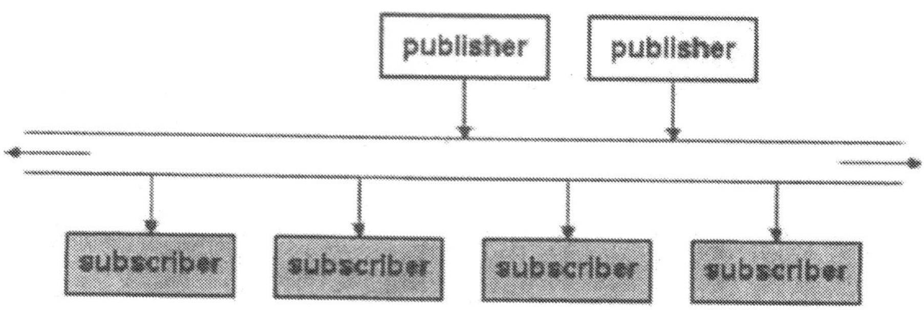

Figure 8.3. The Teknekron Information Bus.

Distributed content-based systems were developed and replaced the first centralized implementations. SIENA [10], Gryphon [3], and the latest version of Elvin [25], all use a distributed network of brokers that is used to perform "smart" matching and event distribution (routing). Events are generally sent only toward destinations that have subscriptions matching those events, thus saving network bandwidth. All three systems use slightly different methods to achieve this. SIENA uses advertisements to setup routing paths and filters events before sending them to subscribers. Elvin's publishers are informed of subscriptions so they may "quench" events (not generate events) for which there are no subscribers. Gryphon uses a very efficient *link-matching* algorithm which matches events to subscriptions in a distributed fashion. All brokers in the network are arranged in a *neighbor* relation (each broker is assigned a small set of "neighboring" brokers). Each broker performs only enough matching steps to determine which of its neighbors should receive a given event. This algorithm proves to be more efficient and scalable than the matching algorithms of SIENA and Elvin.

The Gryphon, SIENA, and Elvin projects show that, despite past doubts, content-based systems can scale well both in terms of event matching and event distribution. To achieve such high scalability, all three systems had to restrict the subscription language to one similar to the attribute-value pair event schema described in section 1.2.2. There is a trade-off between scalability and expressiveness of subscriptions and systems like Ready [13], which use a more expressive subscription language, do not scale to large distributed systems.

The next section describes important issues in the design of content-based systems in more detail.

3. Design Issues in Content-Based Systems

There are three major issues that need to be resolved when designing a content-based publish subscribe middleware:

- event matching
- event delivery
- security guarantees

Event matching is the process of matching published events to existing subscriptions. This process determines which subscribers are interested in a given event and where the events should be forwarded next. In a large, high volume system, the performance of the matching algorithm is very important to the overall system throughput and latency. Matching algorithms are directly tied to the type of subscription language and their performance is adversely affected by the expressiveness of the language.

Event delivery is the process of delivering events from publishers to interested subscribers across a network. Delivery mechanisms are closely tied to the matching algorithm. The type of delivery mechanism often determines the scalability and performance characteristics (throughput and latency) of a publish subscribe system. Event delivery mechanisms can be divided into two categories:

- *centralized* - all publishers and subscribers connect to the same server (*broker*), which receives events from publishers and forwards them to interested subscribers
- *distributed* - multiple brokers are connected over the network. Events are forwarded from publishers through distributed brokers to subscribers

Centralized systems are easier to design and implement but the demands of today's applications on the middleware layer makes them useful for only small application domains. Distributed design of content-based systems is necessary for scalability and performance required by today's applications. Section 4 describes a number of distributed delivery mechanisms based on IP multicast.

Security is a very important issue in distributed applications. Security issues in content-based systems are more complex than in subject-based (or group communication) systems. The main problems related to security in content-based systems are as follows:

- *authorization and access control* - who is allowed to subscribe to different information, who is allowed to publish messages, what are required credentials to perform specific actions, is distribution of events restricted, who has access to subscriptions of different subscribers
- *policy control* - who is allowed to change the security policy
- *trust* - which parts of the pub-sub system are trusted (which brokers/servers are trusted)

The three issues are described in more detail in the following two subsections. Section 5 presents a more detailed discussion of current research in the area of secure delivery of events.

3.1 Event matching and delivery

Event matching and event delivery are closely tied together and we will discuss both issues in this section. The strength of the subject-based approach is a simple matching algorithm and the potential to easily leverage group-based multicast techniques. Each subject can be assigned to a multicast group to

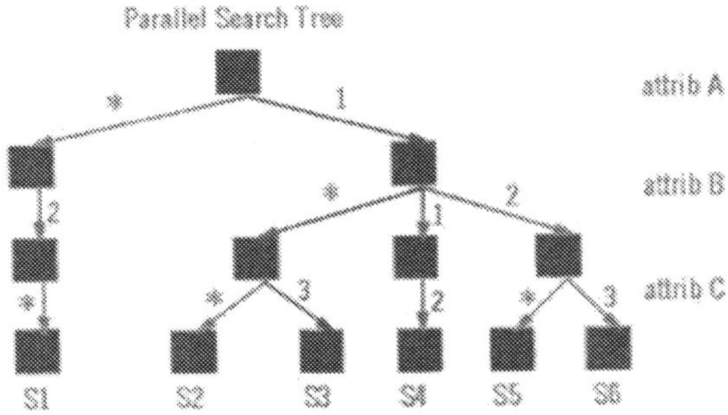

Figure 8.4. A matching tree.

provide scalability and performance. In fact, commercial subject-based products take advantage of these strengths to provide high performance publish subscribe middleware.

While content-based systems are more general and flexible, it has proved challenging to provide efficient and scalable implementations of the content-based paradigm. Content-based systems have gone through a number of *evolutionary* steps.

Centralized systems, where all publishers and subscribers are connected to the same broker, are the easiest to implement. The broker matches events to subscriptions and forwards events to interested subscribers. Matching algorithms are often tree-based where subscriptions are arranged in a tree and a tree parsing algorithm is used to search for matching subscriptions for each published event. A tree in figure 8.4 represents six subscriptions (S1 ... S6) with three integer attributes (A, B, C) per subscription. This simplified tree allows only equality checks.

S3 in figure 8.4, for example, corresponds to the following subscription:

$$[A = 1, B = *, C = 3]$$

The attribute B in this case is a *wildcard* or a *don't-care*, which means that any event with A = 1 and C = 3 will match subscription S3 (the value of B doesn't matter in this case).

Centralized systems do not scale well beyond local area networks and more scalable solutions were needed to show that content-based systems could be used as *real* middleware platforms. Gryphon and SIENA adopted a *broker network* approach. This approach consists of a number of distributed brokers connected to each other using network links. Events travel from publishers, through intermediate brokers to interested subscribers. Figure 8.5 shows an example broker network.

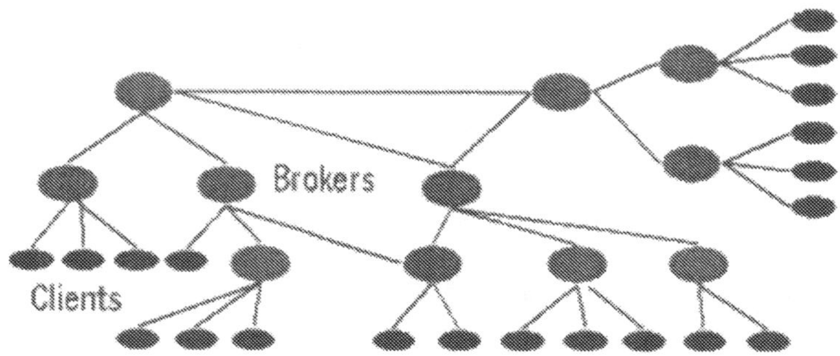

Figure 8.5. Broker network.

To achieve scalability, new event matching and routing techniques had to be developed. SIENA [10] uses publisher advertisements to establish routes between brokers. Advertisements are compared to existing subscriptions to establish routes for events. When actual events are published, the matching and routing is simplified by the existence of those routes.

Gryphon project moved one step further. A distributed matching algorithm called *link matching* was developed. The algorithm is based on subscription tree approach (like in figure 8.4) but each broker performs only enough matching to determine which of its neighbors should receive a particular event. As an event arrives at a broker, only part of the subscription tree is searched to determine where to forward the event next (see [3] for details).

In an attempt to increase message throughput and overall performance of a publish subscribe system, a number of event distribution approaches utilizing IP multicast have been explored in [21]. Using group multicast in content-based systems is non-trivial because there are no subjects that can be easily matched to multicast groups. We describe and compare a number of possible multicast algorithms in section 4.

3.2 Security in content-based systems

Messaging technology has been introduced to create more flexible and scalable distributed systems. Publishers publish messages (events), and the publish-subscribe system delivers them only to the interested subscribers. Publishers are often decoupled from subscribers, creating more scalable solutions. Figure 8.1 in section 1.1 shows a typical publish-subscribe system. Events may be delivered via intermediate brokers, who determine the set of subscribers to whom an event should be delivered. Decoupling of publishers and subscribers works well for increasing scalability but, as we will see, makes it difficult to develop secure solutions.

The application domains mentioned at the beginning of this chapter require different security guarantees. For example, an application distributing premium stock reports may require confidentiality to make sure that only authorized (paying) subscribers can access the data. Integrity may be required to ensure that reports have not been modified in transit from publishers to subscribers and sender authenticity to make sure that fake reports are not sent by third parties. The lack of those security guarantees in content-based systems has prevented their wider use even in applications that could greatly benefit from content-based subscriptions.

The fact that in content-based systems every event can potentially go to a different subset of subscribers makes efficient implementation of certain security guarantees difficult. There are 2^N possible subsets, where N is the number of subscribers. With thousands (tens of thousands or hundreds of thousands) of subscribers it is infeasible to setup static security groups for every possible subset. Even the use of a limited number of intermediate trusted servers/brokers to reduce the complexity can leave each broker with hundreds or thousands of subscribers, making the number of possible groups too large.

Different security requirements place varying constraints on the design of content-based middleware. The security models that need to be supported by content-based middleware and the design of a system that implements those models are active areas of research in the security and publish subscribe communities. Some possible security models are listed below:

- *no control* - this model provides no access control. Subscribers are free to make any subscriptions and the publish subscribe system delivers events to all interested subscribers. Certain security guarantees like confidentiality, integrity, and sender authentication may be desirable, depending on application.

- *subscription control* - access control is done at subscription time. When a new subscription is submitted, the system checks whether the subscriber is allowed to make this particular subscription. Once a subscription is determined to be valid, the subscriber receives matching events until the subscription expires, at which time the subscription must be reauthorized. Depending on application, guarantees like confidentiality, integrity, and sender authentication are also provided.

- *event control* - access control is done on a per-event basis. Applications with strict security requirements may require that recipients for each event are authorized at the time of publishing. A publisher must authorize each subscriber that matches an event or it must produce a list of authorized subscribers that can be used by the publish subscribe system to eliminate unauthorized subscribers. The per-event access control is much more powerful and flexible but it suffers from difficult implementation and possible performance penalties. Again, confidentiality, integrity and possibly sender authentication should most likely be included in this model.

The above three example security models show that when designing a publish subscribe system, one cannot provide only one security model but rather a range of possible models to support multiple application domains. Section 5 describes in more detail a specific problem of secure delivery of events from a broker to its subscribers.

4. Event Multicast in Content-Based Systems

Subject-based systems can easily leverage IP multicast by mapping a multicast group to each subject. The use of IP multicast in content-based systems, on the other hand, is non-trivial. There is no easy way to map groups of subscribers to multicast groups. The naive use of group-based multicast for implementing content-based publish subscribe may require as many as 2^N groups where N is the number of communication end-points. Rather than treating each subscribing client as a communications end-point, we assume that the communication end-points are *brokers*, which are servers that manage client connection and event distribution. Brokers reduce the complexity of routing events by reducing the total number of end-points known to the distribution system. Each end-point broker performs a local matching operation before forwarding the event to the subscribing clients. The local matching operation determines the set of interested clients connected to that broker. An implementation of matching for content-based subscription is described in [2] and shown to take time sub-linear in the number of subscriptions.

The current IPv4 specification for IP Multicast provides a maximum of 2^{24} locally scoped multicast addresses. The practical limit is smaller, since routing table space in backbone routers is a scarce resource [8]. Thus, it is important to reduce the number of groups needed. However, multicast technology is evolving rapidly, so it is difficult to know how few is "few enough." Rather than setting an arbitrary limit (other than the architected limit of 2^{24}), we examine ways to reduce the number of groups needed for a given number of brokers, favoring approaches that use fewer groups over those that use greater numbers. Three general approaches to reducing the number of groups needed are presented below.

1 *Reduce group precision.* In this approach, events are sent to multicast groups that may contain brokers that do not have subscriptions for the event. In the extreme case, messages are sent to all brokers (Sect. 4.2).

Another way to reduce precision is to combine groups to form larger groups until the number of groups is within an acceptable limit. This approach is explored in Sect. 4.5.

2 *Send multiple multicasts.* In this approach, the set of end-points is divided into mutually exclusive subsets, thereby reducing the total number of required groups. For example, if N end-points are divided into two equal subsets, the number of groups required in each subset is $2^{N/2}$, and the total number of groups required is $2 \times 2^{N/2}$. However, each event must be sent to two groups in this case. This approach is explored by the algorithm in Sect. 4.3.

3 *Send over multiple hops.* In this approach, each publisher sends to a small subset of neighboring brokers, which in turn forward the event to their neighbors, and so on. This approach is explored in Sect. 4.4.

Hybrid approaches that combine more than one of the above approaches are also possible.

4.1 The Ideal Algorithm

In an environment where we could have as many groups as we need, we could assign a multicast group to every required subset of the set of brokers. Every such group may be reached using a single multicast, and every event published is always sent to the group which contains exactly those brokers subscribing to the event. We call this the *Ideal* algorithm.

Of course, for any system with non-trivial size, the ideal algorithm requires an impractical number of multicast groups. This makes the ideal algorithm useless in practice. Nonetheless, the ideal algorithm provides a useful benchmark for evaluation – we expect the ideal algorithm to provide a lower bound on the performance of each of our multicast strategies.

4.2 Flooding

A simple solution to the problem of content-based routing is to send every published event to all brokers. In this approach, only one multicast group is needed consisting of all the brokers in the system.

A simple optimization to avoid sending events that do not match any subscribers is to first perform a matching operation at the publishing broker against all subscriptions. The additional overhead of this matching step (on the order of 100 microseconds for 10,000 subscriptions) is not significant relative to overall network latencies (on the order of a hundred milliseconds).

4.3 Clustered Group Multicast (CGM)

The CGM algorithm is based on the use of *clusters*: mutually exclusive subsets of brokers where each subset has its own set of multicast groups. We observe that if we divide N end-points into 2 clusters, we reduce the number of groups in each cluster to $2^{N/2}$ groups, and the total number of groups to $2 \times 2^{N/2}$. The cost of this approach, however, is that it may be necessary to multicast an event twice: once to a group in each cluster. In general, if we divide N into C clusters, the total number of groups needed is given by $g = C * 2^{N/C}$. Table 8.1 shows, for a given number of groups and number of clusters, the number of end-points that can be supported. For example, if we have 2^{13} multicast groups available, we can support 80 broker end-points by dividing them into 8 clusters of 10 brokers each. Since the groups within a cluster enumerate all possible combinations of brokers, each broker must join half these groups (those that include the broker) at system configuration time.

Max Groups	*# Clusters*	*# Endpoints*
2^{24}	8	168
2^{24}	4	88
2^{17}	8	112
2^{17}	4	60
2^{13}	8	80
2^{13}	4	44

Table 8.1. Number of end-points supported by CGM.

Each broker contains an instance of the subscription matching engine with entries for all client subscriptions in the system. When an event is published, the publisher's broker matches the event against all subscriptions, and sorts the resulting list of brokers by cluster. It then looks up the group in each cluster that contains exactly those brokers destined to receive the event. The publisher's broker then performs up to C multicasts, where C is the number of clusters. Some clusters may have no matching brokers and are therefore skipped.

The choice of cluster assignment has a significant impact on performance. For example, brokers that match a single subscription, but that are spread over multiple clusters will require multiple multicasts. One approach builds clusters by grouping brokers with similar subscription sets. Another uses geographic (or network) location data to group brokers into clusters. The algorithms described here use the latter approach.

A variation of the CGM scheme, the Threshold Clustered Group Multicast (TCGM), which behaves like the CGM algorithm or the flooding algorithm based on the number of of brokers matching a particular event, is described in [21].

4.4 The Neighbor Matching Algorithm

The neighbor matching algorithm is derived from [4]. In this approach, each broker designates a number of nearby brokers as "neighbors." Each broker performs just enough tests of the event content to determine which subset of its neighbors are on the next hop to a final destination broker.

There is one major difference between the earlier work and neighbor matching: In the earlier work, the authors assumed a point-to-point link to each neighbor (which is why in that work, the algorithm was named "link matching"). In this paper, we are assuming that there is a multicast group for each possible combination of neighbor brokers. When an event arrives at a broker, the broker computes the set of brokers on the "next hop" and forwards the event to the corresponding group.

There are a number of potential advantages of this approach. First of all, it is more scalable as the number of brokers in the system grows. Each broker has to know about only its immediate neighbors, not about all the brokers. For k neighbors, a broker can have a maximum of 2^k groups. Furthermore, the knowledge of those group names does not need to be widely disseminated; only neighbors need to subscribe to a group.

The disadvantages of the approach are that there is extra processing required on brokers, extra bandwidth required on the links between brokers and the network, and potential extra delay from publisher to subscriber because of the extra hops required.

4.5 Group Approximation Algorithm

[21] introduces a new algorithm based on group approximation techniques. The group approximation algorithm is a single multicast approach which reduces the number of groups required by combining actual groups to approximate groups. This approach reduces precision because an approximate group often contains a superset of the brokers which match an event. That is, some brokers may receive *waste* events: events which do not match any subscription held by a broker. The volume of waste events can affect system performance. Thus, an important aspect of this technique is to construct approximate groups which minimize the volume of waste events received by each broker, given a fixed number of multicast groups.

One way to choose approximate groups is to make use of the information contained in the subscriptions stored at each broker. In particular, we may require a separate group for each disjoint *matching set* entailed by a collection of subscriptions. The matching set of a subscription is the set of events which

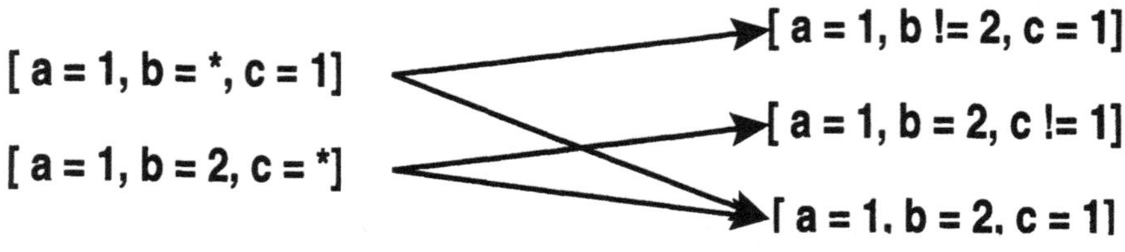

Figure 8.6. Example subscriptions (on the left) and their corresponding disjoint matching sets (on the right) for the simple event schema `[a: integer, b: integer, c: integer]`. The events in the first set match the first subscription, events in the second set match the second subscription, and events in the third set match both.

satisfy the constraints of the subscription. Figure 8.6 gives example subscriptions and corresponding matching sets. Note that the ideal algorithm creates a multicast group for each disjoint matching set.

The intuition behind the group approximation algorithm is the observation that, in systems with large event schemas, many groups have a relatively low probability of receiving events. Therefore, combinations of such groups also have a relatively low probability of introducing wasted events.

The group approximation algorithm operates as follows. Let g be the desired number of groups, then:

1 Determine the set of required multicast groups and their probability of receiving an event.

2 Combine pairs of groups until there are no more than g groups in the system.

The choice of groups to combine at each step has a significant impact on the waste generated by approximation. For more details about group approximation algorithm, as well as the other algorithms described here, see [21].

4.6 Summary of Multicast Algorithms

Each of the algorithms described above makes specific tradeoffs in order to exploit multicast. While a common goal is to reduce the number of groups, there are several other criteria by which these algorithms may be categorized:

- *Precision:* Defined as the ratio $\frac{\#\ matched\ brokers}{\#\ receiving\ brokers}$. Precision is one measure of the amount of waste in the system.

- *Number of Multicasts:* The number of multicast sends required in order to distribute an event.

	Ideal	Flood	CGM	Neighbor	Approx
Precision	1	$\mathcal{P}_b$	1	1	$1 - \frac{W}{N}$
# Mcasts	1	1	$1...C$	1 per hop	1
Groups	2^N	1	$C \times 2^B$	$N \times 2^k$	Configurable
Grps/Broker	2^{N-1}	1	2^{B-1}	$k \times 2^{k-1}$	Variable
Config.	Static	Static	Static	Static	Dynamic
Manag.	Hard++	Trivial	Moderate	Moderate	Hard

Table 8.2. Summary of event distribution algorithms where N is the number of brokers, k is the average number of neighbors of each broker, $\mathcal{P}_b$ is the probability that an arbitrary broker will match an event, C is the number of clusters, B is the number of brokers in each cluster, T is the threshold value for TCGM, and W is the total waste induced by the group approximation algorithm.

- *Total Number of Groups:* A bound on the total number of groups required in the system.
- *Groups Per Broker:* A bound on the number of groups each broker is required to join.
- *Configuration:* The stage at which multicast groups must be created. "Static" means that groups can be created before the subscription set is known.
- *Manageability:* An indication of the complexity and ease of management of a particular algorithm.

Table 8.2 summarizes the characteristics of each of the algorithms under consideration. Note that the ideal algorithm is infeasible to implement in most systems and is only presented for comparison purposes.

[21] describes simulations testing each of the algorithms and presents detailed analysis of the results. The potential choice of a multicast algorithm depends on the workload in the content-based system (distribution of subscriptions and frequency with which events are published).

5. Secure end-point delivery

A very important issue in content-based systems is secure delivery of events from a broker to its subscribers. A messaging system routes events from a publisher to *end-point brokers*. The brokers then distribute those events to their subscribers. In content-based systems, each message could potentially go to a different set of subscribers. For example, figure 8.7 shows two events (E1 and E2) delivered by the delivery system to the broker. Each event is then sent to a different subset of subscribers connected to the broker.

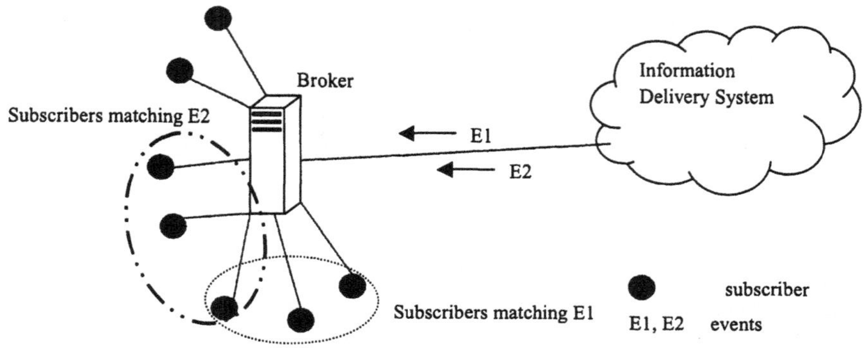

Figure 8.7. Secure end-point delivery problem.

We want to add certain security guarantees to content-based systems. In this section, we focus on efficient implementation of security guarantees like confidentiality. The system must guarantee that only authorized subscribers can read an event. Data must be protected from other (not authorized) subscribers as well as other malicious users on the network.

Security in content-based systems hasn't received much attention until recently but there are a number of related research areas. Specifically, work on secure group communication is relevant to this research. Secure group communication systems are usually meant to provide a secure channel for the exchange of data between group members. Secure groups are often identified by a *session key*, known to all group members, which is used to encrypt all data sent to the group. Key management services are used to facilitate member joins and leaves (including expulsions) as well as periodic re-keying to ensure validity of the session key.

Approaches based on *logical key hierarchies (LKH)* [31, 32, 33] provide an efficient approach to achieving scalable, secure key distribution on membership changes in group communication systems. A logical key hierarchy (LKH) is a singly rooted d-ary tree of cryptographic keys (where d is usually 2, but can be arbitrary). A trusted session leader assigns the interior node keys, and each leaf node key is a secret key shared between the session leader and a single member. Once the group has been established, each member knows all the keys along the path from their leaf node key and the root. As changes in membership occur, re-keying is performed by replacing only those keys known (required) by the leaving (joining) member.

The LKH-based key management algorithms are closely related to the secure end-point delivery problem discussed in this section and we compare our dynamic caching algorithms to an LKH-based solution. See [22] for more details on LKH and other related secure key management approaches.

Another related approach is *broadcast encryption*. It was first introduced by Fiat and Naor [12] in context of pay-TV. The authors presented methods for securely broadcasting information such that only a selected subset of users can decrypt the information while coalitions of up to k unprivileged users learn nothing. Unfortunately, the schemes presented in [12] as well as in extensions found in [6, 7, 27] require large numbers of keys to be stored at the receivers or large broadcast messages. Another problem in the context of secure content-based systems is that a coalition of more than k unprivileged users can decrypt the information.

This section describes issues and solutions for only a subset of the complex security problem in an entire publish-subscribe system. To provide event confidentiality, we assume that events are protected on their way from a publisher, through the delivery system, to the end-point brokers. We focus on the data security on the last leg from end-point brokers to subscribers. We assume that all brokers are trusted and that all subscribers and publishers are properly authenticated to the system. All subscribers and publishers also have an individual symmetric pair-key shared only with their broker (generated during the authentication process). The issues of security in transit between publishers and end-point brokers as well as the issue of broker trust are the subjects of current work.

The publish subscribe system allows dynamic access control to the events. This means that a predicate can be used at event publish time to check the set of subscribers who are authorized to receive the event. It is possible for a subscriber to be interested in an event (have a subscription for a particular event) but not be authorized to read that event (because of restrictions from a publisher). To simplify discussion, we assume that all interested subscribers are also authorized subscribers, i.e., each subscriber is authorized to read all events to which it subscribes.

Given the above restrictions and assumptions, providing confidentiality in content-based systems in an efficient way is non-trivial. Since every event can potentially go to a different subset of subscribers, it is infeasible (for large numbers of clients) to set-up static security groups. The simplest solution would be to encrypt each event separately for every subscriber receiving the event (using individual subscriber keys). For large systems where each broker has thousands of subscribers, this could mean hundreds or thousands of encryptions per event. An additional performance hit involves changing keys for each of the encryptions, which drastically slows down encryption algorithms like DES [23].

In short, the problem presented here is to preserve confidentiality using a small number of encryptions while distributing events from an end-point broker to its subscribers. Due to lack of good workloads in this area, we consider two extreme scenarios — (1) where events go to random subgroups of subscribers and (2) where there are some popular subgroups and some unpopular subgroups. We present research exploring a number of dynamic caching approaches to reduce the number of encryptions and to increase message throughput.

5.1 Group key caching

Our main goal is to reduce the number of encryptions required to ensure confidentiality when sending events from end-point brokers to subscribers. This reduction in number of encryptions in turn increases event throughput necessary for large and scalable systems. Due to the complexity of the problem - there are 2^N possible subgroups (where N is the number of subscribers) and every event can potentially go to a different subgroup - we explore only dynamic caching algorithms.

Our algorithms are compared to the naive solution (described below) and the number of encryptions required by the naive solution is our upper bound. Our algorithms must use less encryptions than the naive algorithm without introducing other performance overheads. The naive approach is described below.

1 new event E arrives at a broker and is matched to a set of subscribers G = [S1, S2, ..., SL] (where S1, ..., SL indicate interested and authorized subscribers)

2 generate a new key K_G

3 create a message $[\{E\}_{K_G}, \{K_G\}_{K_{S1}}, \{K_G\}_{K_{S2}}, ..., \{K_G\}_{K_{SL}}]$ and send it to subscribers S1 through SL

For every event arriving at an end-point broker and matching to K subscribers, the naive approach needs K encryptions. With each event matching a random subset of subscribers, for a broker with N subscribers, an average event goes to N/2 subscribers. This means that with 1000 subscribers per broker, a broker must on average perform 500 encryptions per event.

The approaches presented here aim to improve on that number. All of our dynamic caching algorithms require the broker and subscriber to keep a certain size cache. In general, the cache stores keys for most popular groups. Cache entries have the following format: $< G, K_G >$

G is a bit vector identifying which subscribers belong to the group and K_G is the key associated with the group. K_G is used to encrypt events going to this particular set of subscribers. We assume that all subscribers have enough resources to cache all necessary keys. Subscriber S_1 must cache every entry $< G_X, K_X >$ cached at the broker such that $S_1 \in G$. In practice, subscribers with limited resources may have smaller caches. This means that such subscribers may not have all the appropriate keys cached at the broker. A secure protocol for key request/exchange has to be developed in order for our caching algorithms to support subscribers with limited resources. Currently, we assume that every subscriber S has all the cache entries $\{< G, K > | S \in G\}$ cached at the broker.

The next few sections describe each of the algorithms in detail. We present theoretical analysis and simulation results comparing each of our approaches and the naive solution. We also present simulation results for LKH-based solution for comparison. We derive approximate expressions for the average number of encryptions for two different distributions of groups:

Random - each arriving event goes to a random subset of subscribers. Every one of the possible 2^N groups has the same probability of occurrence.

Popular Set - there is a set of groups that occur more often than others. An event has a higher probability of matching a group from the *popular group set* S. The distribution has the following parameters:

> $| S |$: the size of set S (number of groups in the popular set)
>
> p: probability that an event matches a group from S
>
> Every event matches a random group from S with probability p. Every event matches a totally random group with probability $(1 - p)$.

To enable simpler derivation, we assume the cache to be *smart*. This means that it caches only groups from set S (if cache size is less than or equal to $| S |$). In practice, it is possible to closely approximate this behavior by using a *frequency-based* cache. By caching groups that occur more frequently, this approach would cache groups from set S after long enough time (since groups from S occur more frequently than other random groups).

We introduce a measure of average number of encryptions per message E. Due to uniform distribution of subscribers in a group, the average number of subscribers in a group is N/2, so:

$$E_{naive} = \frac{N}{2} \tag{8.1}$$

5.1.1 Simple caching. The simplest solution to reducing the number of encryptions in the system is to use a plain caching scheme. This approach assumes that, based on customer subscriptions, many events will go to the same subset of subscribers. Creating and caching a separate key for those groups would take advantage of *repeating groups* and it would reduce the number of encryptions performed at the broker. The major parameters affecting the performance of this approach are the number of clients, cache size, and the distribution of groups. The basic algorithm works as follows:

1 new event E arrives at a broker and is matched to a set of subscribers $G = [S_1 ... S_N]$

2 search the current cache

(a) if an entry $< G, K_G >$ is found in cache

- send $\{E\}_{K_G}$ to all subscribers in G

(b) if entry $< G, K_G >$ is not found in cache

- generate a new key K_G
- create the following message and send it to subscribers in G:

$$\{E\}_{K_G}, \{K_G\}_{K_{S1}}, ..., \{K_G\}_{K_{SN}}$$

- add new entry $< G, K_G >$ to cache

This approach works well if many events need to be delivered to the exact same set of subscribers. We present approximate formulas for average numbers of encryptions:

Random distribution: we know that if there is a cache hit (incoming event matches a group stored in cache), the event only needs to be encrypted once with the stored key. If there is a cache miss, then the number of encryptions is the same as in the naive case. The probability of a cache hit is:

$$P_{ch} = \frac{C}{2^N} \tag{8.2}$$

Where C is the cache size and N is the number of subscribers. The probability of a cache miss is:

$$P_{cm} = \frac{2^N - C}{2^N} \tag{8.3}$$

From equations 8.2 and 8.3 we can get the expression for the average number of encryptions per message:

$$E = \left[\left(\frac{C}{2^N} \times 1\right) + \left(\frac{2^N - C}{2^N} \times \frac{N}{2}\right)\right] \qquad C < 2^N \tag{8.4}$$

We can see that as the number of subscribers gets larger, the simple cache will perform only slightly better than the naive approach. For 100 subscribers and a cache of size 10,000, this approach averages 50 encryptions per message (just like the naive solution).

Popular-set distribution: there are 2 cases: whether the group comes from the *popular set* or not (since the *smart* cache only caches groups from the popular set). If the group does not come from the popular set (with probability $(1 - p)$) then the average number of encryptions is:

$$E_{np} = \frac{N}{2} \tag{8.5}$$

If the group comes from the popular set, then the number of encryptions depends on whether there was a cache hit or not. Similarly to the simple cache with random groups, the average number of encryptions for groups coming from the popular set is:

$$E_p = \left[\left(\frac{C}{\mid S \mid} \times 1\right) + \left(\frac{\mid S \mid - C}{\mid S \mid} \times \frac{N}{2}\right)\right] \qquad C <\mid S \mid \tag{8.6}$$

Combining equations 8.5 and 8.6, we get the expression for average number of encryptions with simple cache and popular set distribution:

$$E = (1-p)\,\frac{N}{2} + p\left[\left(\frac{C}{\mid S \mid} \times 1\right) + \left(\frac{\mid S \mid - C}{\mid S \mid} \times \frac{N}{2}\right)\right] \qquad C <\mid S \mid \tag{8.7}$$

For the same parameters as in the random distribution case and the popular set of size 10,000 and probability $p = .9$, simple cache requires only 6 encryptions on average.

5.1.2 Build-up cache. This algorithm extends the simple caching approach. The base of this solution is an observation that many groups are subsets of other, larger groups. In other words, larger groups can often be constructed by combining smaller groups. The algorithm takes advantage of this observation by performing *partial cache matching*. It searches for subsets of a given group G trying to reconstruct G using a number of smaller groups already in the cache. Figure 8.8 shows the basic idea behind the algorithm. Optimally, the algorithm would try to find a minimum number of groups from cache that completely *cover* group G. This would make the algorithm intractable. We use a heuristic that reduces the complexity of the search process to O(C), where C is the size of the cache. We search the cache from largest to smallest entries based on the assumption that finding a larger matching group will lead to a better overall match. The details are as follows:

1 new event E arrives at a broker and is matched to a set of subscribers G

2 search the cache for subgroups of G, starting from largest size groups

   (a) if a full match is found (entry $< G, K_G >$ is found in cache)
      - send $\{E\}_{K_G}$ to all subscribers in G

   (b) if a partial match is found ($< G1, K_{G1} >$ where $G1 \subset G$ and $G1 \neq G$)

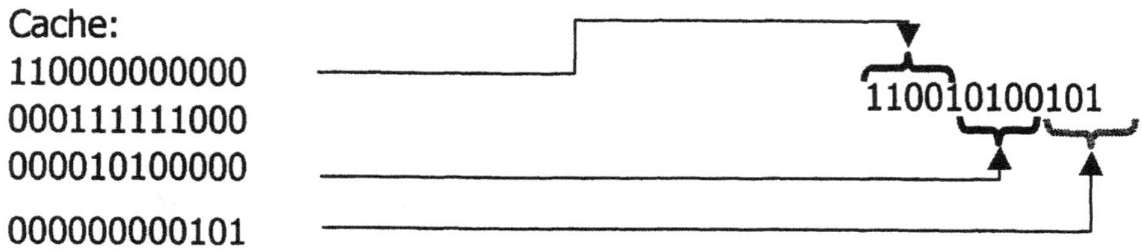

Figure 8.8. Build-up cache.

- store $< G1, K_{G1} >$ in a temporary set S
- take the difference of $G_{new} = G - G1$

3 repeat 2 until $G_{new} = \emptyset$ or there are no more entries in cache to search

4 if $S! = \emptyset$ encrypt and send a message to every subgroup in S

5 if $G_{new}! = \emptyset$

- generate a new key $K_{G_{new}}$ and save the entry $< G_{new}, K_{G_{new}} >$ in cache
- generate a new message (as in step (b) of the simple algorithm) and send it to all subscribers in G_{new}

5.1.3 Clustered cache. A known technique for reducing complexity of certain problems is clustering. There are 2^N possible groups and the above approaches may require very large cache sizes when the number of subscribers grows into thousands. This section describes a new clustered cache technique, which needs a much smaller number of encryptions than the first two algorithms.

For a server with N clients, we divide the set of clients into K clusters. The server has to keep K separate *cluster caches*, but those caches can be much smaller than the caches from section 5.1.1 and 5.1.2. Each cluster cache holds entries of the same format as the simple and build-up caches ($< G, K_G >$), but G only consists of subscribers belonging to the particular cluster. The algorithm works in the following way:

1 new event E arrives at a broker and is matched to a set of subscribers G

2 G is divided into K subsets according to the cluster choices $(G_1, G_2, ..., G_K)$

3 for each cluster

(a) search cluster cache for the appropriate group (one of G_1 through G_K)

- the algorithm works as the simple cache approach for each cluster
- send message to the appropriate group in the cluster
- add new entries to cluster cache as dictated by the simple algorithm

An event has to be encrypted separately for each cluster. Assuming a cache hit in every cluster, an event has to be encrypted K times.

An interesting issue in this algorithm is the choice of clusters. A simple solution is to assign subscribers to clusters randomly (or according to subscriber ids: first X subscribers to cluster 1, next X to cluster 2, etc.). Another approach would be to assign clusters based on subscription similarity. Subscribers with similar subscriptions would be assigned to the same cluster.

We calculate the average number of encryptions per each cluster and multiply it by the number of clusters to get the total. A cluster of size N_K, with individual cache of size C_k, is very much like the simple cache from section 5.1.1. We present an approximate formula for random distribution only. Derived from equation 8.4, the average number of encryptions for clustered cache is:

$$E = \left[\left(\frac{C_K}{2^{N_K}} \times 1 \right) + \left(\frac{2^{N_K} - C_K}{2^{N_K}} \times \frac{N_K}{2} \right) \right] \times K \qquad C_K \leq 2^{N_K} \quad (8.8)$$

To simplify derivation and the resulting formulas we do not account for the fact that some clusters may have no subscribers matching a particular event, therefore reducing the number of encryptions. Equation 8.8 gives us an upper bound on the average number of encryptions for the clustered cache and random group distribution. For the same parameters as in the previous algorithms, the clustered cache requires at most 11 encryptions on average.

Next section describes an algorithm combining clustered and simple caches that combines the advantages of both approaches.

5.1.4 Clustered-popular cache. The clustered cache works well in a generic case of random groups. It doesn't take advantage of the fact that some groups occur more frequently, as in the popular-set distribution. The *clustered-popular* algorithm was designed to enhance clustered cache with support for frequently occurring (popular) groups. The basic idea is to combine clustered cache with simple cache in one. For each event, both caches are checked. If there is no hit in the simple part of the cache, the clustered approach is used to reduce the number of encryptions. The algorithm works as follows:

1 new event E arrives at a broker and is matched to a set of subscribers G

2 search the simple cache

(a) if an entry $< G, K_G >$ is found in cache

- send $\{E\}_{K_G}$ to all subscribers in G

(b) if entry $< G, K_G >$ is not found in cache

- generate new key K_G and add new entry $< G, K_G >$ to cache
- search the clustered cache as in section 5.1.3
- send messages to the appropriate group in each cluster

We omit the derivation of formulas for average number of encryptions for this approach here (see [22] for details).

Table 8.3 shows average numbers of encryptions calculated using the derived formulas for two different sets of parameters. The clustered-popular approach always uses half of the cache for the simple part and second half for the clustered part. The **small set** has the following parameters: there are 100 clients and cache size is 10,000 entries. The clustered approach uses 10 clusters of 10 subscribers each with cluster cache size of 1000. The clustered-popular cache uses 11 clusters of about 9 subscribers each. The popular set distribution uses a set of 10,000 groups and probability $p = .9$. The **large set** has the following parameters: there are 1,000 clients and cache size is 100,000 entries. The clustered cache uses 100 clusters of 10 subscribers. The clustered-popular cache uses 114 clusters of 8 or 9 subscribers each. The popular set has 100,000 groups and probability $p = .8$.

		No cache	Simple cache	Clustered cache	Clustered-popular cache
Small	Uniform groups	50	50	11	17
	Popular groups	50	6	11	6.7
Large	Uniform groups	500	500	109	112
	Popular groups	500	101	109	68

Table 8.3. Calculated average numbers of encryptions.

Clustered-popular cache performs best in the large set and the popular-set group distribution. Next section shows that the clustered-popular approach outperforms other algorithms as the size of the cache increases. The actual numbers for both clustered algorithms should be lower than the theoretical values. This is because we did not account for the fact that for some events, not all clusters will have interested subscribers. This was done to simplify derivation. Also, the values for simple cache should be higher because of our assumption of *smart* cache. We were not able to approach the smart cache because of the size of our sample in the simulations.

5.1.5 Simulations. We ran a number of simulations to confirm our theoretical results as well as to compare the simple, clustered, and clustered-popular approaches to the build-up cache as well as to an LKH-based approach presented for comparison. The LKH-based approach works by creating a hierarchical tree of keys. When an event arrives at broker and is matched to a group of subscribers G, the tree is searched, bottom-up, for keys that are common to as many as possible subscribers from G, but are not known to subscribers not in G. The LKH-based scheme corresponds to a fixed cache size - the key tree only changes with changes in the number of subscribers. We ran a number of simulations testing different parameter settings. Each simulation consisted of 15,000 events. All caches were *warm*. We show results for both random and popular-set group distributions. The number of clusters used by the cluster-based algorithms depends on the number of clients. When solving equation 8.8 for the number of clusters K, we get the optimal number of clusters when $2^{N_K} = C_K$ (N_K is the number of clients per clusters and C_K is the cluster cache size). We chose the largest K which gave $2^{N_K} \geq C_K$. We used the same method to determine number of clusters for the clustered popular approach.

Figure 8.9 shows results for cache size 10,000, and figure 8.10 shows results for cache size 20,000. Figure 8.10 shows results for only the clustered, clustered-popular, and LKH-based approaches.

The clustered-popular is the only algorithm that performs differently with popular-set group distribution. We can see that clustered, clustered-popular, and LKH-based algorithms outperform the simple and build-up caches for random groups. In the case of popular-set distribution, all perform similarly with LKH-based approach being the worst. When the cache size is increased to 20,000, the clustered approaches clearly outperform the LKH-based algorithm. Figure 8.11 shows the effect of increasing cache size on the number of encryptions required by each algorithm for different group distributions. The results (except for the LKH-based approach, which is based on simulation results) in this figure are based on the approximate formulas derived in [22]. As we can see, clustered and clustered-popular approaches are similar with random group distribution, but the clustered-popular algorithm clearly outperforms all other solutions when the popular-set distribution is used.

In order to make judgments about usefulness of any of these algorithms, we need to relate number of encryptions to a performance measure like throughput. We claimed that algorithms that reduce the number of encryptions required are desirable because a large number of encryptions per message reduces message throughput at the broker. We measured throughput of the DES algorithm depending on the number of encryptions per message to show that this is the case. We ran a number of experiments encrypting an 8-byte piece of

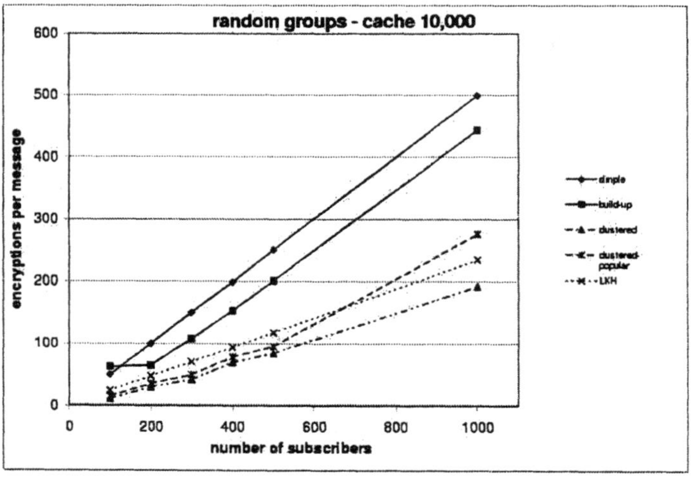

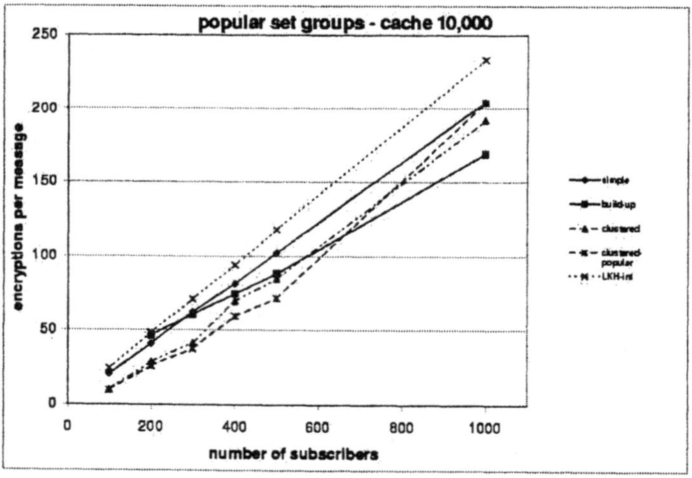

Figure 8.9. Number of encryptions, cache size 10,000.

data. We varied the number of different keys used. We used the DES algorithm on a 550Mhz Pentium III running RedHat Linux operating system.

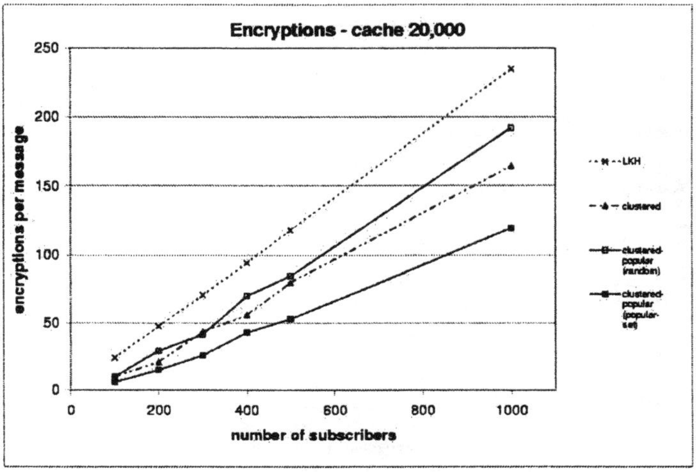

Figure 8.10. Number of encryptions, cache size 20,000.

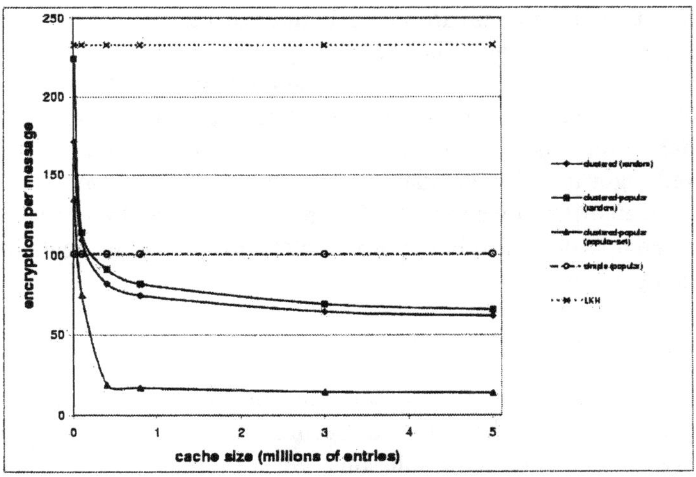

Figure 8.11. Effects of cache size on the number of encryptions.

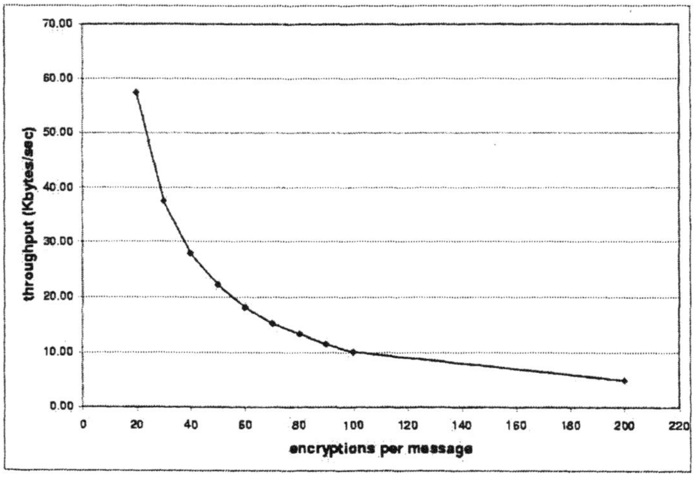

Figure 8.12. Message throughput as a function of the number of encryptions per message

The results of our experiment are shown in figure 8.12. The adverse effect of number of encryptions on throughput is clearly visible. Based on our results, we see that clustered-popular has higher throughput than LKH-based approach by about 100% in the case of popular-set distribution and by about 33% in the case of random group distribution. Clustered-popular is also about 50% better than clustered algorithm for popular-set distribution. It, however, underperforms clustered algorithm by about 10-20% in the case of random distribution. The LKH-based approach also has approximately 40% lower throughput than the clustered algorithms.

5.2 Conclusions

There is a growing need for security solutions for content-based publish subscribe systems. Section 5 identifies the "secure end-point delivery" problem and explores a number of possible solutions. We are concerned with providing confidentiality when sending events from brokers to subscribers. The problem is that in content-based systems, every event can potentially have a different set of interested subscribers. There are 2^N possible subsets, where N is the number of subscribers. With thousands of subscribers it is infeasible to setup static security groups for every possible subset.

A number of key management systems for group communication solve a similar problem but none of them was designed to handle the dynamic nature of content-based event delivery. We explored a number of dynamic caching approaches. A simple solution is to encrypt each event separately for each interested subscriber; however this requires a large number of encryptions for large sets of subscribers. Our main goal is to reduce the number of encryptions required to preserve confidentiality while sending events only to interested subscribers. The number of encryptions is important because it translates directly into message throughput (see figure 8.12).

All of our approaches use a dynamic caching scheme, where the broker and subscribers must cache subgroup keys. Each cache entry has a format $< G, K >$, where G identifies the set of subscribers belonging to a subgroup and K is a key associated with the subgroup. All secure communication intended for all subscribers in G can be encrypted using key K. Through theoretical analysis and simulation results, we show that our clustered and clustered-popular approaches perform better as cache size increases. Both cluster-based algorithms outperform LKH-based solution for most cases. Clustered-popular algorithm performs especially well in the case of the popular-set group distribution.

Our results show that cluster-based algorithms can be a practical solution to the end-point delivery problem. They do not impose heavy overhead as hash tables can be used for cache lookup. The cache size requirement on the subscriber side is also lower than the simple cache or build-up cache algorithms. In the case of the clustered cache, each subscriber only needs $\frac{C_k}{2}$ cache entries, where $C_K = \frac{C}{K}$. C is the total cache size at the broker, K is the number of clusters and C_K is the size of one cluster part of cache. If client resources are limited, a protocol for key request/exchange is needed for a subscriber to request appropriate keys from the broker. [22] presents more detailed discussion and analysis of the results shown here.

6. Summary

Messaging (or publish subscribe) systems have become increasingly popular as the middleware layer of large distributed applications. A number of subject-based commercial solutions currently exist that offer high performance messaging middleware. Many application domains (i.e. supply-chain management, financial applications, e-commerce applications, workflow systems, information distribution systems) require scalable and flexible publish subscribe middleware and could greatly benefit from content-based solutions. Recent research has shown that it is possible to build highly scalable and high performance content-based systems. The lack of security guarantees in content-based systems has prevented their implementation in applications that could greatly benefit from content-based subscriptions. Security is essential to most distributed applications and has come to the forefront of content-based systems research.

This chapter introduced and defined publish subscribe systems. We showed the evolution of systems from early subject-based through subject hierarchies to current content-based prototypes. Section 3 describes issues in content-based event matching, event delivery and security guarantees - all essential to content-based middleware system. Event matching and delivery/routing are mature enough to offer scalable and efficient solutions but security in content-based systems is still a new research area.

Section 4 presents a number of algorithms trying to leverage IP multicast for delivery of events in content-based systems. The mapping of groups of subscribers to multicast groups is non-trivial. The algorithms presented here aim at reducing the number of multicast groups used for delivery of events. Simulation results (presented in [21]) show that the choice of the algorithm depends on the workload in the system and the network conditions.

Section 5 presents a detailed discussion of one of the important research issues in content-based security. An in-depth discussion of the secure end-point delivery problem is presented together with descriptions of possible solutions and their comparisons. We used theoretical as well as experimental data to compare the different approaches.

Content-based systems have a lot of potential in the area of large-scale distributed systems but it won't be realized until acceptable security solutions are developed. The problems discussed here are only a part of the large number of security issues in content-based systems.

References

[1] D. A. Agarwal, P. M. Melliar-Smith, and L. E. Moser. Totem: A protocol for message ordering in a wide area network. In *In Proceedings of the First ISMM International Conference on Computer Communications and Networks*, pages 1–5, June 1992.

[2] Marcos K. Aguilera, Robert E. Strom, Daniel C. Sturman, Mark Astley, and Tushar D. Chandra. Matching Events in a Content-Based Subscription System. In *Proceedings of Principles of Distributed Computing (PODC '99)*, Atlanta, GA, May 1999.

[3] Guruduth Banavar, Tushar Chandra, Bodhi Mukherjee, Jay Nagarajarao, Robert E. Strom, and Daniel C. Sturman. An Efficient Multicast Protocol for Content-Based Publish-Subscribe Systems. In *International Conference on Distributed Computing Systems (ICDCS '99)*, June 1999.

[4] Guruduth Banavar, Tushar Chandra, Bodhi Mukherjee, Jay Nagarajarao, Robert E. Strom, and Daniel C. Sturman. An Efficient Multicast Protocol

for Content-Based Publish-Subscribe Systems. In *International Conference on Distributed Computing Systems (ICDCS '99)*, June 1999.

[5] Kenneth P. Birman. The process group approach to reliable distributed computing. *Communications of the ACM*, 36(12):37–53, December 1993.

[6] Carlo Blundo and Antonella Cresti. Space requirements for broadcast encryption. In *Advances in Cryptology - EUROCRYPTO'93, LNCS 950, pp. 287-298, Springer Verlag 1994.*

[7] Carlo Blundo, Luiz A. Frota Mattos, and Douglas R. Stinson. Generalized beimel-chor schemes for broadcast encryption and interactive key distribution. *Theoretical Computer Science*, 200(1-2):313–334, 1998.

[8] S. Bradner and A. Mankin. *The Recommendation for the IP Next Generation Protocol.* IETF. RFC 1752.

[9] Nicholas Carriero and David Gelertner. Linda in context. *Communications of the ACM*, 32(4):444–458, April 1989.

[10] Antonio Carzaniga. *Architectures for an Event Notification Service Scalable to Wide-area Networks.* PhD thesis, Politecnico di Milano, December 1998. Available from http://www.cs.colorado.edu/~carzanig/papers/.

[11] Erik Fair. Usenet, spanning the globe. *Unix World*, 1:46–49, November 1984.

[12] Amos Fiat and Moni Naor. Broadcast Encryption. In *Advances in Cryptology - CRYPTO'93*, August 1993.

[13] R. Gruber, B Krishnamurthy, and E. Panagos. An Architecture of the READY Event Notification System. In *Proceedings of the Middleware Workshop at the International Conference on Distributed Computing Systems 1999, Austin, TX*, June 1999.

[14] Mark Hayden and Robert van Renesse. Ensemble home page. http://www.cs.cornell.edu/Info/Projects/Ensemble/index.html.

[15] B. Krishnamurthy and D. S. Rosenblum. Yeast: A general purpose event-action system. *IEEE Transactions on Software Engineering*, 21(10):645–857, October 1995.

[16] D. Malki, Y. Amir, D. Dolev, and S. Kramer. The transis approach to high availability cluster communication. Technical Report TR 94-14, The Hebrew University of Jerusalem, June 1994.

[17] M. Mansouri-Samani and M. Sloman. Gem: A generalized event monitoring language for distributed systems. *IEE/IOP/BCS Distributed Systems Engineering Journal*, 4(2):96–108, June 1997.

[18] Shivakant Mishra, Larry L. Peterson, and Richard D. Schlichting. Consul: A Communication Substrate for Fault-Tolerant Distributed Programs.

Technical Report TR 91-32, Dept. of Computer Science, The University of Arizona, November 1991.

[19] Object Management Group. *Notification Service.* http://www.omg.org/cgi-bin/doc?telecom/98-06-15.

[20] B. Oki, M. Pfluegl, A. Siegel, and D. Skeen. The information bus - an architecture for extensible distributed systems. *Operating Systems Review*, 27(5):58–68, December 1993.

[21] L. Opyrchal, A. M. Astley, J. Auerbach, G. Banavar, R. Strom, and D. Sturman. Exploiting ip multicast in content-based publish-subscribe systems. In *In Proceedings of Middleware 2000*, April 2000.

[22] L. Opyrchal and A. Prakash. Secure distribution of events in content-based publish subscribe systems. In *In Proceedings of the 10th USENIX Security Symposium*, pages 281–295, August 2001.

[23] Federal Information Processing Standards Publication. *Data Encryption Standard.* National Bureau of Standards, 1997.

[24] Brian Reid. Usenet readership summary report for june 1994. Monthly posting to news.lists Internet news group, June 1994.

[25] B. Segall, D. Arnold, J. Boot, M. Henderson, and T. Phelps. Content Based Routing with Elvin4. In *Proceedings of AUUG2K*, Canberra, Australia, June 2000.

[26] Bill Segall and David Arnold. Elvin has left the building: A publish/subscribe notification service with quenching. In *Proceedings of AUUG97*, Brisbane, Australia, September 1997.

[27] Douglas R. Stinson and Tran van Trung. Some new results on key distribution patterns and broadcast encryption. *Designs, Codes and Cryptography*, 14(3):261–279, 1998.

[28] Sun Microsystems. *Java Message Service.* http://java.sun.com/products/jms.

[29] Tibco. Tibco randezvous. http://www.tibco.com/products/rv/index.html.

[30] R. van Renesse, R. Cooper K. Birman, B. Glade, and P. Stephenson. Reliable multicast between micro-kernels. In *In Proceedings of the USENIX workshop on Micro-Kernels and Other Kernel Architectures*, pages 269–283, April 1992.

[31] D. Wallner, E. Harder, and R. Agee. Key management for multicast: Issues and architectures (draft). draft-wallner-key-arch-01.txt, September 1998.

[32] C. K. Wong, M. Gouda, and S. S. Lam. Secure group communication using key graphs. In *In Proceedings of ACM SIGCOMM '98*, pages 68–79, September 1998.

[33] Y. Yang and S. Lam. A secure key management protocol communication lower bound. Technical Report TR2000-24, The University of Texas at Austin, Austin, TX, September 2000.

Chapter 9

EXPERIMENTAL STUDY OF SCALABILITY ENHANCEMENT FOR REVERSE LOGISTICS E-COMMERCE

Sarah M. Ryan

K. Jo Min

Sigurdur Olafsson

Abstract: Methods of improving scalability in online auctions include limiting the number of bidding opportunities, providing price information to users, and recommending auctions that may be of interest to the users. We constructed an experimental prototype auction system in the context of reverse logistics for electronics products. Experiments were designed to test the effects of the number of trading opportunities and the amount of previous price and bid information presented to users. The participants' profits improved with the number of trading opportunities but showed mixed effects for increasing price and bid information. The induction of decision trees for an auction recommender is illustrated along with the use of attribute selection to reduce the size of the tree.

Key words: Business-to-business e-commerce, empirical research, recommender systems

1. INTRODUCTION

In recent years, business to business e-commerce has emerged as an important means to enable efficient exchange, reduce marketing costs for sellers and procurement costs for buyers, and unify fragmented markets. Several e-commerce portals have been created for well-defined markets such as automotive parts (Covisint.com, 2001). On the other hand, a fragmented market is characterized by large, dynamic sets of buyers and sellers who are generally unfamiliar with each other. One example is the market for post-retail and surplus items such as used products, returned and refurbished products, and surplus inventories of unsold products nearing obsolescence. In particular, because of their short lifecycles, large quantities manufactured and in use, and potential generation of hazardous waste, electronic products constitute one important category.

More generally, reverse logistics can be defined as the process of planning, implementing and controlling the efficient, cost effective flow of raw materials, in-process inventory, finished goods and related information from the point of consumption to the point of origin for the purpose of recapturing value or proper disposal. It also includes processing merchandise returned because of damage, seasonal inventory, restocking, salvage, recalls and excess inventory. Our focus is on moving goods from their typical final destinations for the purposes of capturing value or proper disposal. Remanufacturing and refurbishing activities as well as recycling for material recovery are common ways of capturing value. Hazardous material content may be of concern also. Kokkinaki et al. (2001a) have identified several ways that electronic commerce could help to unify the markets in reverse logistics. Kokkinaki et al. (2001b) presented a prototype electronic marketplace for PCs including decision support for reuse, remanufacturing or recycling.

The designers of an e-commerce system meant to unify a fragmented market must consider (1) how to encourage participation in the online marketplace, and (2) how to ensure that the system can scale to handle increased traffic as participation grows. This paper focuses on the roles of exchange mechanisms and information delivery in achieving these two related goals. Potential participants will be encouraged to trade in an online marketplace if they perceive that the online system makes exchange easier and/or more profitable than exchanges outside it would be. The system could speed the identification of potential trading partners, provide an efficient exchange mechanism, and deliver useful information about recent prices of items while reducing the need for time-consuming online searches. By performing much of the time-consuming information processing offline or in periods of reduced traffic, these efforts to expand the market could simultaneously enhance the scalability of the system.

The information delivery methods studied in this paper can be better understood by contrast to a popular consumer-oriented online marketplace. Suppose you drop your personal digital assistant (PDA) on the floor and crack the screen. After a call to the manufacturer reveals a fairly steep repair price for your discontinued model, you decide to seek a replacement. Since you own several accessories for your old PDA, you are looking for an identical used or surplus model and, knowing the repair price, you have a ceiling on the amount you are willing to pay. A search for this model on eBay.com yields one thousand records of ongoing auctions, most of which are for accessories rather than the actual PDA. By laboriously examining pages of auctions, you choose several soon-to-end auctions and, by repeated page updates, watch them to learn the settlement prices. Then you choose an auction to enter and start bidding. You can enter your maximum price and let the system bid up for you. In addition, you can ask to be notified when you are outbid. But you soon learn that in order to win your replacement PDA, you must closely monitor the auction at its finish, which generally occurs at an inconvenient time, and repeatedly enter new bids. You are likely to give up and pay the manufacturer's repair price.

Suppose in contrast that you were a regular buyer of used PDAs and that you could participate in a system that proactively notified you when an auction for your preferred model began. This system could inform you of current prices without repeated searches and careful monitoring of auctions on your part. Armed with this reliable price information, you could enter a single bid and then go about your business while the auction was resolved, knowing that other potential buyers were limited to a single bid as well. Assuming you were satisfied with the auction outcomes, such a system would encourage your repeated business and would simultaneously save itself the burden of complying to numerous search requests, page refreshes, and bid updates (both automatic and manual).

We would expect that price information and limits on bidding opportunities would significantly affect buyer and seller behavior, the well-being of participants and the market as a whole. However, these effects have not been studied. In addition, the ability of knowledge discovery methods to reliably classify auctions as interesting to a particular buyer or not has not been rigorously tested. This paper describes an empirical study of a simulated business to business e-commerce system. Experiments were conducted to test the effects of different price information displays and different numbers of bidding opportunities. Decision trees for an auction recommender system were constructed using the data collected. The results of data analysis and data mining for these initial experiments illustrate the promise of the recommender system and suggest several intriguing questions about auction design to explore in further research.

Internet auction has been extensively examined as a way to allocate goods and services effectively and efficiently (see e.g., Kumar and Feldman (1998) or Teich, Wallenius, and Wallenius (1999)). In this paper, we employ an auction mechanism called *sealed bid double auction* (see e.g., Wurman, Walsh, and Wellman, (1998)). Under this auction mechanism, each of multiple potential sellers and multiple potential buyers can submit a bid only once during the auction. Each bid is "sealed" so that only the one that has submitted the bid has any knowledge of the bid throughout the auction process. This auction is, conceptually, one of the simplest auction models in use today. This model relatively (relative to, for example, the English auction model; see e.g., McAfee and McMillan (1987)) induces participants to reveal their true valuation. Considering such factors as the conceptual simplicity, current usage in Internet auction, and relative revelation of true valuation, it is an ideal standard auction mechanism. That is, this mechanism can be used as a benchmark model in the future when other auction mechanisms are comparatively studied.

2. DESCRIPTION OF EXPERIMENTAL PROTOTYPE

For our experiment, we constructed a simulated auction environment. With emphasis on electronics recycling, our business-to-business (B2B) e-commerce participants consist of three Manufacturers, three Demanufacturers, and three Recyclers. Each participant is given a pre-specified level of cash and relevant inventory of items, and each participant's objective is to maximize the level of cash at the end of the auction process. At the end of the auction process, any leftover inventory is worthless.

The auction process consists of several rounds, in each of which there is an open auction for each type of item. In each round, a participant can bid a price and quantity to sell (of items the participant has) and a price and quantity to buy (of items the participant wants). Any cost and price factors that are not explicitly discussed in the next section are negligible.

2.1 Roles of Manufacturer, Demanufacturer, and Recycler

Each Manufacturer produces coffeemakers by using one CPU and one unit of Plastic. The Manufacturer can sell any number of coffeemakers at a fixed unit price of $20. The CPUs and the units of Plastic can be bought in any round of the auction process. At the end of each round, we assume that equal quantities of any CPUs and units of Plastic the Manufacturer has are immediately transformed into coffeemakers. Finally, we assume that the Manufacturer is provided with an initial inventory of obsolete Computers, which can be sold in any round of the auction.

A Demanufacturer can buy obsolete Computers in any round of the auction process. As soon as the Demanufacturer gets them, they are immediately disassembled into memory boards, CPU's and Shells. Each memory board is sold outside the auction system at $4 per unit. Also, if the Demanufacturer does not manage to obtain a fixed quota of memory boards, then the Demanufacturer will have to pay a cash penalty of $2 for each board the Demanufacturer is short after the last round. We assume that the Demanufacturer is provided with an initial inventory of CPUs and Shells. Such inventory as well as the CPUs and Shells from the disassembly of the obsolete Computers can be sold in any round of the auction.

Finally, a Recycler can buy Shells in any round of the auction process. As soon as the Recycler gets them, they are melted down instantaneously into units of recycled Plastic. Each Recycler is provided with an initial inventory of Plastic. Again, we will assume that the initial inventory of units of Plastic as well as the units of recycled Plastic can be sold in any round of the auction process.

Figure 1 depicts transformations of products and materials by the three types of participants. Figure 2 illustrates the exchange of items within the online marketplace and externally.

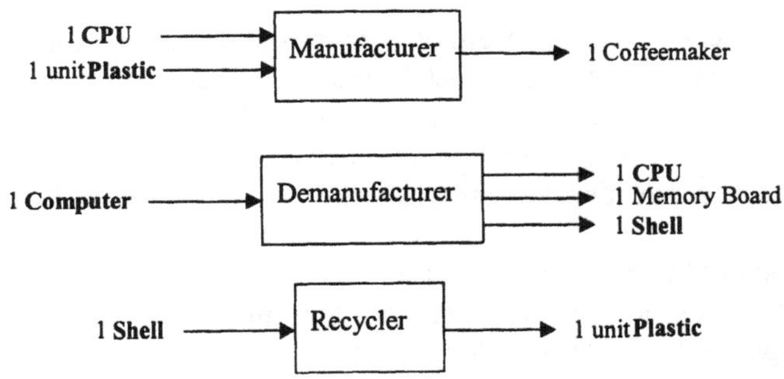

Figure 1. Relationships among products traded online (bold) and sold externally.

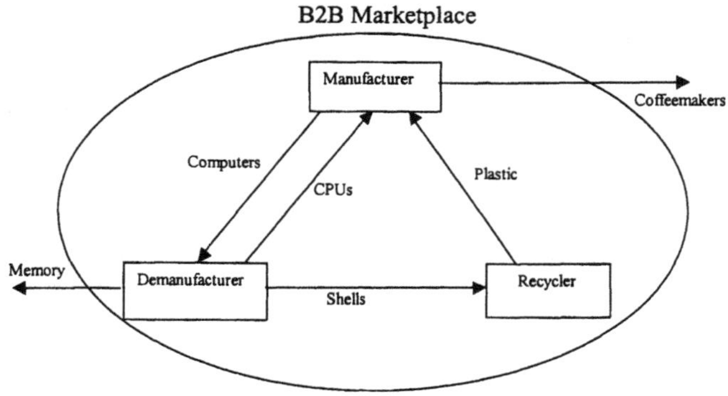

Figure 2. Relationships among participants in the online marketplace

2.2 Sealed Bid Double Auction (SBDA) Mechanism

The auction process consists of either 3 rounds ("short" process) or 6 rounds of SBDA ("long" process). In each round, for a pre-specified kind of product, each potential seller or buyer can submit a bid consisting of the desired quantity and the corresponding price that the potential buyer is maximally willing to pay or the potential seller is minimally willing to accept.

At the conclusion of a pre-specified bid submission period for each round, the auction processing unit rank orders all buy bids according to the maximally acceptable buying price (MABP) and all sell bids according to the minimally acceptable selling price (MASP). By matching rank ordered MABP's and MASP's as well as their respective bid quantities, the minimum MABP and the maximum MASP that maximizes the total transaction quantity can be found.

The settlement price is the average of the minimum MABP and the maximum MASP. If the total quantity demanded with the minimum MABP or higher is not equal to the total quantity supplied with the maximum MASP or lower, which should be the typical case, the following priority is given: Among buyers, the buyers with higher than the minimum MABP have the priority. Among sellers, the sellers with lower than the maximum MASP have the priority. It is quite possible that a buyer or a seller without priority may not be able to buy or sell all the quantity specified in the bid. We will assume that such a buyer or seller is willing to buy or sell less than the exact quantity specified in the bid. Finally, if two or more buyers submit the same MABP or two or more sellers submit the same MASP, then the buyer or seller who submitted the bid earlier has the higher priority.

According to the settlement price and the regulations just described, the transaction records are revealed to all the sellers and buyers with winning bids. The auction process terminates after three rounds (short process) or six rounds (long process) of such SBDA.

3. EXPERIMENTAL DESIGN

We designed the experiments to test the effects of two factors: the information level and the number of rounds in the auction. Each experiment consisted of one auction process. The information factor, INFO, had three levels. The MIN level displayed the settlement price of each auction only to the winning buyer and seller. The MED level displayed the settlement price of each auction to all participants in the auction. The MAX level showed each participant in an auction not only the settlement price but also the price bids entered by his/her competitors (for a buyer, the other buy bids and for a seller, the other sell bids). The number of rounds (K) in each experiment had two levels: $K = 3$ for a short process or 6 for a long process. The level of 3 was chosen as the minimum number of rounds necessary for Computers sold by the Manufacturers to be transformed into Plastic available for their purchase. Six rounds were deemed to be long enough for the market to make maximum use of the starting inventories.

The subjects for the experiments were undergraduate engineering students at Iowa State University. We recruited 27 participants and randomly divided them into three groups, each of which participated in five experiments over two sessions. The first session for each group began with training and some practice rounds. Each subject adopted a different role (M, D, or R) for each experiment conducted in a given session. We conducted four experiments of each of the factor combinations 3*MED, 6*MIN, 6*MAX and one experiment of each of the factor combinations 3*MIN, 3*MAX, 6*MED. The combinations of factor levels were randomly assigned to subject groups and sessions.

3.1 Data Analysis

Online auctions provide a wealth of data from which we can study the behavior of participants, the evolution of prices, consensus about prices or the lack of it, and the overall economic outcomes for participants and the market as a whole. Let i be the index for participants where $i \in M = \{1,2,3\}$

for the Manufacturers, $i \in D = \{4,5,6\}$ for the Demanufacturers and $i \in R = \{7,8,9\}$ for the Recyclers. The item type is denoted by j = 1 for Computers, j = 2 for CPUs, j = 3 for Shells and j = 4 for Plastic. The index l = 1, …, L identifies the experiment (process) and k = 1, …, K_l refers to the auction round, where K_l is either 3 or 6.

For each experiment $l = 1,\ldots,L$, we observed values for the following variables:

p^l_{ijk} = sell price bid by participant i for item j in round k

($p_{ijk} > 0$ if sell bid, < 0 if buy bid)

q^l_{ijk} = sell quantity bid by participant i for item j in round k

($q_{ijk} > 0$ if sell bid, < 0 if buy bid)

r^l_{ijk} = quantity sold by participant i of item j in round k

($r_{ijk} > 0$ if sold, < 0 if bought)

t^l_{jk} = quantity that changes hands of item j in round k

s^l_{jk} = settlement price of item j in round k

($s^l_{jk} > 0$ if the auction was successful, $= 0$ if the auction failed)

d^l_{ik} = cash gain for participant i in round k (loss if negative), which includes revenue from sales of coffeemakers and memory boards. In the final round, any penalties for Demanufacturers are applied also.

From these recorded values we computed aggregate measures of the auction outcomes. In the following, $I[\cdot]$ is an indicator function that equals 1 if the quantity in brackets is true and 0 otherwise.

$T^l_j = \sum_{k=1}^{K} t^l_{jk}$ is the total quantity of item j exchanged

$F^l = \sum_{j=1}^{4}\sum_{k=1}^{K} I\left[s^l_{jk} = 0\right]$ is the number of auctions that failed

$$B_i^l = \sum_{k=1}^{K} I\left[\text{brokering by participant } i \text{ in round } k \text{ of expt. } l\right], i =$$

1,…,9. Brokering attempts were defined according to Table 1.

$$B_T^l = \sum_{i=1}^{9} B_i^L \text{ is the total number of brokering attempts by all}$$

participants.

Let I_{i0} be the initial cash held by participant i. Based on experience with preliminary tests, we attempted to equalize the expected gain in wealth for the three participant types by setting I_{i0} at successively higher values for Manufacturers, Demanufacturers and Recyclers, respectively. In addition, each participant started with some initial stocks of items: for $j = 1, \ldots, 4, I_{ij}$ is the initial stock of item j held by participant i. We valued these initial inventories using estimates of their value over all experiments. The value of item j was estimated as $w_j = \sum_{l=1}^{L}\sum_{k=1}^{K_l} s_{jk}^l \Big/ \sum_{l=1}^{L}\sum_{k=1}^{K_l} I\left[s_{jk}^l \neq 0\right]$. Then the net gain (loss if negative) for participant i in the lth experiment was computed as

$$G_i^l = \sum_{k=1}^{K_l} d_{ik}^l - \sum_{j=1}^{4} w_j I_{ij} - I_{i0} .$$

In order to study the degree of consensus about item prices, we also computed the squared deviation of the bids about the settlement price in the successful auctions:

$$V_{jk}^l = \sum_{i:p_{ijk} \neq 0} \left(\left|p_{ijk}^l\right| - s_{jk}^l\right)^2 \text{ if } s_{jk}^l \neq 0 .$$

Table 1. Identification of suspected brokering behavior by participant and item type

	$j=1$ Computer	$j=2$ CPU	$j=3$ Shell	j=4 Plastic
$i \in M$	$p>0$: normal $p<0$: broker	$p>0$:broker $p<0$: normal	$p \neq 0$: broker	$p>0$: broker $p<0$: normal
$i \in D$	$p>0$: broker $p<0$: normal	$p>0$: normal $p<0$: broker	$p>0$: normal $p<0$: broker	$p \neq 0$: broker
$i \in R$	$p \neq 0$: broker	$p \neq 0$: broker	$p>0$: broker $p<0$: normal	$p>0$: normal $p<0$: broker

Because we observed some qualitative differences in the three subject groups (for instance, one group appeared quiet and studious, while another was talkative), we first checked for any statistically significant differences in the outcomes. First we analyzed the overall outcomes of the experiments.

For each of the response variables T_1, T_2, T_3, T_4, $Q_T = \sum_{j=1}^{4} T_j$, B_T, and F, we tested the null hypothesis that the subject groups were identical using a nonparametric Kruskal-Wallis test (n = 15). The subject group showed a significant effect only on Q_T, the total quantity of material traded. Next we examined outcomes for individual participants. We analyzed the net gain or loss for each participant according to their type, G_M, G_D, G_R; as well as the number of brokering attempts by each individual participant. Of these six response variables, with 45 observations, the subject group membership was significant only for G_M, the gain by Manufacturers. Despite these two warnings that the subject groups were not completely homogeneous, we felt that since homogeneity could not be rejected for most of the performance measures, we could group all the responses across subject groups for the remainder of the analysis in reasonable safety.

In this preliminary study, the sizes of the data sets were relatively small. The sample sizes were L = 15 experiments for the responses T_j, F, and B_T and n = 45 for G_P and B_P, where $P = M$, D, or R (15 experiments with 3 players of each type). Because assumptions of normality seemed untenable in most cases, we relied on nonparametric tests as much as possible. To analyze settlement prices and bid variation, we excluded any rounds in which the settlement price was zero, i.e., the auction had failed. We looked at the effect of the round number and found that if we also excluded the last round, in which all participants had the incentive to "dump" excess inventory and no one was motivated to buy Shells, the hypotheses that the round number had no effect on either settlement price or bid variation could not be rejected. Finally, in the bid variation analysis we eliminated two outlier bids (in two separate experiments) that were an order of magnitude larger than all the other bids. In the end, we had between 50 and 56 rounds of auctions, depending on the item, over which to analyze prices and bids. More details of the statistical analysis, including p-values for hypothesis tests, can be found in Ryan, Min and Olafsson (2002).

3.2 Results

The number of trading opportunities (auction rounds in each experiment) could be expected to affect the wellbeing of participants as well as the health of the market. Based on the profit (gain) for each player i, we make the following observations. When we compare the profits from the three round auction with those from the six round auction, for all three groups of players (Manufacturers, Demanufacturers, and Recyclers), the average profit is substantially higher from the six round auction. We believe that this result occurred because of the substantial expansion of the set of feasible strategies in the case of the six round auction. The longer experiments allowed more material to change hands and also appeared to encourage brokering behavior, particularly by Recyclers. However, we observed that, although some brokering attempts were made by all groups of participants, in no case did a player manage to sell items in a later round for a higher price than he had paid for the same items earlier.

With a couple of exceptions, the number of rounds had little effect on settlement prices or bid variation. The mean settlement price for CPUs was significantly larger for K=3 than for K=6 rounds. The mean variation of bids around the settlement price was larger for CPUs but smaller for Shells in 3 rather than 6 rounds.

The settlement price information revealed to traders and would-be traders could have effects on their behaviors and their eventual economic wellbeing. A trader who was able to use this information effectively might profit by brokering items she did not want or simply by bidding more effectively on items she did want. We originally considered the MED level of information to be the typical amount of price information one would expect when participating in an auction (for instance, this is the information revealed by eBay). Surprisingly, we found that by some measures, participants tended to be worse off when presented with this information than when they received either less or more information about past settlement prices and bids. Generally speaking, the settlement prices and bid variations were similar under MIN and MAX information but *different* under the MED level of information. Due to the small sample sizes, not all these effects were statistically significant. Figure 3 illustrates the effect of information on settlement prices and squared deviations of bids about the settlement price for two different items.

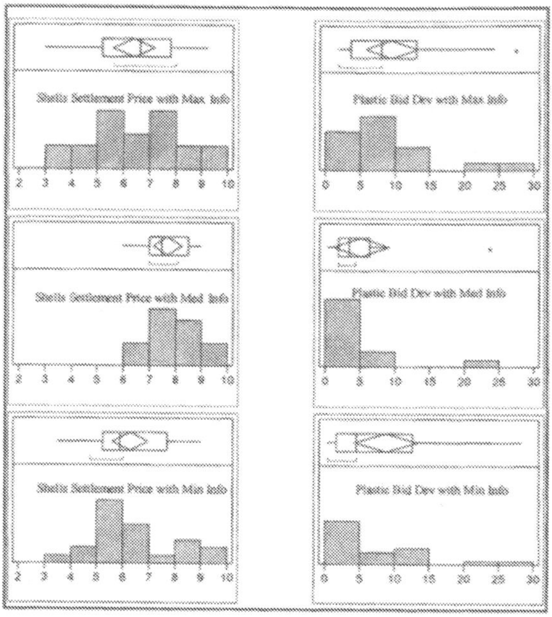

Figure 3. Histograms of settlement prices of Shells and variation in bids for Plastic for different information levels

In particular, the mean settlement price for Computers took its smallest value under MED information, while the settlement prices for the other items were larger under MED information than under either MIN or MAX information. For each item, the mean bid variation was smallest under MED information. These results seem to suggest that by giving all players information about the settlement price, but not competitors' bids, the auction administrator could steer prices for items either higher or lower. Higher prices could be explained by a conjecture that, in contrast to MIN information, MED information motivates a potential buyer who failed to obtain items previously to bid higher than the known settlement price. Subsequent bids are steered to a narrow range just above the previous settlement price. However, if this buyer also learns the range of buy bids in the previous auction, he may feel more comfortable bidding somewhat lower than the previous settlement and, in any case, will not be steered to such a tight interval. Sellers would be motivated to bid in the opposite direction; however, in our setup there is more pressure for Manufacturers and Demanufacturers to buy parts and materials than there is to sell.

When we compare participant gain across the amount of information, the MIN information results in the highest average profit for all three groups.

For the Manufacturer group and the Recycler group, the MAX information results in the second highest average profit. On the other hand, for the Demanufacturer group, the average profits under MAX and MED information differ by only 2.7%. Finally, we observe that the total average profit of all three groups under MIN information is substantially higher than that under MAX information. The total average profit of all three groups under MAX information, in turn, is substantially higher than that under MED information.

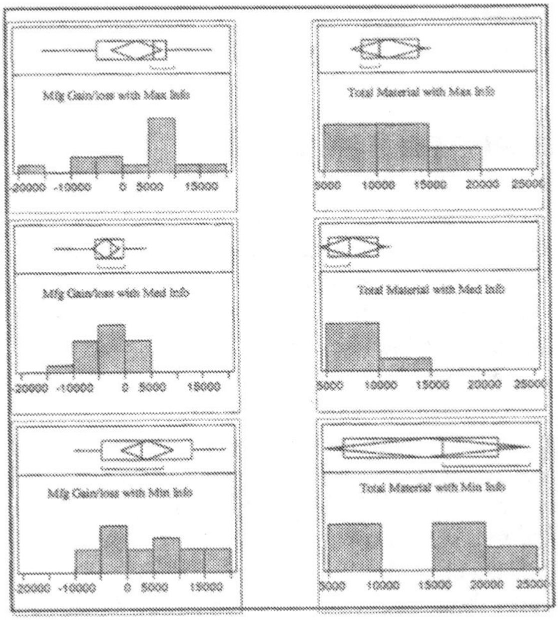

Figure 4. Histograms of gain/loss by Manufacturers and total material traded by information level.

These observations seem to suggest that the MED information case not only leads to a relatively narrow range of settlement prices, but also results in a relatively small amount of profit for all groups of players. If the auction administrator is paid by a fraction of the profit, these results imply that it should strive to provide either MIN or MAX information, and to avoid providing MED information (though providing MAX information may be necessary especially if potential sellers and buyers demand it). However, the

MIN information level may be seen as more consistent with the sealed bid auction environment.

The total amount of material that changes hands during the auction process is smallest under MED information. Figure 4 illustrates the gain/loss by Manufacturers and the total material traded by information level. Finally, the amount of brokering by all participants and particularly by Recyclers (who had the least regular trading to do) was least under MED information. These outcomes suggest that settlement price information alone somehow discourages trading and does not inspire confidence of traders in their ability to successfully broker items.

4. AUCTION RECOMMENDER

In this section we consider what can be learned about the preferences of the auction participants using knowledge discovery and data mining techniques. In particular, we are interested in what auctions might interest each user and should thus be identified or recommended. A **recommender** is a system that can identify which products and hence auctions are of interest to a particular user, and display the information that is relevant for the user to participate in such an auction. Such systems have received considerable attention in the literature with early recommender systems generally based on simple nearest-neighbor collaborative filtering algorithms (Resnick et al., 1994; Shardanand and Maes, 1995). Later versions have used other learning methods, including Bayesian networks (Breese et al., 1998), classification methods (Basu et al., 1998), association rules (Lin et al., 2000), regression (Vucetic and Obradovic, 2000), and intelligent agents (Good, 1999). The key to the successful implementation of such systems is knowing the participants. The more intimate this knowledge, the more likely it is that the system will be able to recommend the correct products.

To gain as much knowledge as possible about the behavior of the auction participants, a data set is constructed that describes the details of their behavior for every auction as well as the characteristic of the auction itself. The following attributes, which are shown along with their possible values, are used in this set:

<u>Product Description</u>

1) Item {Computer,CPU,Shell,Plastic}
2) ItemContainsCPU {T,F}
3) ItemContainsShell {T,F}
4) ItemContainsPlastic {T,F}

Round Information

5) Round	{1,2,3,4,5,6}

Information on Last Transactions to Occur

6) PriceForLastComputerAuction	numeric
7) QuantityForLastComputerAuction	numeric
8) PriceForLastCpuAuction	numeric
9) QuantityForLastCpuAuction	numeric
10) PriceForLastShellAuction	numeric
11) QuantityForLastShellAuction	numeric
12) PriceForLastPlasticAuction	numeric
13) QuantityForLastPlasticAuction	numeric

Information on Last Transaction with Participants Involvement

14) TimeOfParticipantsLastComputerAuction	numeric
15) PriceForParticipantsLastComputerAuction	numeric
16) QuantityForParticipantsLastComputerAuction	numeric
17) CategoryOfParticipantsLastComputerAuction	{buy,sell,none}
18) TimeOfParticipantsLastCpuAuction	numeric
19) PriceForParticipantsLastCpuAuction	numeric
20) QuantityForParticipantsLastCpuAuction	numeric
21) CategoryOfParticipantsLastCpuAuction	{buy,sell,none}
22) TimeOfParticipantsLastShellAuction	numeric
23) PriceForParticipantsLastShellAuction	numeric
24) QuantityForParticipantsLastShellAuction	numeric
25) CategoryOfParticipantsLastShellAuction	{buy,sell,none}
26) TimeOfParticipantsLastPlasticAuction	numeric
27) PriceForParticipantsLastPlasticAuction	numeric
28) QuantityForParticipantsLastPlasticAuction	numeric
29) CategoryOfParticipantsLastPlasticAuction	{buy,sell,none}

Information on Participant

30)ParticipantName	{M1,M2,M3,D1,D2,D3,R1,R2,R3}
31)ParticipantType	{Manufacturer,Demanufacturer, Recycler}

Class Attribute

32) ParticipateInAuction	{yes,no}

From the experiments, 2592 data objects or instances were obtained, one for each user for each auction of every round. Each of these instances has a value for all of the 32 above mentioned attributes, including the class attributes, that is equal to yes if the user participated in the auction and no otherwise. The knowledge discovery task of the recommender system is to relate the class attribute with the others, all of which are observable before an auction must be recommended.

4.1 Constructing a Recommendation

Before using data mining to construct a model for recommendation, let's consider how a very simplistic model might be constructed. As the basic roles of each type of participant are known, for example a Manufacturer needs to sell Computers and buy CPUs and Plastic to make coffeemakers, one might be tempted to always recommend Computer, CPU, and Plastic auctions to a Manufacturer and never recommend any of the Shell auctions. However, this simplistic model ignores the facts that a Manufacturer's interest in Computers, CPUs and Plastic changes as the auction evolves, and that the Manufacturer might want to act as a broker for Shells as well as Computers. Since past behavior of the users is available, data mining techniques can be used to construct a more intelligent recommender.

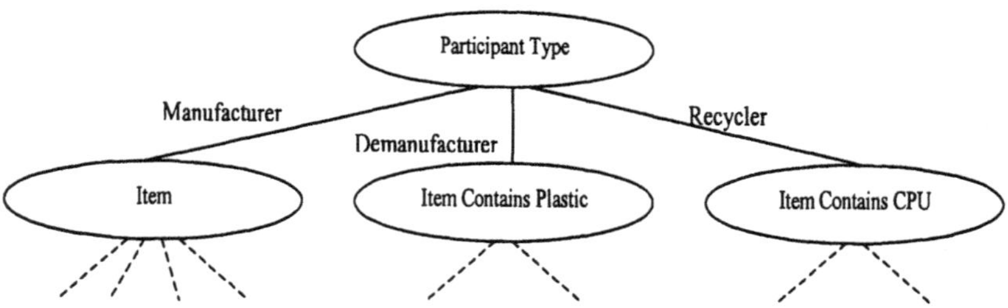

Figure 5. Top of the recommender decision tree

The relevant data mining problem for this context is **classification**. Classification is the process of building a model that describes *a priori* specified classes or concepts based on data objects with known classifications, and then using this model to classify new data objects where the class is unknown. Here the concept to be learned is whether or not a user wants to participate in an auction. The primary performance measure of classification models is its accuracy, but its scalability in terms of induction speed and use of memory is also a critical issue. The most common approach for classification is decision tree induction using algorithms such as C4.5

(Quinlan, 1993) and CART (Breiman, Friedman, Olshen, and Stone, 1984). To build a recommender for the auction system a decision tree model is induced using the C4.5 decision tree algorithm. At the top the decision tree splits according to participant type (see Figure 5), so the three branches can be thought of as decision trees for a Manufacturer, Demanufacturer and a Recycler. Let's consider the tree in Figure 6, which shows the branch or decision tree for a Manufacturer. Clearly, this decision tree implies the need for a much more complicated recommendation model than the simplistic model above. As an example, let's read off the tree when a Computer auction should be recommended to a Manufacturer. The following rules can be directly inferred from the tree:

- If the time since the Manufacturer last participated in a Computer auction is less than or equal to one, then **recommend** the auction.
- Otherwise, if the time since the Manufacturer last participated in a Computer auction is more than one, and the Manufacturer bought the Computers last time, then **recommend** the auction.
- Otherwise, do **not recommend** the auction.

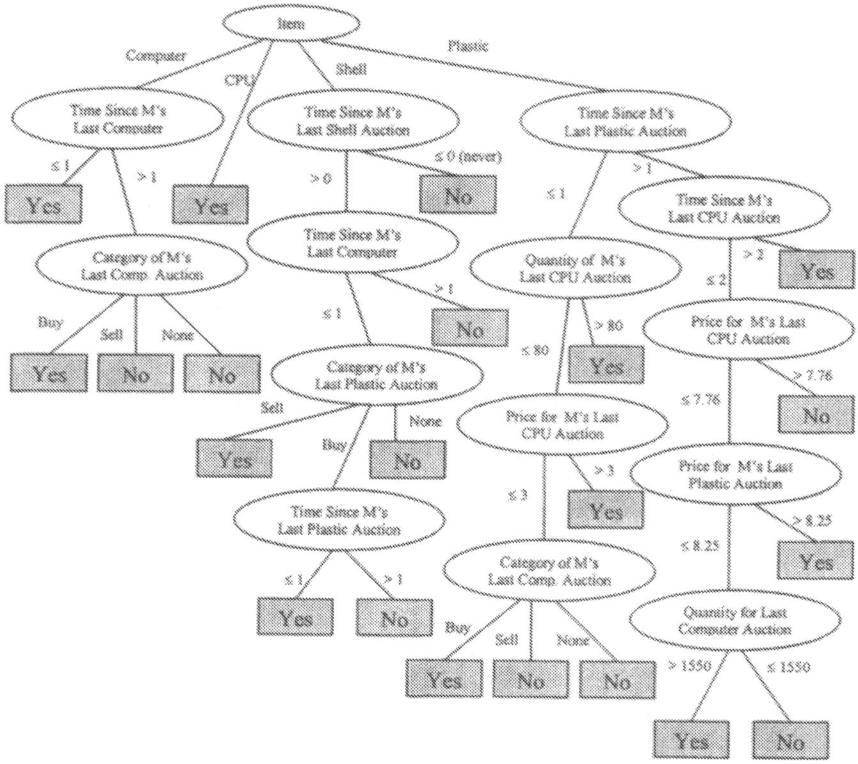

Figure 6. Branch of decision tree for a Manufacturer

Intuitively, one might explain these rules as follows. If the Manufacturer participated in a Computer auction last time (or has never participated), this Manufacturer is likely to be still trying to sell Computers and thus the auction should be recommended. If the Manufacturer did not participate last time, this indicates that they have already sold all of their Computers. The auction is only recommended if they acted as brokers and bought Computers before, in which case they are likely to want to continue in their broker roles. In this case, the recommender discovers two interesting nuggets of information:

1. A Manufacturer may become disinterested in Computer auctions after all Computers have been sold and thus these auctions should no longer be recommended.

2. However, if a Manufacturer wishes to act as a broker and also buy Computers, then Computer auctions should continue to be recommended for future rounds.

This is an example of discovering structural knowledge. The decision tree not only provides a predictive model but also adds new insights into the decision making process of Manufacturers. Finally, note that these rules give a classification for each auction being either recommended or not recommended, but they do not give any confidence assessment of the quality of the recommendation.

Now let's consider when a Shell auction should be recommended to this Manufacturer. The following rules can be read off the tree:

- If this Manufacturer has never participated in Shell auctions, then do **not recommend** the auction.
- Otherwise, if there has been more than one round since the last participation in a Computer auction, do **not recommend** the auction.
- Otherwise, if the Manufacturer's last participation in a Plastic auction was to sell Plastic, then do **recommend** the auction.
- Otherwise, if the time since the Manufacturer's last participation in a Plastic auction is less than one round, **recommend** the auction.
- Otherwise, do **not recommend** the auction.

A Manufacturer will only participate in a Shell auction to act as a broker, and these rules clearly attempt to detect brokering behavior. For example, unless this Manufacturer has brokered Shells before, Shell auctions are not recommended. Furthermore, other behavior such as the Manufacturer selling Plastic is taken as an indication of a wish to serve as a broker and thus Shell auctions are recommended. Finally, note that determining the interest in a Plastic auction is the most complicated of all and requires considering the quantity, price, and timing of previous CPU auctions, brokering behavior as indicated by buying Computers, settlement price of previous Plastic auctions, and the amount of Computers sold most recently. In particular, note the importance of the detailed behavior regarding CPU auctions, which appears to reflect the fact that CPUs and Plastic combine to make coffeemakers.

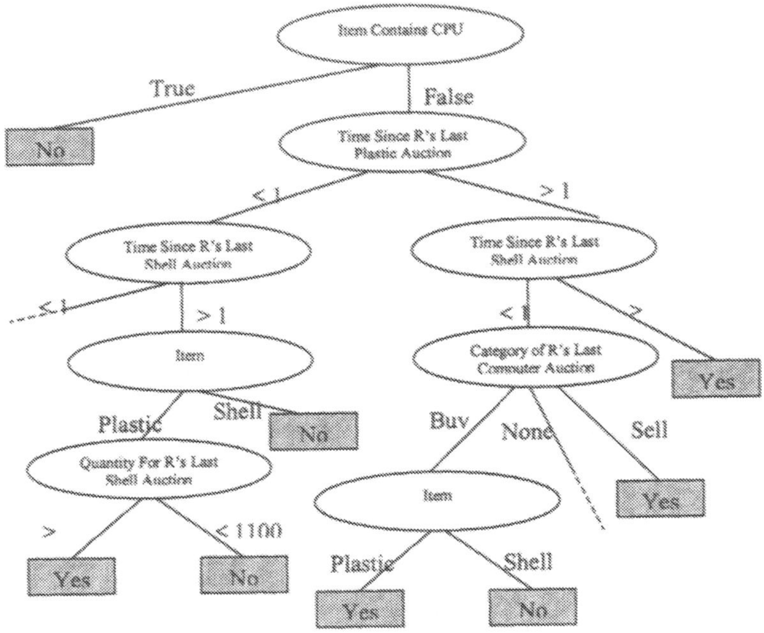

Figure 7. Partial branch for Recycler

Now let's consider the decision of recommending an auction to a Recycler. A partial decision tree for this decision is shown in Figure 7. Note that the tree first splits on the question of whether the item contains CPU or not. If the item contains CPU, that is the item is CPU or Computer, then the auction is not recommended, which implies that an auction participation solely for the purpose of being a broker is never recommended. If the item does not contain CPU, that is the item is Plastic or Shell, then the auction may or may not be recommended and the next two nodes split by determining the time since the last Plastic and Shell auction in which the Recycler participated. For example, if the Recycler participated in a Plastic auction last time but did not participate in a Shell auction last time then the auction is not recommended if it is for Shells and is only recommended for Plastic if the Shell quantity of the Recycler obtained from the last auction is large (1100 units). These rules appear to be discovering the situation that the Recycler is no longer interested in buying more Shells and is only trying to sell the current inventory of Plastic. We finally note that the branch corresponding to Demanufacturer is similar in its discovery and, as indicated by Figure 5, starts by splitting according to whether the item contains Plastic, that is, if the item is Plastic in which case it would only be of interest for brokering, or if it is Computer, CPU, or Shell, in which case it would be of primary interest.

Attribute Selection

In this simple system the number of attributes is only 32 and is thus quite manageable for any standard induction algorithm. However, it is clear that this number of attributes will increase very rapidly. Let's first consider how many transactional attributes there are regarding a single item, such as a Computer:

PriceForLastComputerAuction
QuantityForLastComputerAuction
TimeOfParticipantsLastComputerAuction
PriceForParticipantsLastComputerAuction
QuantityForParticipantsLastComputerAuction
CategoryOfParticipantsLastComputerAuction

Thus, for every new item there are six new transactional attributes, resulting in a very rapid explosion of the number of attributes as the number of items handled by the system increases. Next, let's consider the disassembly information:

ItemContainsCPU
ItemContainsShell
ItemContainsPlastic

Here there are only three attributes regarding components of the items. However, this is a function of the extreme simplification of the prototype system. When dealing with realistic products the number of component related attributes can be expected to grow very rapidly.

When the number of attributes becomes large, the scalability of the system may be compromised as the knowledge discovery algorithms do not always scale up in a satisfactory manner. Thus, it becomes imperative to use attribute selection to determine which attributes are essential to inducing a high quality recommender model. Such attribute selection is commonly used as a preliminary step preceding a learning algorithm such as decision tree induction and has numerous benefits. The resulting tree may be simpler, which often makes it easier to interpret and thus more useful in practice. It is also often the case that simple trees generalize better when they are used for prediction. Thus, a smaller tree is likely to score higher on interestingness measures and may score higher in accuracy. Finally, discovering which attributes should be kept often provides valuable structural information and is therefore important in its own right.

The literature on attribute selection is extensive, and some of the methods applied for this problem in the past include genetic algorithms (Yang and Honavar, 1998), various sequential search algorithms (see e.g., Aha and Bankert, 1996; Caruna and Freitag, 1994), correlation-based algorithms (Hall, 2000), evolutionary search (Kim, Street, and Menczer, 2000), rough sets theory (Modrzejewski, 1993), randomized search (Skalak, 1994), branch-and-bound (Naranda and Fukunaga, 1977), and the nested partitions method (Olafsson and Yang, 2001). These and other attribute selection methods are typically classified as either filtering or wrapper methods. Filtering methods produce a ranking of all attributes before the learning algorithm is applied. Wrapper methods use the learning algorithm to evaluate subsets of attributes. As a general rule, filtering methods are faster whereas wrapper methods usually produce subsets that result in more accurate models.

Using the NP-Wrapper attribute selection method developed by Olafsson and Yang (2001), the following eleven attributes were selected for the entire decision tree (with C4.5 used to evaluate the attribute subsets):

1) Item
5) Round
11) QuantityForLastShellAuction
13) QuantityForLastPlasticAuction
18) TimeOfParticipantsLastCpuAuction
21) CategoryOfParticipantsLastCpuAuction
24) QuantityForParticipantsLastShellAuction
25) CategoryOfParticipantsLastShellAuction
27) PriceForParticipantsLastPlasticAuction
30) ParticipantName
31) ParticipantType

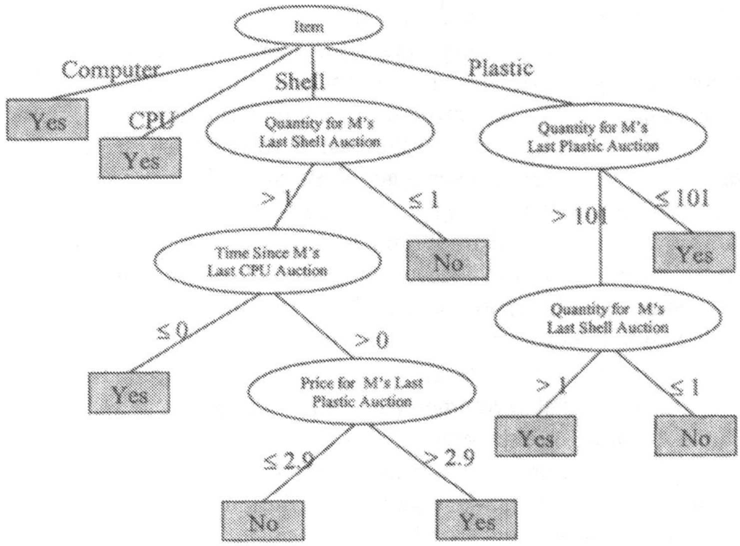

Figure 8. Decision Tree for Manufacturer after Attribute Selection

Note that attributes selected are from a variety of categories and the total number of attributes is reduced to about a third of the original size. With these attributes a new decision tree can be constructed. The branch corresponding to recommending auctions for a Manufacturer is shown in Figure 8. This tree is clearly much simpler than the one constructed before (6 split nodes versus 15 nodes before), yet its accuracy is about the same or 84.5% versus 84.7% estimated using 10-fold cross validation. The induction time for this model is also much shorter, or 1.3 seconds versus 4.3 seconds using all attributes, and when the model is applied for classification in real time its simpler structure will result in a faster response. The trade-off with respect to time is that considerable computational time is expended to select the attributes. However, this can be done off-line and therefore need not affect the scalability of the system adversely.

5. CONCLUSION

In order to build and sustain participation in an online auction site while allowing the system to scale robustly to a large number of users, site administrators might consider providing users with decision support, such as information about past prices and recommendations of current and upcoming auctions. Scalability can also be enhanced by limiting the number of bidding opportunities for any given item. The results in this paper, based on limited experimentation with a simple prototype auction system, suggest that some apparently innocuous types of decision support can have unintended consequences for user behavior and profits, prices of items traded online, and the quantities of material traded. If the goal is to unify a fragmented market or generally to encourage exchange of items such as end-of-life electronics products, the methods used to achieve that goal may be self-defeating. To follow up this initial study, much further research needs to be done to discover more about the effects of price information displays, auction mechanisms (other than sealed bid double auction), and extensiveness of auction opportunities.

From the experimental data, we were able to illustrate the induction of decision trees for an intelligent recommender and to trim the set of relevant auction attributes it uses. The smaller attribute set provides greater scalability. We did not attempt to study the impacts of the recommender system on the outcomes for participants, prices or the market as a whole. Our findings on price information suggest that it is dangerously easy to lead the market in directions that benefit neither it nor its participants. Much further research, both analytical and experimental, is necessary to help auction site administrators design effective decision support systems that achieve the goals of their sites.

***Acknowledgment**: This work was supported by the National Science Foundation under grant DMI-0075575.*

REFERENCES

Aha, D. W., and Bankert, R. L. (1996). A comparative evaluation of sequential feature selection algorithms. In D. Fisher & J.-H. Lenz (Eds.), *Artificial Intelligence and Statistics V*. New York: Springer-Verlag.

Basu, C., Hirch, H., and Cohen, W. (1998). Recommendation as classification: using social and content based information for recommendation. In *Proceedings of the National Conference on Artificial Intelligence*.

Breese, J., Heckerman, D., and Kadie, C. (1998). Empirical analysis of predictive algorithms for collaborative filtering. In *Proceedings of the 14th Conference on Uncertainty in Artificial Intelligence*.

Breiman, L., Friedman, J., Olshen, R. and Stone, C. (1984). *Classification and Regression Trees*. Wadsworth International Group, Monterey, CA.

Caruana, R., and Freitag, D. (1994). Greedy attribute selection. In *Proceedings of the Eleventh International Conference on Machine Learning*, pp. 28-36. New Brunswick, NJ: Morgan Kaufmann.

Covisint.com (2001). Accelerating the pace of business. http://www.covisint.com/downloads/print/overview.pdf (October 29, 2001).

Good, N., Schafer, J.B., Konstan, J.A., Borchers, A., and Sarwar, B. (1999). Combining collaborative filtering with personal agents for better recommendations. In *Proceedings of the National Conference on Artificial Intelligence.*

Hall, M. A. (2000). Correlation-based feature selection for discrete and numeric class machine learning. *Proceedings of the Seventeenth International Conference on Machine Learning*, Stanford University, CA. Morgan Kaufmann Publishers.

Kim, Y. S., Street, W. N, and Menczer, F. (2000). Feature selection in unsupervised learning via evolutionary search. In *Proceedings of the 6th ACM SIGKDD International Conference on Knowledge Discovery and Data Mining.*

Kokkinaki, A. I., Dekker, R., de Koster, M. B. M., Pappis, C. and Verbeke, W. (2001a). From e-trash to e-treasure: how value can be created by the new e-business models for reverse logistics. In *EURO 2001, the Operational Research Conference*, Rotterdam.

Kokkinaki, A. I., Dekker, R., Lee, R., Pappis, C. (2001b). An electronic marketplace for PCs in reverse logistics networks. In *2nd European Conference E-COMM-LINE,* Bucharest.

Kumar, M. and Feldman, S. (1998). Internet auctions. In *Proceedings of the Third USENIX Workshop on Electronic Commerce*, Boston, MA.

Lin, W., Alvarez, S.A. and Ruiz, C.. (2000). Collaborative recommendation via adaptive association rule mining. In *Proceedings of ACM WEBKDD 2000.*

Liu, H. and Motoda, H. (1998). *Feature Selection for Knowledge Discovery and Data Mining.* Kluwer, Boston.

McAfee, R. and McMillan, J. (1987). Auction and bidding. *Journal of Economic Literature*, **25**, 699-738.

Modrzejewski, M. (1993). Feature selection using rough sets theory. In P.B. Brazdil, editor, *Proceedings of the European Conference on Machine Learning*, pp. 213-226.

Narendra, P. M., and Fukunaga, K. (1977). A branch and bound algorithm for feature subset selection. *IEEE Transactions on Computers*, **26**(9), 917-922.

Olafsson, S. and Yang, J. (2001). Intelligent partitioning for feature relevance analysis. Working Paper, Industrial Engineering Department, Iowa State University, Ames, IA.

Quinlan, J. R. (1993). *C4.5: Programs for Machine Learning.* Morgan-Kaufmann, San Mateo, CA.

Resnick, P., Iacovou, N., Suckak, M., Bergstrom, P., and Riedl, J. (1994). Grouplens: an open architecture for collaborative filtering of netnews. In *Proceedings of ACM CSCW'94 Conference on Computer-Supported Cooperative Work*, 175-186.

Ryan, S., Min, J. and Olafsson, S. (2002). Experimental study of reverse logistics e-commerce. In *Proceedings of IEEE International Symposium on Electronics and the Environment*, San Francisco, 218-223.

Shardanan, U. and Maes, P. (1995). Social information filtering: algorithms for automating 'word of mouth'. In *Proceedings of ACM CHI'95 Conference on Human Factors in Computing Systems*, 210-17.

Skalak, D. (1994). Prototype and feature selection by sampling and random mutation hill climbing algorithms. *Proceedings of the Eleventh International Machine Learning Conference*, pp. 293-301, New Brunswick, NJ: Morgan Kauffmann.

Teich, J., Wallenius, H. and Wallenius, J. (1999). Multiple-issue auction and market algorithms for the World Wide Web. *Decision Support Systems*, **26**, 49-66.

Vucetic, S. and Obradovic, Z.. (2000). A regression-based approach for scaling-up personalized recommender systems in e-commerce. In *Proceedings of ACM WEBKDD 2000.*

Wurman, P., Walsh, W. and Wellman, M. (1998). Flexible double auctions for electronic commerce: theory and implementation. *Decision Support Systems*, **24**, 17-27.
Yang, J., and Honavar, V. (1998). Feature subset selection using a genetic algorithm. In H. Motada and H. Liu, editors, *Feature Selection, Construction, and Subset Selection: A Data Mining Perspective*, Kluwer, New York.

Chapter 10

WEB-BASED DISTRIBUTED MULTI-AGENT ARCHITECTURE FOR IMPLEMENTING VALUE NETS

Soundar R. T. Kumara*, Chad Dodd

Department of Industrial and Manufacturing Engineering
The Pennsylvania State University, University Park, PA 16802, USA

Taioun Kim
Department of Industrial Engineering
Kyungsung University, Nam-gu, Busan, Korea

Abstract: The value net creates a network of suppliers, and distributors such that the product becomes customer centric. In order to implement the value net model, information processing and distribution need to occur in real time. Software agent technology is becoming popular due to the inherent characteristics of autonomy, distributedness and modularity. In this reserach, we adopt agent technology to handle all real time decision processes, making the value net model a complex multi-agent network of decision makers. For the agents to properly coordinate their respective activities we develop Valcomm, a Web-based multi-agent language grounded in the XML and Java. Such a system will enhance the overall performance of all the participating organizations. Valcomm will truly implement extended enterprises.

Keywords: value net, multi-agent, XML, supply chain

---

* Corresponding author: skumara@psu.edu

1. INTRODUCTION

Performance enhancement of organizations can be accomplished by embracing information technology. When enterprises interact with other enterprises forming extended enterprises information exchange becomes critical for sustained operation. Supply chains are an example of implementing extended enterprises. Supply chains rely on forecasts and implement these forecasts. Value nets, on the other hand which are digital implementation of supply chains, respond to customer requests on-line and in real-time. This leads to the enhanced overall performance of the organization.

Over the past few years' enormous technological advances have changed the way many organizations conduct business. During the current digital age, companies are expected to be agile, flexible, and customer oriented. They're required to provide cost quotes, delivery times, delivery tracking, and production updates to the customers in real time. In response, organizations are reengineering and streamlining various processes and business practices to improve customer relations and productivity. One process that businesses in all sectors of industry are finding to improve customer relations, productivity, and efficiency is the supply chain.

The supply chain is a process that spans the entire enterprise integrating processes based on flawless delivery of basic and customized services. Therefore the objective of Supply Chain Management (SCM) is to optimize information and product flow from the receipt of the order to the purchase of raw materials, to delivery and consumption of the finished product [Fox et al, 1993]. For the supply chain to operate efficiently during the digital era, all the functions and processes must be integrated within a network providing a rapid and quality response to any demand or informational request from the customer. Therefore the supply chain becomes a digitalized and dynamic, high performance network of customer/supplier partnerships and information flows. This new model of the supply chain is known as a value net [Bovet and Martha, 2000]. This gives the facility to truly integrate the internal operations of other organizations forming extended value nets that represent extended enterprises.

The value net model creates an informational infrastructure that facilitates the distribution of information from a customer to the various manufacturing sites and suppliers. Therefore the problem is one to fulfill the customer order with high quality products meeting delivery requirements while minimizing the cost of the parts, manufacturing, and transportation. To make these decision-making processes efficient and hence make the value net reliable, it is imperative that we integrate and automate all the decision-making processes to solve the various different value net problems. Within the old supply chain model the automation of some decision processes led to Enterprise Resource Planning (ERP). However the solution to the value net problem lies within a network of autonomous and intelligent software agents that react to the environment and to the different nodes of the net. Each agent would represent a part of the decision making process within the value net. The agents thus create an agile network of decision makers that react to the customer in real time as opposed to a rigid process flow that is decided before the customer places an order.

Intelligent software must be utilized to create real time decisions distributed across a worldwide network (World Wide Web – WWW) to make the value net instantaneous. Thus intelligent agents by virtue of their adaptability, autonomy, and social ability make them a viable technology in the implementation of a value net. Each intelligent agent within the network would represent a part of the decision making process and would contain information specific to their process. In the Internet world, agents would communicate with each other to realize the value net.

We consider the value net of a manufacturing enterprise as a world-wide network of suppliers, factories, warehouses, distribution centers, and retailers through which information is processed and raw materials are acquired, transformed into products, and then distributed to customers. Therefore in this chapter we will model the value net with heterogeneous systems and intelligent agents distributed across the Internet. The focus of this work is more on modeling the communication in the value net.

Problem Definition

The current supply chain problem within many organizations is two fold. The first problem concerns the way orders are proposed by the customer, quoted by the manufacturer, and agreed upon by both the customer and the manufacturer. The second problem concerns the actual decision processes and methodology used from the time an order is placed to the delivery of the final product to the customer. Though these two problems may seem to be different, both deal with integrating information from different sources. This similarity leads to the generation of a common solution strategy. In this chapter we look at an agent based framework for implementing such a solution strategy.

The first part of the current supply chain problem lies within the company organization and process flow. A typical manufacturing organization has many different facilities, customers and suppliers interconnected by a supply chain. Many companies use an ERP system to help with the decision making process. This type of ERP system begins with a demand forecast, then considers the manufacturing facility, and then works gradually down to each supplier through Electronic Data Interchange (EDI). Smaller organizations have the same supply chain model, however they utilize more archaic and manual methods to make the necessary decisions. These types of supply chains are slow, inefficient, and often, but not always, conform to a fixed product line. Therefore a multi-agent value net needs to be implemented, so that all information given to the company and driven by the customer will be distributed to the network of departments and suppliers in real time. The idea of value net is shown in Figure 1.

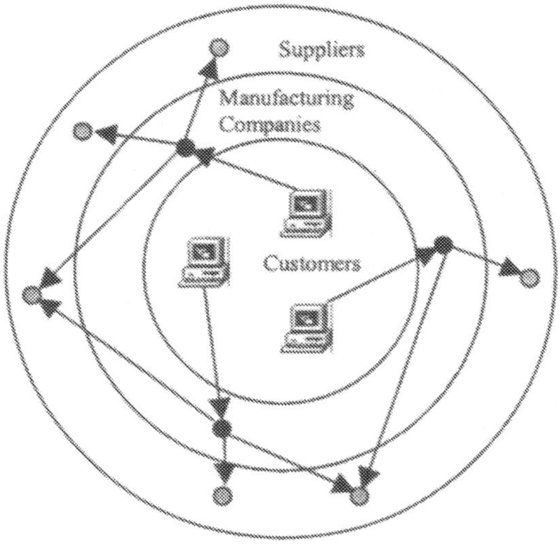

A value net is centered around the customer, where decisions are made and transmitted in real time to the manufacturing facilities and suppliers within the network. The first tier of the network is the customer, then the manufacturing facility, and finally in the third tier the suppliers

Figure 1. A Value Net [Bovet and Martha, 2000]

The other problem of the current supply chain model involves the decision-making processes. This problem is particularly complex since the input to one manufacturing facility is the output from another facility within the network. Thus each facility must make decisions based upon its suppliers' decisions and consequently making the value net as strong as the decision-making ability of each facility. Therefore, a facility needs to make accurate decisions in real time concerning its inventory levels, manufacturing schedules, and total cost and then relay that information to its respective customers as well as other facilities in the extended entreprise.

Current supply chain models and multi-agent supply chain models have many limitations that prevent them being customer centric. The customer will drive future orders, and the decisions within the value net, and therefore these decisions will have to be made instantly and accurately. Hence to make any organization truly efficient, agile, and customized a technology-based multi-agent value net must be implemented. We therefore propose a multi-agent system to implement a customer centric value net in the Internet driven eMarkets.

2. LITERATURE REVIEW

Presently the automation of the supply chain model has been researched using intelligent agents, however a web-based value net using multi-agents has not been investigated. In the traditional supply chain manufactures produce products and push them through the distribution channels to the customer. In contrast, a value net is centered around the customer, and it captures "their real choices in real time and transmits them digitally to all net participants" [Bovet and Martha, 2000].

Various multi-agent systems have been researched and created in order to generate models, however Fox et. al. [1993] were probably the first ones to propose organizing the supply chain as a network of intelligent agents. In 1996 Kalakota et al. [1996] generated a model that created more generalized agents; agents that represented the distribution centers, manufacturing facilities, inventory, and loading. Therefore only the flow of material through the supply chain was modeled in this multi-agent design. The real time supply chain optimization system generated by this team utilized the Internet and EDI to transfer information. In addition, the work completed by Hinkkanen et al. in 1998 has suggested various distributed decision support systems for real time supply chain management that are completely managed by agents. They suggest that an agent can be modeled as a human and thus agents can drive the supply chain [Hinkkanen et al, 1998]. Like the work completed by Kalakota, et al., this proposal suggests using EDI as the medium to exchange information. Yung and Yang [1999] proposed another multi-agent supply chain model. This model was designed to solve the constraint satisfaction problem within the supply chain and therefore consisted of three types of agents; propagation, interface, and control agents. Huhns and Stephens [2001] suggested automating the interchange of standard business documents. Enterprise need only describe their supply processes using OAG (Open Application Group) standard business documents and UML (Unified Modeling Language) interaction diagrams. The methodologies and tools convert these UML diagrams into specifications for software agents, which then cooperate in automating the resultant supply chain. Balakrishnan et. al. [1999] discuss several research issues related to product realization in the digital era, along with research focus related to agent research. Considering rapidly expanding B2C or B2B environment, negotiating supplier contracts between suppliers and customers is a complicated process. Collins et. al. [2001] has proposed multi-agent test-bed designed to support multiple agents in negotiating contracts for tasks with temporal and precedence constraints.

3. MULTI-AGENT ARCHITECTURE

The Multi-Agent Value Net (MAVN) proposed in this paper consists of multiple agents that communicate across the Internet and Intranets of each business enterprise involved in the network. Each business within the network is considered as an agent society, a group of agents working together to achieve a goal.

Utilizing the hierarchical architecture and the message performatives, the communication complexity of the multi-agent value net will grow controllably as the system increases in size. Typically each departmental agent has a set of non-redundant tasks, however as the degree of coordination decreases, departmental agents will have to share tasks as to reduce resource contention and increase agent response time. Since the coordinators are responsible for the group of departmental agents, the amount of work a coordinator oversees and distributes will grow with an increase in the departmental agents. Hence the coordination problem shifts from the departmental level to the coordinator level. After the expansion of the departmental agents, the coordinator agents will begin to increase workload and decrease resource availability. To alleviate their workload the coordinator agents will need to allocate their required tasks to other agents. The two-tier multi-agent architecture will then expand into a three-tier architecture with two levels of coordinator agents, and an expanding departmental tier. The three-tier architecture generates a departmental coordinator for each department that will be responsible for coordinating all the activity within the respective departments. The department coordinator will then respond to the coordinator of the entire society of agents. The three-tier architecture and responsibility distribution is illustrated in Figure 2.

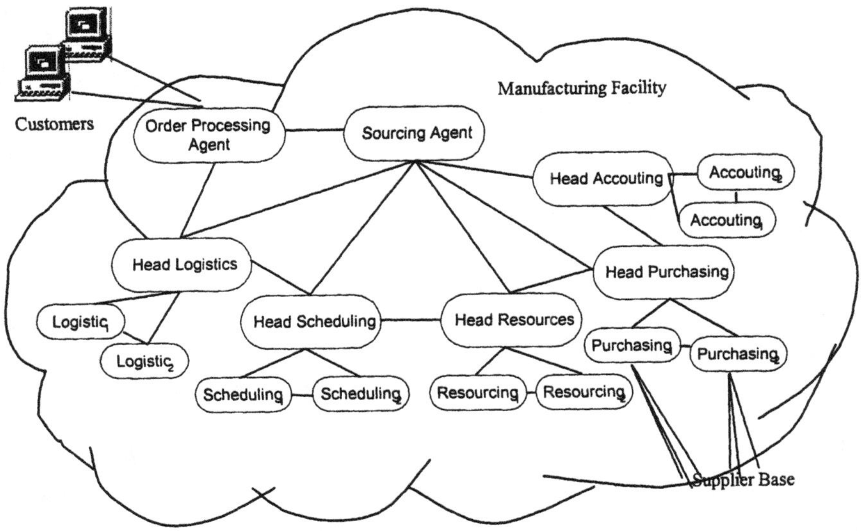

Figure 2. Multi-Agent Value Net (MAVN) Architecture

The hierarchical architecture of MAVN is centered around the roles and responsibilities of individual agents. In this specific design, we are utilizing a variation of the master-slave organizational structure and we have developed three types of agents, the master agents (coordinator agents), the slave agents (departmental agents), and agents which will have duties as both a master and a slave (service agents). The departmental agents will be responsible to the coordinator agents and the coordinator agents are responsible to the customer. Service agents will have a unique dual role; they will be responsible to a coordinator but will also have coordinator responsibilities.

The master-slave-service architecture defines the specific roles each agent has within the dialogue of a conversation. Depending on an agents' responsibilities, within the value net architecture, an agent will either have a passive or an active role in a conversation.

The passive agent accepts a query from an external source and then reacts toward solving the given problem from their decision algorithms, beliefs, and / or environmental variables. After analyzing the problem or request and generating a solution, the passive agent will inform the speaker of a recommended solution. The active agent will then issue the queries and also make the necessary assertions to departmental and service agents based on customer requests and environmental conditions. The active agent does not have the ability to form queries but can dictate and distribute them to departmental and service agents. Therefore within MAVN, the coordinators will have an active role, the departmental agents will have a passive role, and the service agents will have both a passive and an active role depending on the request.

4. AGENT COMMUNICATION

Coordination is the process of managing dependencies between activities. Within MAVN the activities are managed through messages and interactions. Interactions occur when two or more agents are brought into a dynamic relationship through a set of activities, which have a strong influence on the future of the environment. In this research we have designed and implemented a communication language and protocol, which generates successful social interactions between agents and coordinates all activities within the multi-agent system. This language is called the Multi-Agent Value Net Communication Language (Valcomm). Valcomm will allow individual agents to understand the goals and activities of other agents through social interactions and therefore manage the relationships and communications passed within the system. Finally, Valcomm will define a protocol needed to achieve the global goals within the multi-agent system.

Valcomm is a message based communication language. The messages will be sent across the Internet or Intranets, depending on the location of both the sender and the receiver, using TCP/IP supported transport services. Since an agent is considered a programmable object, it will be able to exchange messages within a structured set of conversations between other agents or humans across the Internet.

4.1 Information/workflow within the multi-agent value net

The workflow and information flow of an order through the multi-agent value net can be split into two parts: information/workflow for an order proposal and information/workflow for an order. The information workflow for a proposal is shown in Figure 3.

After the proposals are agreed upon by both the manufacturing facility and the customers, the following steps are accomplished.

1) After the customer has agreed upon an order, the proposal, constraints, activities, and final order are all passed to the Sourcing Agent.

2) The Sourcing Agent will store and distribute all the information regarding a final order throughout the life cycle of that order.

3) Each departmental agent will determine if the necessary activities are still feasible. If the activities are no longer feasible, a message is returned to the Sourcing Agent describing the problems.

4) If the Sourcing Agent receives a reject message, the Sourcing Agent will decide if an alternative plan can be developed. If a plan cannot be developed the customer is notified of the problems.

5) If the Sourcing Agent receives an Accept message from all the departmental agents, then the order information is sent to the Accounting Agent, and the Accounting Agent takes care of all the billings.

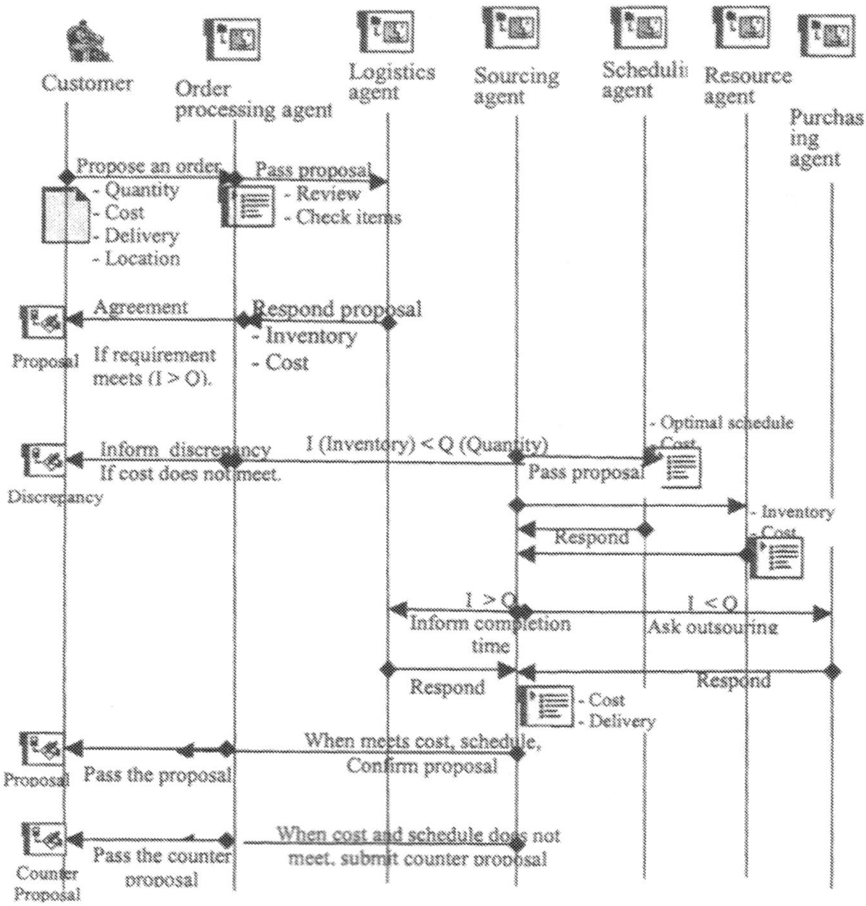

Figure 3. Information / workflow process for the processing of a proposal

4.2 BDI model

The BDI (belief, desire, and intention) logic model is used to describe an agents' cognitive state. An agent is continually updating its beliefs based on perceptions, using its beliefs to reason about possible plans, committing to certain intentions based on its beliefs and desires, and realizing these intentions by acting. We will use the predicate BEL_s to describe the beliefs of the speaker, BEL_L to describe the beliefs of the listener, and f to denote a ground formula as in first order logic. The belief set (*BEL*) of an agent is a set of ground formulas that the agent believes to be true. Due to an application dependent set of rules we are able to illustrate the revision of an agents' beliefs as in equation (1) [Satapathy and Kumara, 1999].

$$Bel_L + BEL_S(f) \rightarrow BEL_S' \text{ (Belief Addition)} \quad (1)$$

$$Bel_L - BEL_S(f) \rightarrow BEL_S' \text{ (Belief Extraction)}$$

We also need to introduce another notation *BEL* tautologically implies f,

$$BEL \models f \text{ iff } f \in (AD) \quad (2)$$

if and only if every truth about *BEL* also satisfies f and if and only if f exists within the algorithmic decisions of the agent. In other words the belief set of an agent contains the results of algorithmic decision created by the same agent. The algorithmic decision is generated from the agents' current belief set and the current set of constraints given to the agent by a speaker.

Desires are goals that an agent is trying to achieve. We will use the predicate *GOAL* to describe the set of goals for a specific agent. Each goal within the set is considered as 'g' and $g \in GOAL$. Typically, 'g' is a ground formula that the agent is trying to achieve and therefore 'g' does not exist in the agents' current set of beliefs. However the agent believes that 'g' will exist in a future set of beliefs [Rao and Georgeff, 1995]. When the goal becomes a belief then 'g' will not exist in the set of goals but in the agents set of beliefs. We can add and extract goals within a set just as we had done with the belief set.

$$GOAL + g \rightarrow GOAL' \text{ (Goal Addition)} \quad (3)$$

$$GOAL - g \rightarrow GOAL' \text{ (Goal Extraction)}$$

Intentions are the set of plans or actions arranged in some order to accomplish the goals or desires of the agent. The resources agent will intend to use all of one product before ordering more of the same product. This will help the resources agent achieve the desire of low inventory for all raw material and subcomponents in the future by doing inventory distribution in the current state.

Let *INTENT* denote a set of plans and each plan has to contain a set of actions *ACT* that when performed results in achieving a goal. Each action, designated by a within the set of actions *ACT* generates a set of expected results. Therefore we can designate a plan as a collection of actions on a time line by *PLAN(BEL, BEL')*. *PLAN(BEL, BEL')* indicates that if an agent believes *BEL* then the actions in the plan can be executed and if the actions are successfully executed then the agent believes in *BEL'*. Assuming that a plan is an element of the intentions of an agent, we can denote an action a to be an intended action by $INTENT \models a$, stated as follows [Satapathy and Kumara, 1999]:

$$INTENT \models a \text{ iff } (\exists P(BEL, BEL') \in INTENT) \text{ s.t. } (a \in P(BEL, BEL')) \quad (4)$$

Similar to the addition and retraction of beliefs and goals, we can also define the addition and extraction of intentions as follows:

$$INTENT + PLAN(BEL, BEL') \rightarrow INTENT' \text{ (Intent Addition)} \quad (5)$$

$$INTENT - PLAN(BEL, BEL') \rightarrow INTENT' \text{ (Intent Extraction)}$$

4.3 Semantics of the Valcomm Performatives

The semantics of the performatives used in the agent based value net communication language (Valcomm) are important for the customer-coordinator, coordinator-coordinator, service-coordinator, service-departmental, and coordinator-departmental interactions. The general Performatives are summarized in the Tables 1,2 and 3. These are based on Rao and Georgeff [1995].

Table 1. Performatives interacting with the knowledge base of another agent

Performatives	Issued for the cases	Expected actions of listeners
Ask	The listener will assert the truth based on their knowledge base and / or algorithmic decisions.	$[BEL_L \models f] \Rightarrow$ [mess (inform(content, data))]
Inform	Giving the listener information based on the experiences or algorithmic expression generated by the speaker.	$BEL_L + BEL_S(f) \rightarrow Bel_L'$
Retract	Which information (*f*), previously given by the speaker, should be disregarded.	$BEL_L - BEL_S(f) \rightarrow BEL_L'$
Subscribe	A form of the ask performative, which is only initiated once and is continued until the unsubscribe performative has been issued.	$[[BEL_L^{t_0} \models f_{t_0}] \Rightarrow$ [mess (inform(content, data))]] $\wedge [BEL_L^{t_0} + BEL_S(f) \rightarrow BEL_L^{t_0}] \wedge [[\forall t > t_0, BEL_L^t \models BEL_S(f_t)] \wedge \Rightarrow$ [mess (inform(content, data))]
Unsubscribe	To retract a current subscribe message.	$BEL_L - BEL_S(f) \rightarrow Bel_L'$

Table 2. Performatives used to bind other agents

Performa-ives	Issued for the cases	Expected actions of listeners
Commit	To commit the listener to perform a future activity (*f*).	$[GOAL_L + g_S(f)] \wedge [INTENT_L \models ACT] \Rightarrow$ [mess(Accept(content, data))[$[GOAL_L + \neg g_S(f)]$ $[INTENT_L \models \neg ACT] \Rightarrow$ [mess(Reject(content, data))]
Uncommit	To retract a previously sent commit message.	$[GOAL_L - g_S(f)] \wedge [INTENT_L - PLAN(BEL(f), BEL') \rightarrow INTENT_L'] \Rightarrow$ [mess(Accept(content, data))]
Request	When the speaker believes that a future activity (*f*) will be needed to successfully achieve a future goal.	$TRUST_L \Rightarrow [GOAL_L + g_S(f)] \wedge [INTENT_L \models ACT]$
Unrequest	To retract a current request	$TRUST_L \Rightarrow [GOAL_L - g_S(f)] \wedge [INTENT_L - PLAN(BEL(f), BEL') \rightarrow INTENT_L']$

Table 3. Performatives regarding acceptance or rejection

Performa-ives	Issued for the cases	Expected actions of listeners
Accept	After receiving a subscribe, unsubscribe, commit, and uncommit performative from the speaker.	$BEL_L + BEL_S(f) \rightarrow Bel_L'$
Reject	From the listener of a previously sent subscribe, unsubscribe, commit, and uncommit performative.	$[[BEL_L - BEL_S(f) \rightarrow Bel_L] \wedge [BEL_L \models f] \Rightarrow$ [mess(Inform(content, data))]

4.4 Multi-Agent interactions

The interactions within MAVN are defined by the messages sent between any of the agents or between the customer and an agent. Since the message classes are defined above and the information and work flow follows a set of sequenced operations, all of the interactions within MAVN can be categorized into three general types of interactions; coordinator/customer-coordinator (link A), coordinator-departmental (link B), and then departmental-external coordinator (link C). These links are illustrated in Figure 4.

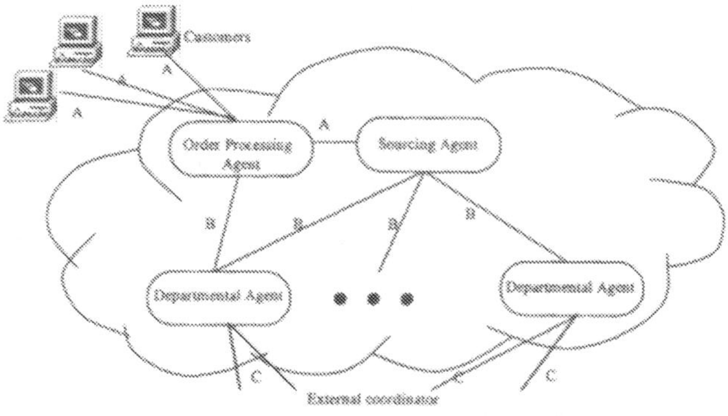

Figure 4. Valcomm Communication Links

4.4.1 Customer / Coordinator – Coordinator Interactions (Link A)

A customer will issue a proposal for an order to the Order Processing Agent across the Internet. This specific customer wants to know if the manufacturing organization will be able to fulfill the order within an appropriate amount of time, and therefore the message class is defined as **ask**. Sometimes the Order Processing Agent will pass the information onto the Sourcing Agent and the message will again contain the **ask** performative. The content of the message is stated as follows (illustrated in predicate logic):

$$\left(\Pr op_{rj}(i) \mid x_{rj},\ \forall i \in I\right) \tag{6}$$

The predicate Prop_{rj} denotes the proposal number 'r' from customer 'j' on product 'i', with respect to the constraint set x_{rj}. The constraint set x_{rj} can be expressed as

$$x_{rj} = \langle q,k,t,c \rangle$$

Where 'q' is the order quantity, 'k' is the delivery location, 't' is the desired delivery time, and 'c' is the cost constraint. The Valcomm for this message is described below:

Valcomm Message

```
<Body>
        <Content>? Proposal</Content>
        <PropNum>1</PropNum>
        <Data>
                <ProductInfo> " ... " </ProductInfo>
                <ShippingInfo> " ... " </ShippingInfo>
                <BillingInfo> " ... " </BillingInfo>
        <Data>
</Body>
```

All the links from A.1 to A.8 are summarized in Table 4.

4.4.2 Coordinator – Department (Link B)

After a proposal is received from the customer, the Order Processing Agent and / or the Sourcing Agent will need to send an **ask** performative to the respective departmental agents inquiring about the cost and the time associated with completing their respective activity. The links related to the Coordinator-Departmental interactions are summarized in Table 5.

4.4.3 Internal Departmental-External Coordinator Interactions (Link C)

The interactions between the Purchasing and Logistics Agents to external organizations will be the same as those described in the previous section. However the Purchasing and Logistics Agents would be the customer and therefore be placing the proposals as opposed to reacting to a proposal.

Table 4. Customer-Coordinator Interactions

Link	Speaker	Listener	Performative	Valcomm Content Section	Content Description
A.1	Customer / Order Processing Agent	Order Processing Agent / Sourcing Agent	Ask	?Proposal	Submit a proposal.
A.2	Order Processing Agent / Sourcing Agent	Customer / Order Processing Agent	Inform	Accept	Confirm the proposal.
A.3	Order Processing Agent / Sourcing Agent	Customer / Order Processing Agent	Inform	CounterProposal {*Area(s) of Dispute*}	A counterproposal.
A.4	Customer / Order Processing Agent	Order Processing Agent / Sourcing Agent	Commit	Execute Proposal	Commit to an order.
A.5	Order Processing Agent / Sourcing Agent	Customer / Order Processing Agent	Accept/Reject	Acknowledge / Problem {*reason*}	Accept or Reject the commitment.
A.6	Order Processing Agent / Sourcing Agent	Customer / Order Processing Agent	Retract and Inform	Deficient {*Area(s) of deficiency*}	Deviation from a current order.
A.7	Customer / Order Processing Agent	Order Processing Agent / Sourcing Agent	Uncommit	HaltExecution	Cancel an agreed upon order.
A.8	Customer	Order Processing Agent	Subscribe / Unsubscribe	Subscribe{*to what, how often*} or Unsubscribe {*to what*	Subscribe/Unsubscribe to the status of an order.

Table5. Coordinator-Departmental Interactions

Link	Speaker	Listener	Performative	Valcomm Content Section	Content Description
B.1	Sourcing Agent / Order Processing Agent	Departmental Agent	Ask	SolveProp{(?cost)(?Active)(?Start)(?Com)}	Submit a proposal.
B.2	Departmental Agent	Sourcing Agent / Order Processing Agent	Inform	Accept or CounterProp{*reason*}	Inform about completion of the proposal.
B.3	Sourcing Agent	Departmental Agent	Commit	Execute Proposal	Commit to an order.
B.4	Sourcing Agent	Departmental Agent	Uncommit	HaltExecution {*Problem*}	Cancel an agreed upon order.
B.5	Departmental Agent	Sourcing Agent	Retract and Inform	Deficient {*Area(s) of deficiency*}	Deviation from a current agreed upon order.
B.6	Departmental Agent	Sourcing Agent	Accept / Reject	Acknowledge or Problem {*reason*}	Accept or Reject the commitment

5. IMPLEMENTATION OF MULTI-AGENTS FOR VALUE NETS

An object-oriented Java and XML approaches were employed in the development of Valcomm, and thus could be used within many multi-agent systems currently deployed or under development and can handle data intensive messaging. Simple Object Access Protocol (SOAP) was adopted as a XML parser. SOAP was designed in 1999 to use valid XML as a wire protocol, or packet layout language, that can be transported to a remote system, typically using HTTP. There are other software packages necessary for the development of Valcomm. These packages are not as critical, however are necessary for development. The three other software packages utilized were the Apache web server, Xerces Java Parser 1.3.1, and Xalan. The protocol stack adopted to implement Valcomm is shown in Figure 5.

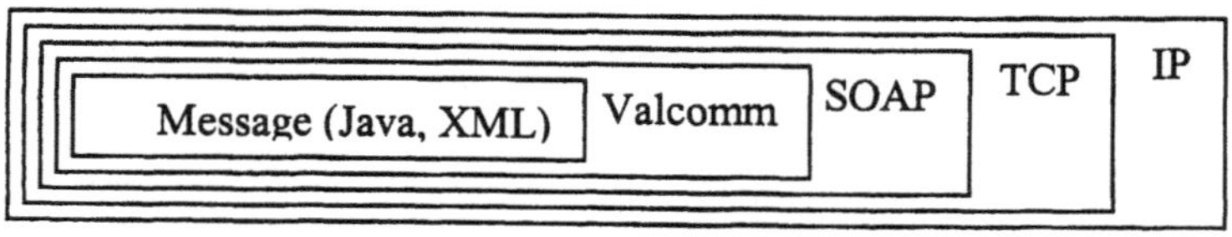

Figure 5. Layers of Valcomm protocol

Valcomm is a message based agent communication language, where XML is being passed as the content through the Internet. Within the Valcomm package, there are many operations and programs that have been developed in order to receive a message, parse the message, place the message into a queue, and then send a message to another agent. In addition Valcomm has been designed such that messages can be sent to an agent from an HTML based web page. Therefore Valcomm performs header processing, error and exception handling, routing, invocation, and transformation. The specifics of each program, and their functions are described in Figure 6.

The ProcessOrderForm class is designed specifically for HTML access to the multi-agent value net. This is a servlet that accepts HTTP POST messages from an HTML form. The servlet extracts information from the HTML form, and then places the pertinent information into an XML document.

The Client class is the central class for sending messages. This class creates the header and an envelope that wraps around the message content the agent would like to send. The header contains the information concerning the receiving agent, the sending agent, and the performative of the message. The body of the message varies depending on the sending and receiving agent and the performative, however the body of the message is encapsulated by XML.

Once the Client class creates the header and the body, the Valcomm message is wrapped by SOAP. SOAP then creates its respective header and body around the Valcomm message. After the Valcomm message is wrapped by SOAP, the entire package is sent to the respective agents via HTTP, which uses TCP and IP as the transport layer. When a message is sent from one agent to another, the Exchange class initially receives the message. Currently Exchange.java accepts the HTTP transport protocol; however can be easily altered to support various transport protocols, such as SMTP, JMS, FTP, etc. A majority of the operations performed on a message prior to reaching an agent are done in the Broker class. After the message is considered authentic, the Broker class determines where the message should be forwarded based on the performative and content. The Archive class is designed to illustrate an agent, and the reception of a message.

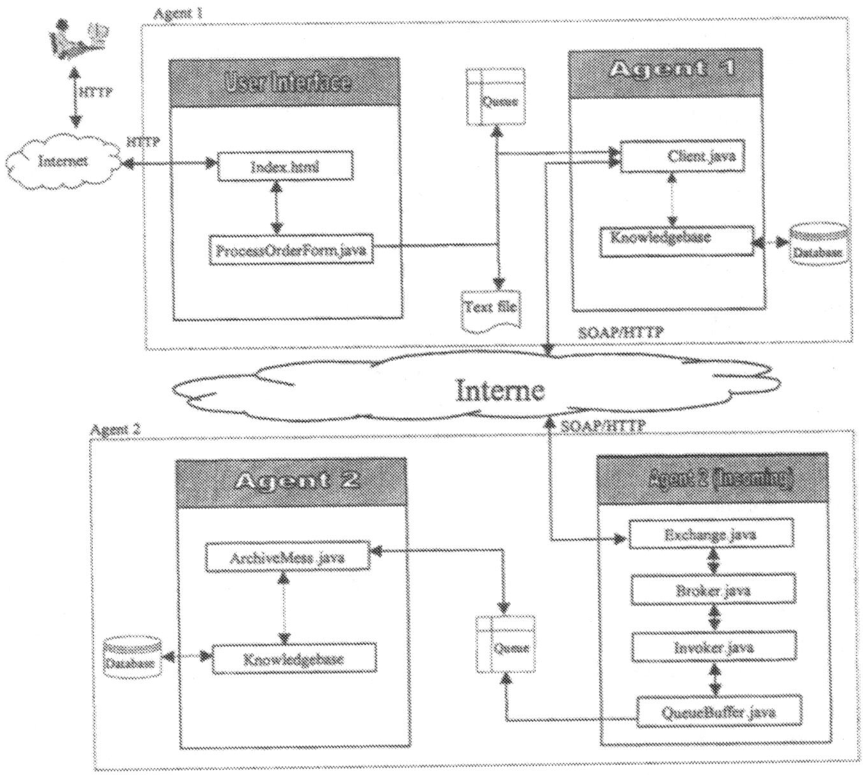

Figure 6. Valcomm Implementation in the Web environment

6. CONCLUSIONS

With the availability of advanced information technology, considerable research has been conducted on generating a flow of information between suppliers, manufacturing facilities, and distribution centers in real time. However this solution only enhances information distribution and not information processing. Therefore in this research, our objectives was to create a network of intelligent real time decision makers that are customer centric. In this chapter, we described the solution to this new business problem as a multi-agent value net model. Value net model exists in the literature and in the spirit of this book we took the reader through a tutorial like implementation schema. We described the multi-agent architecture, and communication language necessary for a customer centric value net implementation.

Traditionally, the communication amongst distributed agents is handled by variations of KQML. However KQML has many limitations when applied to heterogeneous systems that are distributed across the Internet. This problem can be avoided by creating a communication language that utilizes Internet technologies with KQML logic structure. Therefore we need to solve the agent communication problem within the value net model. This language must be able to integrate with customers, other agents, legacy systems, and databases anywhere in the world. The Valcomm language is one such language that was motivated by this specific problem. Valcomm was developed with Java, XML, and HTTP technologies and KQML logic structure can be incorporated into the message system.

In this chapter we have outlined the conceptual and implementation details of multi-agent value nets. If we need to enhance the performance of manufacturing organizations, it is critical to exploit IT, and multi-agent value nets, we believe will be a perfect solution.

7. FUTURE WORK

A full-scale implementation of a multi-agent value net requires future research and development. This research can be split into three areas: agent development, agent communication and general implementation issues. With regard to the agent development, all agents should have the knowledge base, the ability to learn, and the algorithms necessary to make educated decisions. For communication between agents, the language should contain a directory service, network security and message delivery. The third issue involves solving time constraint and wireless capabilities. A customer centric value net needs to react to the customer in real time, and the requirements of wireless implementation are endless.

In addition, we are experimenting with DARPA Cougaar (Cognitive Agent Architecture) platform to implement the multi-agent value net. Details of Cougaar can be found at www.cougaar.org.

ACKNOWLEDGEMENTS

Professor Kumara acknowledges the National Science Foundation for their generous support (under DMI-0075584) for making this research feasible.

REFERENCES

[1] Bovet, D., and J. Martha. *Value Nets: Breaking the supply chain to unlock hidden profits.* John Wiley and Sons, Inc: New York, New York, 2000.

[2] Balakrishnan, A., S. Kumara, and S. Sundaresan. Manufacturing in the digital age: Exploiting information technologies for product realization. *Information systems frontier.* Vol. 1, No. 1, 25-50. July 1999.

[3] Collins, J., C. Billot, and B. Mobasher. Decision processes in agent-based automated contracting. *IEEE Internet computing.* March-April. 61-72. 2001.

[4] Fox, M.S., J.F. Chionglo, and M. Barbuceanu. The integrated supply chain management system. http://www.eil.utoronto.ca/ Department of Industrial Engineering, University of Toronto, Toronto, Ontario, Canada, 1993.

[5] Hinkkanen, A., R. Kalkota, P. Saengharoenrat, J. Stallaert, and A.B. Whinston. Distributed decision support systems for real time supply chain management using agent technologies, *Working Paper, University of Texas*, Austin, Texas, 1998.

[6] Kalakota, R., J. Stallaert, and A.B. Whinston. *Implementing real-time supply chain optimization systems.* 1996.

[7] Huhns, M.N., and L.M. Stephens. Automating supply chain. *IEEE Internet computing*. July-August. 90-93. 2001.

[8] Rao, A.S. and M. Georgeff. BDI Agents: from theory to practice. In *Proceedings of the First International Conference on Multi-Agent Systems (ICMAS-95).* pages 312-319, San Francisco, CA, June 1995.

[9] Satapathy, G., and S.R.T. Kumara. Semantics of KQML Messages based on illocutionary acts and BDI model of an agent. *Working Paper Series, Department of Industrial and Manufacturing Engineering, Pennsylvania State University,* University Park, PA, 1999.

[10] Yung, S.K., and C. C. Yang. A new approach to solve supply chain management problem by integrating multi-agent technology and constraint network. In *Proceedings of the 32nd Hawaii International Conference on System Sciences (ICSS),* March, 1999.

INDEX

A

B

C

D

E

F

J

K

M

P

R

S

T

V

X

www.ingramcontent.com/pod-product-compliance
Lightning Source LLC
Chambersburg PA
CBHW060747150926
51501CB00003B/116
* 9 7 8 1 4 0 2 0 7 4 9 1 2 *